MICROSCOPIC IDENTIFICATION OF MINERALS

MICROSCOPIC IDENTIFICATION OF MINERALS

E. Wm. HEINRICH

PROFESSOR OF MINERALOGY
THE UNIVERSITY OF MICHIGAN
ANN ARBOR, MICHIGAN

McGRAW–HILL BOOK COMPANY

New York · St. Louis · San Francisco · Toronto · London · Sydney

MICROSCOPIC IDENTIFICATION OF MINERALS

PREFACE

The immersion method of the identification of nonopaque minerals was pioneered in the United States by the late Professor E. S. Larsen, Jr., from whom the writer became familiar with it. The method now is a standard technique, both in the United States and abroad, recognized as the outstanding method for the rapid identification of a large number of rock-forming minerals and for estimating their compositions from their optical properties.

Owing to the lack of modern compilations of optical data on rock-forming minerals, this book was begun in 1948 when the writer began to offer instruction in the immersion method at The University of Michigan. It has been used successfully in lecture form and as laboratory notes for hundreds of students who had previously completed a one-semester course in the principles of optical crystallography, taught by the late Professor C. B. Slawson and by Professor R. M. Denning.

Whereas other textbooks in this field have emphasized the properties of minerals in thin sections, this book is concerned primarily with their characteristics as immersed microfragments, secondarily with the optical properties of immersed detrital grains. The book also attempts to present the best available charts that relate the variation of optical properties in a series or a group to compositional variation.

Thus the book may be considered as an elementary introduction to optical microscopy in practice and as a companion volume to any of several modern textbooks that deal with the theory of optical microscopy.

The writer gratefully acknowledges the many generous permissions granted by mineralogists to reproduce their original illustrations. These are individually acknowledged with each figure. Permissions for reproductions also are thankfully acknowledged to the *Bulletin of the Geological Society of America, The American Mineralogist, The Mineralogical Magazine,* and the United States Geological Survey. The photomicrographs are the result of the painstaking care and skill of Mr. Fred Anderegg. Edith D. Heinrich assumed most of the drudgery of preparing the final manuscript.

Credit for any significant or useful features that the book may prove to possess should, of course, revert first to the many mineralogists whose researches are herein summarized and compiled; second to my colleagues, especially to Professors R. M. Denning and the late C. B. Slawson for their critical reviews; and third to the many classfulls of students who survived the evolution of its manuscript and cheerfully disclosed any errors and inconsistencies. For those mistakes and omissions that remain to be uncovered the writer is solely responsible.

E. Wm. HEINRICH

CONTENTS

CONTENTS

PART ONE

GENERAL INFORMATION

CHAPTER ONE

INTRODUCTION

The recognition of minerals by means of their optical properties is doubtless the single most widely used and most widely applicable method of mineral identification. Optical methods applied to minerals in thin section have long been used and were developed to a state of considerable refinement before the immersion method became widely known. The first application of the immersion method was by Maschke in 1872. The first tables for mineral determination by means of the technique were published by van der Kolk in 1900. In the United States the first such determinative tables were prepared by Rogers (1906), and three years later a reference book on optical mineralogy, which contained determinative tables and minerals descriptions including available optical data, appeared (Winchell and Winchell, 1909).

Without doubt, however, the use of the immersion method received its greatest impetus through investigations of Larsen, who published his systematically arranged results in 1921. This was subsequently revised and expanded by Larsen and Berman (1934). *U.S. Geological Survey Bulletin 848,* although now long outdated, remains a standard reference work in the field. It is to be hoped that a modernized and expanded edition will be forthcoming eventually.

Other Americans whose studies and publications have contributed materially to the microscopic study of translucent minerals include H. E. Merwin, F. E. Wright, C. S. Ross, P. F. Kerr, E. E. Wahlstrom, R. C. Emmons, A. Johannsen, and G. Tunell.

The immersion method is the only technique available for determining the refractive index or indices of minute, translucent solid particles. In this method the particles are placed in a liquid of known refractive index and by various techniques (p. 39) the unknown index of the solid is compared with the known index of the surrounding medium. The general range for translucent minerals is $n = 1.31-2.7$, but most have refractive indices in the range 1.45–2.0. The minerals in the lower part of this range have various types of nonmetallic lusters; toward the upper end lusters become submetallic as the translucency decreases. Such high-index species may be essentially opaque to all but a very small band of the spectrum. Metallic species, completely opaque in the visible spectrum, can be translucent in infrared radiation.

For the sole purpose of identification, indices of minerals usually need not be measured more accurately than to 0.01. However, for a determination of the approximate composition in a series, closer determination of the index is essential, commonly to 0.005. Most index liquids are prepared in steps of 0.01.

Allen (1954, pp. 19, 20) has summarized the factors that influence the accuracy of refractive index determination of solids by means of the immersion method:

1. Limitations inherent in the mechanical design and optical characteristics of the microscope.
2. The skill, experience, and visual acuity of the observer.
3. The size and shape of the particles, their visibility and their indices —the lower indices can be determined more accurately. Accuracy in birefringent crystals is more difficult to attain.
4. The uniformity of the index throughout the specimen.
5. The precision with which the liquids of a set are calibrated for their indices and the accuracy of the temperature coefficients, combined with the smallness of the intervals between adjacent members of the series.
6. The accuracy with which the actual working temperatures of the liquids are determined.

Routine determinations, made with white light and low-dispersion immersion liquids, can be made to an accuracy of ±0.003. If all the above factors are realized ideally and if the index liquids are in intervals of 0.002, calibrated to 0.0002, then an accuracy of ±0.0005 is possible in the determination of the refractive index.

Generally for the identification of chemical compounds, particularly organic substances, it is necessary to measure refractive indices more accurately than for minerals. If the compound is chemically "pure," i.e., if it shows no isomorphous variation (pp. 33–34) the refractive index (or indices) will have definite, invariant value. Since there is a very large number of compounds the indices of which are very closely spaced, accurate measurements are necessary to single out individual members. In those instances in which an invariant index (or indices) is sufficient to identify a particular chemical substance, then a departure from this value indicates a deviation from the pure state, i.e., isomorphous variation is indicated (Allen, 1954).

Beyond the field of mineralogy a knowledge of the immersion method is of considerable advantage in chemistry, ceramic engineering, biology, archaeology, medicine, criminology, and soil mechanics and in all fields in which nonopaque crystalline substances must be rapidly and accurately identified. A chemical analysis may reveal a substance to be $CaCO_3$, but it does not show whether the substance is present in rhombohedral or orthorhombic form. By means of the microscope this distinction can be made easily. Microscopic examination of all materials to be analyzed chemically will detect the presence of additional phases ("impurities") and greatly increase the significance of the analysis. Likewise, samples prepared for x-ray study are best first examined microscopically to ensure their homogeneity. Glasses, refractories, porcelains, chemical compounds, concrete aggregates, soil, dust, fibers, plastics, resins, body calculi, and many other materials may be studied advantageously by the immersion method.

In fact, the immersion method is applicable to all nonopaque solids both crystalline and amorphous unless they dissolve or react rapidly with the index liquids or decompose rapidly in light.

CHAPTER TWO

MICROSCOPIC METHODS OF STUDY OF MINERALS

THE IMMERSION METHOD

GENERAL   The immersion method offers a rapid, simple, and relatively inexpensive way of identifying most nonopaque minerals and many rocks by means of the petrographic microscope. No large quantities of the unknown substance are required; an exceedingly minute amount of the material normally will suffice; indeed, a single grain can be made to do in some cases. Usually no elaborate methods of purification or separation are required; mixtures of three, four, or even more substances can be identified by the more experienced student as easily as the beginner studies a single mineral.

The immersion technique does not supplant thin-section study of rocks. Rather, the two methods supplement each other. Most minerals are more easily identified if their refractive indices can be measured accurately. This is not possible in thin section but remains the chief advantage of the immersion method. Moreover, the accurate determination of refractive indices usually results in more than mere recognition of a particular species. It may also yield information on the approximate chemical composition of the mineral. From a petrogenetic viewpoint it is significant not only to identify the rock minerals but also to determine their positions in isomorphous series or groups. For some species, this may also be done in thin section, but for most minerals it is accomplished more easily and more quickly in mounts in which the immersion medium can be changed readily.

Student preparation of thin sections often yields results of inferior quality, and many sections today are made by commercial laboratories. Apart from the cost, this may involve a long period before sections are available. It may be essential to learn the nature of the rock as soon as possible, and the immersion method can minimize the period between rock collection and identification.

The thin section samples a very restricted part of a hand specimen, which itself, for the sake of economy or other reasons, may have been selected as "typical" of a larger group of specimens. By means of the immersion method a relatively large number of both typical and aberrant types may be studied. In addition, if only one or two grains of a rare mineral are seen in the preliminary examination of the rock, they may be removed for identification.

Mineral identification is only part of the investigation of rocks. The texture of the aggregate also must be studied; this requires the use of thin sections, in which the shapes and relationships of various components have been preserved. Thus thin sections and immersion mounts are well used in conjunction, the latter chiefly for identification of the components, the former for studying their spatial relationships. The mineral composition and texture of a rock are two of the fundamental features from which its petrogenetic history may be determined. Occurrence and chemical composition are, of course, the others.

IMMERSION MEDIA   The ideal immersion liquid should be colorless, nontoxic (either to smell or touch), chemically inert, and inexpensive. Its volatility and dispersion should be low, but its contact angle with glass should be high in order to prevent rapid migration over the slide. The change in index of refraction with temperature should be small. There are very few liquids that possess this combination of ideal properties throughout the index range of the common nonopaque minerals. Formerly, different liquids were used for each index, and liquids with adjoining indices of refraction commonly were immiscible. Thus intermediate liquids could not be prepared. The present practice is to use a small number of liquids, each of which is completely miscible with its neighbor on the index scale. Adjoining liquids should also have equal vapor pressures at ordinary temperatures, equal temperature coefficients, and equal dispersion. An ideal series is impossible to obtain, but it may be approached if the intermediate members are prepared by mixing two suitable miscible end-member liquids of widely separated refractive indices. Suitable combinations are listed in Tables 2.1 and 2.2 for the lower range of indices.

TABLE 2.1   IMMERSION LIQUIDS SUITABLE FOR THE RANGE $n = 1.411-1.778$

n (22° C)	LIQUIDS
1.411	n-decane
1.415–1.465	n-decane + petroleum oil
1.466	petroleum oil
1.470–1.630	petroleum oil (Nujol) + α-chloronapththalene
1.633	α-chloronaphthalene
1.635–1.735	α-chloronaphthalene + methylene iodide
1.739	methylene iodide
1.740–1.775	methylene iodide + methylene iodide with dissolved sulfur
1.778	methylene iodide saturated with sulfur

TABLE 2.2   ALTERNATIVE IMMERSION LIQUIDS SUITABLE FOR THE RANGE $n = 1.34-1.66$

n	LIQUIDS
1.34–1.42	water + butyl carbitol
1.43	butyl carbitol
1.44–1.63	butyl carbitol + α-chloronaphthalene
	or
1.44–1.655	butyl carbitol + α-bromonaphthalene

The use of the liquids in range 1.34–1.42 is limited to minerals that are not water-soluble; unfortunately many species in this range *are* somewhat water-soluble.

If special liquids in the range 1.35–1.45 are required for use with water-soluble minerals of low index, they may be prepared from the lower petroleum distillates from kerosene and gasoline (Harrington and Buerger, 1931).

A series of stable index liquids of low to intermediate index may be prepared by mixing ethyl cinnamate, $C_6H_5CH:CHCO_2C_2H_5$, $n = 1.558$, with either glyceryl triacetate, $(CH_3CO_2)_3C_3H_5$, $n = 1.429$, or tributyl phosphate, $(C_4H_9O)_3PO$, $n = 1.422$. Such mixtures have distinctly higher dispersion (Fig. 2.1) than have most liquids of the same index range now commonly used. Such a set will supply dispersion colors useful for index comparison and determination (Wilcox, 1964).

Methylene iodide decomposes and discolors upon exposure to light, but a little metallic copper or tin in the bottles will prevent the color change. The methylene iodide–sulfur solutions tend to precipitate some sulfur and are the least stable of the group.

For indices above 1.78 various combinations have been used (see, for example, Larsen and Berman, 1934, pp. 13–17, and Meyrowitz, 1955, 1956).

A new series of high-index liquids that avoids some of these difficulties has recently been described (Table 2.3).

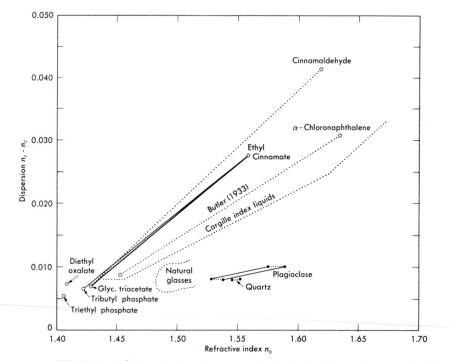

FIGURE 2.1 Relations of dispersion and refractive index of some immersion liquids and typical solids (Wilcox, 1964).

TABLE 2.3   HIGH–INDEX LIQUIDS (MEYROWITZ AND LARSEN, 1951; MEYROWITZ, 1952)

n	LIQUIDS
1.655–1.814	α-bromonaphthalene + 10% S in $AsBr_3$
1.814–2.00	10% S in $AsBr_3$ + 20% S and 20% AsS_2 in 60% $AsBr_3$
2.01–2.15	Se in $AsBr_3$

The color varies from light yellow to amber. The mixing curve for the lower range is not a straight line; for the higher range the curve is a straight line. These liquids and their vapors are poisonous and corrosive, being absorbed upon contact through the skin. Exposure should be minimized; the liquids should be used in well-ventilated areas; immersions should be made rapidly and covered immediately; bottles should be kept tightly closed to avoid hydrolysis.

The indices of the liquids are checked by means of any of several types of refractometers for the lower ranges (Abbé type to about $n = 1.72$; Pulfrich type to about $n = 1.86$), for the higher ranges by means of the method of minimum deviation with a hollow prism mounted either on a one-circle goniometer or on a two-circle goniometer.

The liquids are best stored in small dark bottles with individual glass applicator rods attached to screw caps. The square-sectioned bottles can be set in closely fitting, shallow depressions in narrow wooden frames that will hold about ten bottles. Such trays prevent the bottle from rotating, and the cap with attached rod and adhering liquid may be removed with one hand and the liquid transferred to the slide held in the other hand. Bottles should be kept tightly stoppered when not in use and stored at room temperatures, away from radiators and direct sunlight. The indices should be checked at least once a year.

For liquids used in petrographic work, an increase in temperature causes a decrease in refractive index. For liquids in the range 1.400 to about 1.660, this change is approximately 0.0004 decrease for each degree centigrade. Liquids with methylene iodide have rates of change of greater magnitude, to as much as 0.0007 per degree centigrade. Liquids are standardized at room temperature, and unless they are used under abnormal temperature conditions, no correction is usually necessary.

PREPARATION OF MOUNTS The selection of the material for immersion study is usually best accomplished under a binocular microscope from which the fixed stage has been removed to permit the handling of large specimens. Needle vises or dental picks are useful tools for removing small amounts of the mineral from the specimen. If the amount of material is small, care should be taken to record the exact place of its removal. The selected mineral or rock is crushed but not ground. For relatively large pieces of a hard mineral or rock, it is best to reduce the fragments first in a steel mortar before final crushing in an agate mortar. Normally, sizing of the powder is unnecessary, but in the preparation of large amounts for classroom study, sizing to −80 mesh will produce pieces suitable both for index measurement and 2V determination. Micaceous minerals are difficult to reduce to small sizes. Initial shredding of coarse flakes with a scissors is helpful, or grinding with a small amount of powdered glass, followed by screening, may also assist.

The powdered mineral or rock is transferred to a glass slide from the mortar or container by means of a small spatula or knife blade. Slides 50 by 25 mm in size and 1 to 1.1 mm thick are especially suitable, for longer slides tend to shift position upon stage rotation, and thicker slides may prevent securing interference figures. Usually, several mounts may be made on a single slide, provided the size of the mount is kept at a minimum. Glass cover slips 8 mm square and 0.13 to 0.17 mm thick are well adapted for immersion purposes. Thicker cover slips are unsuitable, but larger ones may be required for many mounts. Larger cover slips may be ruled by means of a diamond point and broken into squares of the desired size.

The beginner should exercise caution in gauging the amount of powder needed. The use of too much powder results in an overcrowded field, with overlapping grains and tremendous visual confusion. Actually, only an exceedingly minute quantity is needed; three or possibly four dozen grains, well distributed over the slide, are adequate for a single substance or when various components are present in approximately equal proportions. A tap of the spatula against the edge of the slide ensures good particle distribution. The applicator rod with a moderate-sized drop should not be applied directly to the fragments. After the cover slip is placed over the immersion, small slivers of filter or blotting paper may be used to remove any excess liquid. Excess liquid results in a loose cover slip that tends to shift grains with rotation of the stage. Too little liquid generally results in a poor distribution of particles, with crowding of pieces toward the edge of the liquid field. Additional liquid may be added by touching the applicator rod to the edge of the slip.

The position of the cleavages governs the orientation of the particle and consequently restricts the measurement of the optical constants. This becomes a characteristic property of the mineral and thus an aid to its identification.

Micaceous minerals may be sprinkled on slides coated with gelatin before being covered with immersion fluid (Fairbairn, 1943; Marshall and Jeffries, 1945). A simple rotating device also is helpful in working with minerals whose perfect cleavages do not coincide with optical directions (Rosenfeld, 1950).

If only a single grain or one well-formed crystal is available, the liquid may be withdrawn from under the cover slip by means of a sliver of blotting paper. Liquid of the desired index should then be used as a wash, withdrawn, and renewed. Always withdraw from one side and add from the other. Grain transfer is thus unnecessary. If grain transfer is necessary, a capillary pipette fitted with a bulb will be helpful.

THIN SECTIONS

GENERAL Sections of standard thickness (between 0.02 to 0.03 mm) may be prepared not only of rocks and other mineral aggregates but also of ceramic materials and other laboratory products consisting of aggregates of solid crystalline or amorphous substances. Not only do such sections provide undisturbed two-dimensional samples in which the original spatial relations and shapes (texture) of grains may be determined, but they also usually permit measurement of a sufficient number of optical properties that the more common species may be identified. In addition, they display the mineral with its usual associates (paragenesis). No other device has been as extensively employed in petrographic studies, but under some circumstances thin-section studies may yield but incomplete information (Heinrich, 1956):

1. If the rock is very coarse-grained and mineralogically variable, as in the case for some conglomerates, breccias, agglomerates, and pegmatites.
2. If the mineral constituents are sufficiently unusual that their refractive indices must be measured to aid in their identification.
3. Generally if it is required to determine the position of a species in an isomorphous series or group. This can normally be done more quickly and more accurately by means of optical data determined in immersion.
4. If the rock contains a very small percentage of genetically significant accessory constituents, only a very few grains of which might be cut by a thin section. This applies particularly to the insoluble residue minerals of carbonate and saline sedimentary rocks and to the accessory detrital and authigenic minerals, both heavy and light fractions, in sandstones and related rocks.

Thus, for many mineralogical studies the optical methods, thin sections and immersions (either crushed fragments or released grains), supplement one another.

Today most thin sections are prepared commercially, although it is good practice to require the beginning student in microscopic petrography to persevere until he succeeds in producing a usable section he himself has manufactured. Not only does this teach him the difficulties and pitfalls involved in such preparations, but it helps to instill a measure of care in handling thin sections beyond that absorbed from the warnings and maledictions of the instructor. Techniques for the production of thin sections in quantity vary. For descriptions of equipment and techniques reference should be had to Meyer (1946), Reed and Mergner (1953), Heinrich (1956), and Baumann (1957).

The mounting medium is commonly either Canada balsam ($n = 1.535$–1.540) or a thermoplastic cement, Lakeside No. 70 ($n = 1.536$).

Older sections or those hastily or improperly prepared may show a variety of defects some of which may sorely tax the ingenuity of students attempting to assign them mineralogical identities:

1. Grains of grinding compound; usually silicon carbide.
2. Holes, especially in coarse-grained minerals showing two or more cleavage directions.
3. Organic fibers.
4. Impregnating media for binding porous or friable rocks; e.g., Bakelite varnish, $n = 1.63$; or Lucite (methyl methacrylate), $n = 1.417$; aroclor 4465, $n = 1.66+$.
5. Crystallized Canada balsam. This may take the form of minute spherulitic bundles of fibers or single fibers randomly oriented, both showing moderate birefringence.

Students should remember that contacts between adjacent grains in thin section are not usually normal to the plane of the section. If the contact-section plane angle is very small, extensive grain overlap occurs with anomalous optical effects in the zone common to both. Sections of rock rich in rhombohedral carbonate minerals may be sprinkled by minute carbonate dust (cleavage particles) randomly oriented with respect to the underlying grains. Sections of graphitic rocks also are bordered by graphite dust plucked from the larger graphite plates during section grinding.

STAINING   Selective staining of minerals to facilitate identification is a widespread procedure, particularly in thin sections prior to modal (quantitative mineralogical) analysis either by means of measuring linear intercepts (Rosiwal method) or by means of point counting (Glagolev-Chayes method) (see Heinrich, 1956, pp. 8–12). Under some circumstances staining of crushed grains may also be advantageous, particularly if grain counts are to be made. Both various silicate rocks and carbonate rocks may be stained, particularly for distinction among nepheline–potash feldspar–plagioclase and between calcite and dolomite. For example, potash feldspar "hidden" interstitially in the matrix of trachybasalts or syenogabbros may be revealed readily; minute interstitial nepheline in the groundmass of phonolites can be detected; and tiny blebs of dolomite exsolved from the calcite of carbonatites will appear conspicuously.

PROCEDURE FOR NEPHELINE

1. Spread, with glass rod, a thin film of sirupy phosphoric acid over section; allow to remain 3 min.
2. Wash by dipping gently in water.
3. Immerse slide in 0.25% solution of methylene blue 1 min.
4. Wash off excess dye by dipping gently in water.

By this technique nepheline, sodalite, and analcite are stained deep blue and melilite a light blue; potash feldspar, plagioclase, and leucite are unaffected. Zonal structure in nepheline may be revealed, the unstained zones being richer in SiO_2.

PROCEDURE FOR NEPHELINE AND POTASH FELDSPAR

1. Spread, with pipette, concentrated HCl on slide surface; allow to remain 4 min.
2. Wash by dipping gently in water.
3. Spread, with pipette, solution of malachite green (1 g in 200 ml of distilled H_2O) over wet slide; allow to stand for 50 to 60 sec. Nepheline is stained strong green.
4. Dry in air 24 hr.
5. Place face down in plastic box over HF. Expose to HF fumes at room temperature for 15 sec. The green nepheline stain is temporarily lost but restored in the next step.
6. Stain with sodium cobaltinitride (ca. 60 g per 100 ml H_2O) by immersing sections for 15 to 20 sec in a Coplin jar filled with this solution. Potash feldspar is stained yellow; nepheline green is restored; plagioclase is left colorless. On high-Na potash feldspars the stain is of poor quality.
7. Wash immediately in a gentle flow of cold water and allow to dry thoroughly before covering.

If potash feldspar alone is to be stained, use steps 5 to 7 only. Plagioclase, except pure albite, may be stained an intense red by etching with HF as above, dipping in $BaCl_2$ solution, rinsing and treating with potassium rhodizonate solution (Bailey and Stevens, 1960). This may be combined with the cobaltinitride stain for potash feldspar.

PROCEDURE FOR BRUCITE-SERPENTINE

1. Immerse slide in dilute HCl in which a few crystals of potassium ferrocyanide have been dissolved. Watch development of stain; immersion time varies. Brucite stains blue; serpentine, pale green.
2. Wash gently, dry thoroughly.

PROCEDURE FOR CALCITE-DOLOMITE AND ASSOCIATED SPECIES. The recommended procedure is that shown in Table 2.4.

BINOCULAR MICROSCOPIC EXAMINATION

The preliminary examination of specimens under the binocular non-polarizing microscope prior to immersion examination is essentially standard procedure in order to (1) obtain some information on the homogeneity

TABLE 2.4   STAINING OF CARBONATE MINERALS (FRIEDMAN, 1959)

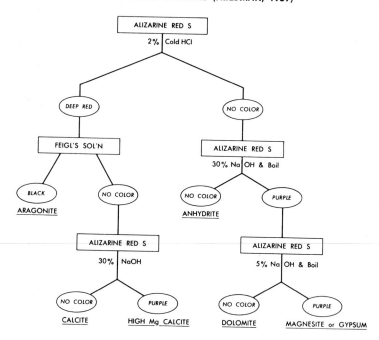

of the specimen and (2) help select particles, crystals, or portions for crushing or direct immersion. In addition, the binocular nonpolarizing microscope is used extensively for:

1. Study of insoluble residues, both from carbonate rocks or evaporite rocks (see Heinrich, 1956, pp. 108, 109).
2. Study of detrital grains for form and other general characteristics (see p. 26), for measurement of length-breadth ratios, and for grain counts. The last is facilitated with a grid background.
3. Estimation of and description of common rock-type fragments from drill cuttings, in preparation of well logs (see Heinrich, 1956, p. 13).

VARIATION IN APPEARANCE OF SPECIES

For the beginning student difficulties in identification of a species may arise as he switches among the various optical methods, because of which superficial differences in appearance are apparently enhanced. For example, in thin section, in which the thickness remains constant, a restricted and therefore diagnostic range of interference colors characterizes a par-

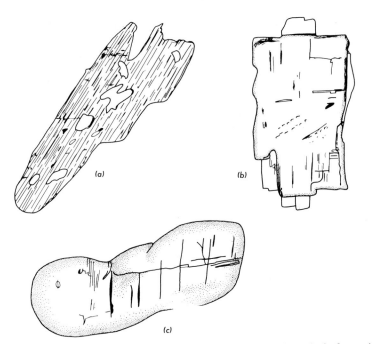

FIGURE 2.2 Variation in appearance of a single species with method of examination. (a) Kyanite in thin section, matrix of quartz grains, $\times 10$. (b) Kyanite, crushed fragment, liquid $n = 1.65$, $\times 80$. (c) Detrital grain of kyanite, liquid $n = 1.54$, $\times 80$.

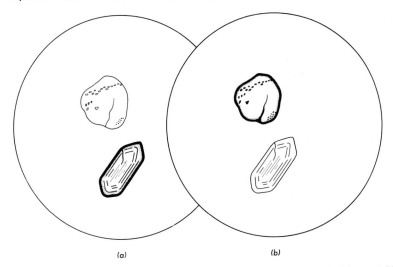

FIGURE 2.3 Relief of quartz (upper) and zircon (lower) grains in (a) liquid, $n = 1.56$, and (b) liquid, $n = 1.80$.

ticular species. In crushed grains of the same mineral the range of inter-ference colors is greater; the colors are generally of higher orders and may vary within a single grain. Similarly, with increased thicknesses for crushed material and detrital grains, mineral colors become deeper and pleochroism is intensified. In thin sections and for detrital grains *original* shapes are significant features; in crushed material a *secondary* shape is produced to facilitate recognition. These and other apparent differences are illustrated in Fig. 2.2. Variations in optical relief also provide marked but superficial differences in appearance (Fig. 2.3).

CHAPTER THREE

OPTICAL PROPERTIES OF MINERALS

GENERAL

Minerals can be divided optically into three groups: isotropic, uniaxial, and biaxial. Amorphous and most isometric minerals are isotropic; tetragonal and hexagonal minerals are uniaxial; orthorhombic, monoclinic, and triclinic minerals are biaxial. As the symmetry of minerals decreases, the minerals become optically more complex and are characterized by an increasing number of measurable optical constants. Theoretically, therefore, it is easier to identify biaxial minerals than isotropic ones, because the latter have few optical characteristics to differentiate them one from another. However, in practice uniaxial minerals generally are the easiest, being of intermediate complexity and fewer in number than the biaxials.

For minerals of all the groups the following general properties may be recorded.

1. Shape, outline
2. Size
3. Cleavage, parting, fracture
4. Twinning, zoning
5. Inclusions, intergrowths, alterations

For opaque minerals the following properties may be observed:

1. Transparency of thin edges or along fractures
2. Color and luster in reflected light

3. Surface grooves or grinding striations on soft opaque minerals in thin section

The additional properties of isotropic minerals that may be recorded are:

1. Color
2. Index of refraction (n)

For uniaxial minerals:

1. Color and pleochroism
2. Indices of refraction (ϵ and ω) and birefringence
3. Optical sign, ($+$) or ($-$)
4. Character of elongation or cleavage direction

For biaxial minerals:

1. Color and pleochroism (absorption formula)
2. Indices of refraction (α, β, γ) and birefringence
3. Size of 2V and dispersion of optic axes
4. Orientation (relation of optical directions, α, β, γ, to crystallographic directions, a, b, c)
    a. Optical character or sign, ($+$) or ($-$)
    b. Type of extinction and size of extinction angles (except in orthorhombic)
    c. Character of elongation or cleavage; position of optic plane

In thin section where it is not possible to determine the refractive indices, the indices are compared with the index of Canada balsam ($n =$ ca. 1.54). If they are lower, the relief can be said to be negative; if higher, the relief is positive.

For crushed fragments of minerals that have good cleavages, the optical properties of specific cleavage pieces will also be diagnostic (e.g., the rhombohedral carbonates, the plagioclase feldspars).

ABNORMAL OPTICAL PROPERTIES

GENERAL Not uncommonly, some rock-forming minerals will exhibit optical properties that apparently are inconsistent with their crystal form or crystal structure. Thus, for example, outwardly isometric species may display birefringence, and some monoclinic minerals may appear uniaxial. The causes of such misalliances of characteristics are:

1. *Strain.* The small 2V's of some calcite and of some quartz, particularly those of metamorphic rocks, result from the mechanical distortion of the crystal structure.

2. *Variations in chemical composition.* The change from normally uni-axial to biaxial character of vesuvianite is a consequence of relatively large compositional variations.
3. *Inversion.* Leucite, which is isometric at high temperatures, inverts upon cooling to a structure of lower symmetry. This transformation is accompanied by multilamellar inversion twinning, and the result is a weakly birefringent pseudomorph of twinned low-temperature leucite that has inherited the trapezohedral form of its high-tempera-ture, isometric ancestral polymorph.

Examples of the various types of relationships are:

1. Minerals with isometric form but which display birefringence: *leucite, perovskite, grossularite, spessartite.*
2. Minerals with tetragonal or hexagonal form (uniaxial) but which may be biaxial: *calcite, apatite, quartz, vesuvianite.*
3. Minerals with orthorhombic, monoclinic, or triclinic structure (bi-axial) but which may be uniaxial or appear essentially uniaxial: *phlogopite, biotite, muscovite,* some *chlorites, stilpnomelane, pigeon-ite, sanidine.*

It should be recalled that in minerals of very low birefringence a rela-tively small change in the value of one of the refractive indices can produce a large change in the size of 2V.

A few mineral species can occur in such fine-grained particles that aggregates of these *appear* isotropic: apatite as collophane, serpentine (var. serpophite), and in some cases, various clay minerals and gibbsite-boehmite mixtures (cliachite) in bauxite. Fine-grained aggregates of antigorite ($2V = 30-50°$) may show $2V = 0°$ to small because of overlap-ping platelets with a and b axes randomly oriented.

ABNORMAL INTERFERENCE COLORS Not all species display interfer-ence colors or sequences of such colors in the order of the color chart. Minerals that are themselves colored (e.g., biotite, hornblende, tourmaline) under plane-polarized light, will, under crossed polars, display interfer-ence tints that result from the combination of the body color (due to absorption) and the interference color of the transmitted rays.

In a few species the birefringence shows noticeable dispersion, i.e., it is unequal for different ends of the spectrum. Therefore, if the birefringence is nil for one λ and appreciable for all other colors, the color of the nil-birefringent λ will be absent in the interference color sequence, without regard to the thickness. For example, melilite, which is essentially isotropic for yellow light, shows a deep blue interference color, commonly called

"Berlin blue," with $\Delta = 100$ to 200 mμ. Another abnormal shade is a deep brown ("tobacco brown"). In biaxial species abnormal interference colors also appear in species showing dispersion of the optic axes, as in titanian augite. These abnormal tints ("anomalous" interference colors) are restricted to relatively few common minerals (Table II).

OPTICAL PROPERTIES OF SUBMICROSCOPIC AGGREGATES Certain substances, which may be demonstrated to consist of mixtures of two (or more) phases by means of x-ray diffraction analysis or by means of the electron microscope, may behave optically in the aggregate as if they were a single, homogeneous species. This phenomenon occurs when the elementary particles constituting the mixture have dimensions no greater than one-half the wavelength of light, a condition prerequisite to the application of Wiener's formulae of *form birefringence* (Donnay, 1945).

Such aggregates, of the proper grain size, are uncommon:

"Vaterite A"	spherulitic calcite + water
Chalcedony	fibrous quartz + water
Opal	cristobalite + water
Iddingsite	goethite + hydrous Mg silicate
Nemalite	fibrous brucite + magnetite
Ishkyldite	fibrous chrysotile + water

In the case of nemalite (Donnay, 1945) the ω index of the fibers is much higher than ω of brucite, and the birefringence of the fibers is much lower, both values depending on the content of magnetite. In opal and chalcedony, the refractive indices decrease with increasing water content.

Anomalous optical properties can also be obtained from mixed-layer structures, such as occur with the clay mineral group (p. 313).

COLOR AND PLEOCHROISM

Generally only those minerals that are strongly colored megascopically will show marked coloration microscopically under conditions of greatly reduced thicknesses. However, color, when observable microscopically, is generally a more characteristic feature than in hand specimen. Colored isotropic minerals show no change in color when their crystallographic orientation is changed under polarized light. In contrast, in colored anisotropic species certain light wavelengths are absorbed differentially with crystal direction, giving rise to different colors for the grain in different orientations under plane-polarized light. Colored uniaxial crystals show two colors and are said to be *dichroic*; colored biaxial crystals may show three different colors and are termed *trichroic;* the color change phenomenon is *pleochroism.*

Usually in biaxial species the optical indicatrix and the absorption indicatrix coincide, so that for each of the three vibration directions there is a corresponding absorption axis. Each axis may show a different degree of color absorption, which can be expressed by means of an *absorption formula.* For example, if light vibrating parallel with the γ direction is most strongly absorbed and that along β is intermediate, the absorption formula may be expressed: $\gamma > \beta > \alpha$. In some minerals two axes may have essentially the same absorption, as detectable by eye. Then the absorption formula might be, for example, $\gamma > \beta = \alpha$.

In some minerals the directions of minimum, intermediate, and maximum absorption do not coincide with the axes of the indicatrix, for example, in some pyroxenes and in axinite.

SHAPE AND FORM

GENERAL In the microscopic examination of minerals shape of grains or crystals must be considered for three different categories:

1. Undisturbed grains in thin section
2. Material crushed and immersed in oils
3. Detrital grains liberated from clastic rocks and immersed in oils or mounted in more permanent media

IN THIN SECTION Thin sections of igneous rocks show grains that may be wholly, partly, or without crystal-face boundaries. Such grains are respectively termed euhedral, subhedral, and anhedral. In glassy rocks where crystallization may have ceased abruptly through chilling, minerals may show only embryonic development as:

1. Crystallites. Tiny skeletal forms of varying shape (Fig. 3.1), identifiable with difficulty and then only with oil-immersion lenses and extra-strong illumination. These grade into:
2. Microlites. Small crystals, identifiable as to species under ordinary high-power magnification.

The habit as well as the degree of crystallization of a species governs the shape of its section in thin section. Common habit-group designations are:

1. Equant, blocky
2. Tabular, lamellar, flaky, micaceous

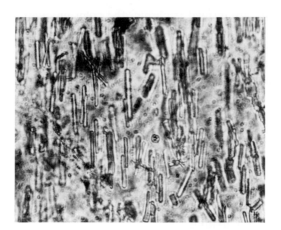

(a)

(b)

FIGURE 3.1 Various types of crystallites (Ross, 1962). (a) Prismatic pyroxene crystal-lites, obsidian, Valles Mountains, New Mexico. Area is 0.1 mm across. (b) Pyroxene trichites, obsidian, Lower Lake, California. Area is 0.1 mm across.

3. Elongated, columnar, prismatic, bladed
4. Acicular, fibrous

It is well to remember that tabular crystals viewed normal to the flatten-ing may look blocky in section and that elongated or bladed cross sections can obtain from both columnar and tabular crystals.

Tabular, columnar, or fibrous minerals in groups or aggregates may be oriented in diverse ways to produce characteristic textures:

1. Parallel arrangement
   a. Foliation in regional metamorphic rocks (parallel planar orientation) to yield slaty cleavage, schistosity, or gneissic structure: chlorite, muscovite, biotite, kyanite, sillimanite, actinolite, hornblende
   b. Lineation in regional metamorphic rocks (parallel linear orientation): sillimanite, kyanite, hornblende
   c. Foliation in intrusive igneous rocks (crude parallel planar orientation): biotite, hornblende
   d. Bedding in pelitic sedimentary rocks (parallel planar orientation): clay minerals, sericite, chlorite
   e. Comb structure in veins (subparallel orientation normal to a surface): quartz, calcite, tourmaline, chrysotile
2. Radial arrangement
   a. Spherulites in glassy volcanic rocks: sanidine plus interstitial cristobalite
   b. Radial groups in vesicles in volcanic rocks: natrolite
   c. Radial groups in metasomatic rocks: tourmaline ("sunbursts"), prehnite ("bowties")
   d. Radial groups in metamorphic rocks: sillimanite ("flamboyant structure"), brucite ("whorls")
   e. Spherulites (recrystallized oolites) in sedimentary rocks: calcite, francolite

Banding resulting from the alternation of layers of different mineral composition may also be observed on the scale of thin sections. Mineralogical banding occurs in metamorphic rocks, in sedimentary rocks, and in veins and cavity fillings. In veins small-scale, scallopy, fine-grained bands give rise to colloform structure. Flow banding in fine-grained or glassy volcanic rocks depends on streaks or layers of contrasting color and/or mineralogy.

Structures of generally circular cross section that appear in thin sections may be either organic or inorganic in origin. Certain sections of some fossils yield circular outlines: foraminifera, radiolaria, corals, crinoid stem plates, etc. Inorganic spherules that consist of concentric layers precipitated about a foreign nucleus are called oolites: calcite, hematite, collophane, chamosite. Some similar spherules are not layered and have been called ovules: collophane, glauconite, bauxite.

Organic structures appear in thin sections of many sedimentary rocks: shells or shell fragments (calcite, aragonite, fine-grained or chalcedonic quartz), diatom tests (opal), bone (dahllite, collophane), silicified wood (fine-grained quartz).

Regular to semiregular intergrowths of two or more species are features common to igneous and metamorphic rocks and to a lesser extent to sedimentary rocks.

1. Graphic, vermicular
    a. Quartz–feldspar: micropegmatite (orthoclase–quartz, microcline–quartz); granophyre (orthoclase–quartz); myrmekite (sodic plagioclase–quartz)
    b. Other species (symplectites—formed by replacement): quartz–muscovite, quartz–tourmaline, muscovite–sillimanite
2. Lamellar, spindle
    a. Feldspars: perthite (orthoclase–sodic plagioclase, microcline–sodic plagioclase)
    b. Pyroxenes: (hypersthene–diopside, hypersthene–augite)
3. Inclusions:
    a. Blebs: antiperthite (orthoclase in plagioclase); glass (in leucite, plagioclase phenocrysts of volcanic rocks, melilite)
    b. Oriented platelets, minute rods: "schiller" inclusions (hypersthene, labradorite); graphite (andalusite)
    c. Irregular, crowded: poikiloblastic or "sieve" structure (staurolite, cordierite, almandite)
    d. Pleochroic haloes: zones of dark coloration developed in iron-bearing silicates (biotite, hornblende) by the oxidation $Fe^{2+} \rightarrow Fe^{3+}$ by means of α-particle bombardment from included radioactive accessory species, especially zircon
4. Overgrowths
    a. Crystallographically oriented: armoring (orthoclase or sanidine on plagioclase); some uralite (hornblende on augite)
    b. Reaction rims: coronas (augite on olivine, hornblende on augite); keliphytic rims (hornblende plus plagioclase on pyrope)
5. Micropseudomorphs
    a. Single-mineral replacements: limonite after pyrite, chlorite after garnet, kaolinite after orthoclase
    b. Multimineral replacements: pinite (sericite and chlorite after cordierite); saussurite (albite and an epidote-group mineral after labradorite); resorbed phenocrysts (magnetite and augite after hornblende); pseudoleucite (orthoclase and nepheline after leucite)

CRUSHED FRAGMENTS Minerals crushed for immersion in oils for identification purposes have shapes and orientations controlled in the main by cleavages or partings, provided that the fragmented pieces are not polygranular. Reduction to −80, +100 mesh will usually produce fragments that are parts of single grains, except for such fine-grained species as some sericite, chlorite, serpentine, sillimanite, glauconite, and clay minerals.

Minerals without cleavage or with but poor cleavage yield, upon crushing, particles with essentially random orientation. As the quality of the cleavage increases, larger percentages of the pieces will be oriented preferentially on cleavage faces. Minerals such as the micas, with perfect and easy basal cleavage, will have essentially all flakes lying on (001). Under crossed polars cleavage faces of anisotropic fragments will display a uniform interference color, owing to the uniform grain thickness. Narrow color bands of decreasing order appear along the steep margins of the grains. If no cleavage is present and the grain is of variable thickness, the interference tints vary over the surface (Fig. 3.2).

When two or more cleavage directions are present, the mineral will tend to lie on the best cleavage, the others appearing as straight boundaries that form either vertical or inclined edges.

Crushed fibrous minerals may yield pieces that are bundles of fibers, in which the c axes of the needles are essentially parallel and the a and b axes are randomly oriented with respect to each other. Such sheaves may yield 0° extinction angles, even if the mineral is monoclinic. The orthorhombic character of such fiber bundles may not be determinable optically.

DETRITAL GRAINS In freeing detrital grains from their aggregates (sandstones, arkoses) it is essential to break down the rock to liberation

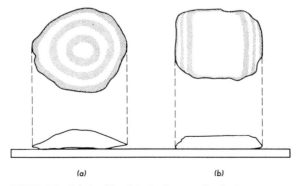

(a)   (b)

FIGURE 3.2 Relationship of grain shape to distribution pattern of interference color bands: (a) no cleavage, (b) with cleavage.

identification purposes such grains may be handled by the immersion technique in the same manner as crushed grains, with the remembrance that orientation owing to cleavage may be a more subordinate feature. The so-called heavy accessory detrital minerals $(G > 2.85)$ are of major importance in reconstructing the provenance of clastic sediments, and for these their detailed characteristics should be determined. These include (Heinrich, 1956):

1. Percentage present
2. Grain size
3. Shape: form and sphericity
4. Color, zoning
5. Inclusions
6. Surface markings: etching, frosting
7. Fluorescence
8. Radioactivity
9. Approximate chemical composition of a member of a series or group through measurement of the appropriate optical properties

Ross (1950) has described the advantages of the dark-field stereoscopic microscope in the examination of detrital grains.

Zoning is a feature that should be noted in both detrital grains and in thin section. In thin-section studies it is particularly significant genetically in such minerals as the plagioclases and pyroxenes. Most zoning is "concentric," i.e., the zones are shells, either distinctly separate or gradational, that are parallel with the crystal outline (tourmaline, fluorite, garnet, plagioclase). Rarely, "hourglass structure" or sector zoning occurs (titanian augite, chloritoid). Zoning becomes apparent variously, by changes in:

1. Color (tourmaline)
2. Extinction angle (plagioclase)
3. 2V (olivine)
4. Birefringence (epidote group)

and by

5. Inclusions (quartz)
6. Selective alteration (plagioclase)

OPAQUE MINERALS IN THIN SECTION

There are a few opaque or nearly opaque minerals, generally of accessory character, that are widespread in rocks. Those that are opaque include magnetite, ilmenite, pyrite, pyrrhotite, and graphite. Chromite is opaque except along cracks or its thinnest edges. A few species of very high relief, such as perovskite, rutile, and cassiterite, may appear to be opaque upon cursory examination, particularly if they are present as minute grains. If a grain appears opaque, it is best to swing into place the substage accessory condenser with its concentrated beam in order to test all parts of the grain for transmission. If no light is passed, the condenser can be swung back, *halfway*, and the grain is then viewed in random reflected daylight for:

1. Color and luster
2. Alteration
3. Cleavage, fractures, and grinding striations (graphite)

OPTICAL ORIENTATION

GENERAL   By *optical orientation* is meant the relationships of the three principal vibration directions to the three crystallographic axes. In orthorhombic species the three vibration directions coincide with the three crystallographic axes; in monoclinic crystals one axis of the indicatrix is coincident with the *b* crystal axis; in triclinic minerals there is no correspondence between any optical and crystallographic directions. In determining the optical orientation of a mineral it is first necessary to recognize what directions in the crystal or grain correspond to *a*, *b*, and *c*. If the mineral is euhedral, this is somewhat simplified. In anhedral units or crushed grains crystallographic directions may be relatable to (1) cleavage planes, (2) parting planes, (3) elongation of fibers, (4) twin planes, (5) growth zones, and (6) oriented inclusions. If the mineral is not known, or is a new species, or if it lacks elongation or one or more particularly well-developed cleavages, certain arbitrary assignments may be required to label those crystal directions that are available. It should be remembered that not all good cleavages are necessarily parallel with one of the three pinacoids, (001), (010), (100).

Normally in acicular or elongated prismatic crystals the elongation axis can be taken as *c*. Crystals of tabular, platy, or flaky habit commonly are flattened parallel with (001).

ORTHORHOMBIC SYSTEM   The six possible combinations of correspondence between optical and crystallographic axes appear in tabular form on the next page.

Thus if the optic plane is parallel with (001), orientation 2 or 3 is possible; if (100) is the optic plane, orientation 4 or 5 may be present; and for

	1	2	3	4	5	6
α	a	a	b	b	c	c
β	b	c	c	a	a	b
γ	c	b	a	c	b	a

(010) as the optic plane 1 or 6 is a possible orientation. A very common orientation is one in which $\beta = b$ and the optic plane is parallel with (010).

Biaxial minerals that show parallel or symmetrical extinction on principal sections; i.e., on faces in zones parallel with at least one crystallographic direction, are orthorhombic. However, if the section cuts all three crystallographic axes, the extinction will be inclined to the cleavage traces, since parallel extinction is in reference to crystal axes and not to cleavage plane traces. Sections parallel with vibration axes (and thus in orthorhombic species with crystal axes) may be recognized by the symmetry of their interference figures.

MONOCLINIC SYSTEM In monoclinic crystals one vibration direction coincides with the b crystal axis; this may be α, β, or γ. Commonly, β corresponds to b, and in this case α and γ can be in any position in the a–c plane, at right angles to each other. Thus with $\beta = b$, the optic plane is parallel with (010).

Since neither α nor γ can coincide with either a or c, they are said to extinguish at an angle to these crystallographic directions. In many species the c direction can be recognized owing to the parallelism of traces of (010), (100), or (110) cleavage directions with c in sections parallel with c. If such a direction of reference can be recognized, the angle between the trace of the vibration direction and the crystallographic axis can be measured.

The extinction angle usually is measured against the trace of the slower of the two rays in the section (that with the larger refractive index). Since for many of the rock-forming minerals that are monoclinic the optic plane is parallel to (010), usually it is desirable to determine the extinction angle $\gamma \wedge c$ on (010). This may be done in two ways:

1. By finding a (010) section by locating a centered flash figure and making a single extinction angle measurement.
2. By choosing, in thin section, a series of sections that show the highest orders of interference colors and measuring the extinction angle on each. If a sufficiently large number of sections is selected, then the largest angle is $\gamma \wedge c$, which is a characteristic optical property of

the mineral. Sections parallel with c but not to (010) will show progressively smaller extinction angles as their orientations depart from (010) toward (100).

Extinction angles are either positive or negative. Two conventions have generally been used for determining the sign:

1. The sense of rotation (clockwise or counterclockwise) of the line generating the angle from the reference direction.
2. Whether the angle generated is toward the positive or negative end of a particular crystallographic axis.

For the plagioclases a number of mineralogists have defined positive extinction angles as those in which the rotation of the line generating the angle turns clockwise from the line of reference.

The mechanics of measurement are concisely described by Tunnell (1953):

Under the microscope one actually turns the stage and crystal with respect to the line, the north-south crosshair, which remains stationary. Thus, if one first sets the trace of the 010-cleavage to coincide with the north-south crosshair and then rotates the stage counterclockwise through an angle less than 90° to a position of extinction, and if the vibration direction which then coincides with the north-south crosshair is the X'-vibration direction (i.e., the faster ray in the section), the sign of the extinction angle is positive. . . . The convention that an extinction angle is considered positive if the line generating the angle is turned clockwise from the reference direction to the X'-vibration direction necessitates that the sign of the extinction angle is reversed when the section is turned over. In conformity with this requirement, in the published stereographic projections the extinction angles at points on the circumference of the projection at opposite ends of a diameter have opposite signs; these two extinction angles, numerically equal, but opposite in sign, of course apply to the same section; the first extinction angle applies with one side of the section up, the second, numerically equal, but opposite in sign applies when the section is turned over.

The second convention can be described with the aid of Fig. 3.3. In Fig. 3.3 the angle $\gamma \wedge c = 30°$ because it has been generated by rotation of the γ direction toward the positive ($+$) end of the axis (front end). The angle $\alpha \wedge a = -18°$, because it is rotated toward the negative ($-$) end of the c axis (downward). In Fig. 3.3 the angle $\gamma \wedge c = -30°$ because it is turned toward the negative ($-$) end of the a axis. Thus in monoclinic crystals in which (010) is the optic plane the extinction angle is negative when it lies in the acute crystallographic angle β (rear), positive if in the obtuse crystallographic angle β (front). With this system the trace of the (001:010) edge must be recognizable via crystal outline, (001) cleavage or parting, zoning, or oriented inclusions.

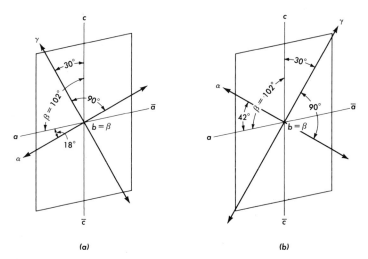

FIGURE 3.3 Extinction angles in monoclinic minerals; section is (010), parallel with the optic plane.

When the extinction angle to be measured is on the front pinacoid (100), as in the plagioclases, the convention is that the sign of the angle is positive if the X'-vibration direction lies in the acute angle between the traces of the two perfect cleavages, (001) and (010). If the (001:100) edge is absent (either through lack of crystal outline or the cleavages), positive and negative angles are not distinguishable. The acute angle is the upper left angle in a section normal to *a*, when seen from the positive end of *a*. For the plagioclases the inconsistency in the two conventions is discussed by Tunell (1953).

In those monoclinic amphiboles and pyroxenes in which the optic plane is parallel with (010), extinction on (010) is inclined, on (100) it is parallel, and on (001) it is symmetrical, bisecting the traces of the (110) cleavages.

TRICLINIC SYSTEM In triclinic crystals the indicatrix may have any orientation relative to the crystal axes; for there is no coincidence between any optical direction and any crystal axis. Thus to represent the optical orientation of triclinic species, the stereographic projection is necessary. The position of the indicatrix is, of course, fixed for a specific wavelength of light and for a particular crystal.

Triclinic minerals show inclined extinction on all vertical sections.

All three genetic types of twinning can be recognized optically in common rock-forming minerals: (1) glide (translation) twinning, calcite; (2) inversion twinning, andradite; (3) growth twinning, plagioclase. Twinning is not recognizable optically in isotropic species.

If twins are contiguous along the composition face, the twins are termed *contact* twins; usually the face is a plane. If the individuals appear to interpenetrate, they are called *penetration* twins, and the composition

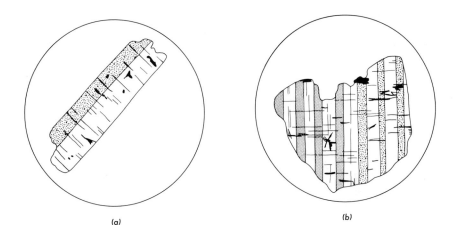

(a)   (b)

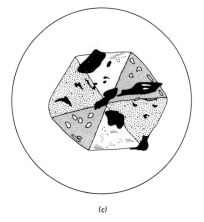

(c)

FIGURE 3.4 Types of twins in minerals. (a) Simple (paired). Carlsbad twins in sanidine phenocryst in trachyte. (b) Multilamellar (polysynthetic) and combined: Carlsbad $+$ albite. Bytownite in gabbro. (c) Cyclic: cordierite porphyroblast in hornfels.

surface is usually irregular. A crystal twinned once is a simple twin (Fig. 3.4*a*); repeated twinning (polysynthetic) may be multilamellar (Fig. 3.4*b*) or cyclic (Fig. 3.4*c*). Combinations of two or more types of twinning occur not uncommonly, as, for example, in the plagioclase feldspars: Carlsbad and albite; albite and pericline; Carlsbad, albite, and pericline.

Twinning is recognizable optically by (1) differences in interference tints and (2) differences in extinction position.

CHAPTER FOUR

MINERAL COMPOSITION AND OPTICAL PROPERTIES

ISOMORPHISM

Relatively few of the common rock-forming minerals are compounds of fixed or invariant composition. Most of them show compositional variations within certain, fixed limits. This chemical variation is due to the fact that structurally equivalent positions in the crystallographic framework may be occupied by ions of different elements. If an ion of one element can thus proxy for or substitute for ("replace") an ion of another element in a crystal structure, the two are said to be isomorphous. The term "isomorphism" was originally introduced to describe the phenomenon in which substances, analogous chemically, also displayed close crystallographic similarities. It may be usefully extended to cover the basis of this similarity, i.e., the abilities of ions or atoms of different elements to inhabit identical structural positions.[1]

In the past, complex variations in the composition of a mineral have commonly been described by considering the substance to represent a mixture of two or more ideal "molecules," or components termed "end members." Some such end members may actually exist as natural extremes

[1] Some mineralogists prefer the term "diadochous" for "isomorphous" (in this sense) and "diadochy" for "isomorphism." The writer makes no claim for this extended use of the term "isomorphism," except to point out it is a logical extension, since the property of substitution is based mainly on "equal form" (ionic radii) and is the basic cause of isomorphism (*sensu strictu*).

of an isomorphous series, but in many cases hypothetical "molecules" must be devised to chart the compositional variations of complex minerals in which three or more isomorphous substitutions are possible. The end-member concept is both useful and effective for expressing the variation in some simple series, but it is basically incorrect in that it imagines the existence of separate characteristic "molecules" as a part of the crystal structure. Such sub-units do not exist as independent entities within the continuous crystal framework. The substitution of an isomorphous ion in the position of another usually is random, but may be periodic (ordered structures).

The extent of the ability of one ion to proxy for another is governed by three factors:

1. The sizes (ionic radii) of the two ions
2. The temperature of crystallization of the mineral
3. The exact nature of the crystal structure

If the radii of two different ions are within about 15% of each other, iso-morphism is possible and normally is extensive at ordinary temperatures. At higher temperatures greater size differences can be tolerated. The type of structure also influences the extent of the isomorphous substitution; for an elemental pair that is completely isomorphous in one structure may have only limited isomorphism in a mineral of a different structure.

In crystals with ionic bonding no restrictions are placed upon the bonds, and thus ions with radii of similar size may substitute one for another. The resulting coordination (i.e., the number of atoms that directly sur-round or are directly attached to a central atom) can be predicated by the radius-ratio rule. In crystals that have covalent bonding this rule does not hold. In such the vectorial properties of the bonds govern both the grouping of atoms and the admission of possible substitute atoms, and thus the degree of isomorphous replacement.

Isomorphism between elements may be complete, as for Mg–Fe^{2+} in olivines, for example, or limited, as in the case of Ca–Mn^{2+} in some epidotes. The substituting ions need not have the same charge or valence, provided that the electrical neutrality of the compound is maintained by means of another substitution; for example:

1. Albite–anorthite
   $NaAlSi_3O_8$—$CaAl(Si_2,Al)O_8$
   Na^{1+}–Si^{4+} for Ca^{2+}–Al^{3+}
2. Anthophyllite
   $(Mg,Fe^{2+})_7Si_8O_{22}(OH)_2$—$(Mg,Fe^{2+})_5Al_2(Si_6Al_2)O_{22}(OH)_2$
   $2(Mg^{2+}$–$Si^{4+})$ for $4Al^{3+}$

Isomorphism of ionic pairs is called *coupled substitution.*
Radicals may substitute for single-element ions, as, for example, OH^{1-} for
O^{2-} or NH_4^{1+} for K^{1+}. In some minerals the number of possible isomorphous substitutions may be very large, so that the complete compositional variation of the species is relatively great, for example:

1. Apatite (fluorapatite), $Ca_{10}(PO_4)_6(F)_2$
   For Ca: Mn^{2+}, Ce, Sr, Na, K
   For PO_4: SiO_4, SO_4, AsO_4, CO_3OH
   For F: OH, Cl, S^{2-}
2. Muscovite, $K_2Al_4(Si_6Al_2)O_{20}(OH)_4$
   For K: Na, Ba, Rb
   For Al: Mg, Fe^{2+}, Fe^{3+}, Mn^{2+}, Cr, V, Li
   For OH: F

Some of the more important isomorphous pairs or groups of elements in the common rock-forming minerals are:

ISOMORPHOUS ELEMENTS	EXAMPLES
Al–Fe^{3+}	epidote group, montmorillonoids, prehnite
Al–Li	lepidolite
Al–Mg	clay minerals, micas
Al–Mn^{3+}	andalusite
Al–Si	micas, feldspars, pyroxenes, amphiboles, clay minerals, zeolites
Ba–Sr	barite, strontianite
Ca–Fe^{2+}	wollastonite
Ca–Mn^{2+}	calcite, apatite, rhodonite
Ca–Na	plagioclase, sodic pyroxenes, sodic amphiboles, zeolites
Ca–Sr	apatite
Ce–Ca	allanite
Ce–Th	monazite
Fe^{2+}–Mg–Mn^{2+}	olivines, pyroxenes, amphiboles, micas, chlorites
Fe^{2+}–Fe^{3+}	oxyhornblende, biotite, glauconite, stilpnomelane, greenalite
Fe^{3+}–Ti^{3+}–Mn^{3+}–Cr	micas, amphiboles
OH–F	micas, amphiboles, apatite, topaz, humite group
OH–O	oxyhornblende, allanite, sphene
K–Ba	sanidine
K–Na	muscovite, alkali feldspars
PO_4–AsO_4–VO_4	many phosphates and arsenates
Si–Be	vesuvianite
Si–Ti^{4+}	dumortierite
SiO_4–PO_4	zircon, apatite, allanite
Zr–Hf–Th	zircon

COMPOSITIONAL–OPTICAL VARIATION

The optical properties of a mineral, as other physical properties, are a function not just of the kinds and proportions of atoms present (i.e., the chemical composition) but also of the exact arrangement of these atoms with respect to each other (i.e., the crystal structure in detail). Nevertheless, the isomorphous substitution of specific ions for certain others brings about some consistent optical changes which are, in general, independent of the crystal structure in which the replacement occurs. This is particularly true of changes in refractive indices. In general, the refractive indices of a mineral will show an increase if ions of the substituting elements have:

1. An increased ionic radius
2. An increased positive charge
3. An increased atomic weight

Thus the indices of refraction of most minerals will increase with isomorphous substitutions of the following kinds: Fe^{2+} for Mg, Ca for Na, Na for K, Ba for K, Fe^{3+} for Al, Fe^{3+} for Fe^{2+}, and OH for F. In some cases these substitutions also bring about an increase in birefringence. Ions of other isomorphous pairs have specific refractive energies that are very similar, and consequently their interchange produces no marked variation in refractive indices: $Fe^{2+}-Mn^{2+}$ and Al–Si.

Other general optical changes also can be assigned to the presence of certain ions: for example, extreme birefringence due to the CO_3 ion in rhombohedral and orthorhombic carbonates of many metals; strong to extreme dispersion due to Ti in such species as sphene, brookite, anatase, rutile, and even titanian augite; the change in 2V [and in optical sign from $(+)$ to $(-)$], with Fe^{2+} for Mg, in olivines, enstatite–hypersthene, augite–aegirine, cummingtonite–grunerite and edenite–hornblende.

Color and composition also may be related closely, with changes in color in some species being sensitive to isomorphous substitutions on a very small scale. Elements capable of causing pronounced color variations are called chromophores. Some of the more common are listed on the facing page.

In minerals of complex composition, particularly some groups of silicates, the number of possible isomorphous replacements may be so large that it becomes difficult to correlate a specific optical effect with the substitution of any one ion. Indeed, some optical changes may be the result of combined substitutions, and one that might normally result from a single-ion replacement may be effectively neutralized by another substitution of opposing effect.

In some minerals another factor, particularly the thermal history, is of

ION	COLOR	MINERAL
Fe^{2+}	green	actinolite, stilpnomelane, hedenbergite, hastingsite, biotite, muscovite
Fe^{3+}	brown, red	oxyhornblende, biotite, stilpnomelane
	greenish yellow	epidote, nontronite
Mn^{2+}	pale rose	rhodochrosite, rhodonite
Mn^{3+}	light red, violet	lepidolite, tourmaline, piedmontite
Ti	deep brown	biotite
	purple	dumortierite, augite
Cr	bright green	augite, fuchsite, margarite, beryl (emerald)
	lavender, violet	clinochlore
Cu	blue, blue green	tourmaline
Ni	yellow green	chlorite

considerable importance in influencing optical properties. This factor is not entirely independent of isomorphism, for the extent of ionic substitution is governed in part by the temperature at which the mineral crystallizes, and the size of exsolved mineral units is a function of the rate at which the crystal cooled. The plagioclases and sanidine exemplify minerals whose optical properties vary with the temperature of formation.

The thermal history may govern optical properties in determining the degree of order-disorder relations between ion pairs such as Al–Si. Order-disorder relations represent one type of polymorphism. Usually optical properties such as refractive indices are but little affected by order-disorder, but the optic angle, crystal system, and extinction angle may vary (sanidine–microcline, indialite–cordierite).

Polytypism is common in many phyllosilicates (layered structures) such as the micas, in which different stacking arrangements can give rise to variations in 2V. In general, refractive indices are not affected. If, as in clay minerals, stacking variations occur between two distinctly different species to form a mixed-layer structure, the structural composite will have composite optics.

The determination of mineral composition by measurement of optical constants will, for most minerals, never be but a substitutive approximation for a good chemical analysis. Thus for mineral groups of very complex composition, the value of the use of a multitude of artificial end members in graphs showing composition vs. optical constants is illusory. For groups of more simple composition this device is satisfactory in many cases.

CHAPTER FIVE

IDENTIFICATION TECHNIQUE

To assist in the identification of minerals in immersion mounts, the following generalized outline of procedure may be employed:

1. Determine whether the mineral is isotropic or anisotropic by crossing the polars to see if the grains show birefringence. If they do not, they are isotropic; if they do, they are either uniaxial or biaxial. To be certain that the fragment is not a surface normal to an optic axis, an interference figure should be sought. If none is obtained, the mineral is isotropic. It is not sufficient merely to look for a dark cross in order to distinguish between isotropic and uniaxial minerals. The accessory plate should be inserted in all cases, for uniaxial minerals of very low birefringence (e.g., apatite and nepheline) will yield interference figures that are diffuse and difficult to detect. With the test plate, however, the opposing quadrants will show color contrasts despite the vague axial cross.

If there is any doubt whether a colorless or weakly colored grain of very low birefringence shows interference colors, the gypsum plate should be inserted and the stage rotated. Any slight departure of the color of the grain from that of the uniformly colored field is then accentuated.

2. If the mineral is truly isotropic, the index of refraction should be measured by successive immersions. In addition, record should be made of color, cleavage, fracture, parting, shape, inclusions, and alteration.

3. If the mineral is birefringent, search should be made for grains on which an interference figure can be obtained. From the figure it can be

determined whether the mineral is uniaxial or biaxial, and the optic sign, $(+)$ or $(-)$ can be noted. If the optic axis falls outside the field, it will be difficult to distinguish between a uniaxial figure or a Bxa figure with a small 2V. A few biaxial minerals may have axial angles so small that the figure appears to be sensibly uniaxial; and conversely, some normally uniaxial minerals may show a small anomalous 2V, owing to strain or twinning.

4. If the mineral is uniaxial positive, the lower index of refraction ω can be measured on any grain. If it is uniaxial negative, the higher index is ω and can be measured on any grain. Randomly oriented grains that show the highest interference colors for the least thickness can be measured for both ω and ϵ. Such grains will yield flash figures.

5. If elongate fragments or needles occur, it is well to note the character of the elongation, i.e., whether the fast or slow ray vibrates parallel with the length or which ray is parallel with any cleavage direction.

6. If the mineral is biaxial, determine the optic sign, $(+)$ or $(-)$, either on the Bxa or optic-axis figure. If the three indices have been measured, the sign generally may be estimated from their values. The mineral is positive if the value of β is closer to that of α and negative if it is closer to γ.

7. Measure 2V (actually 2E) by noting either the spread of the isogyres in the Bxa figure or the degree of curvature of the single isogyre in the optic-axis figure (Fig. 5.1). If the mineral has a large 2V, it is difficult or impossible to distinguish between a Bxa and a Bxo figure. However, if the sign is known either from an optic-axis figure or from the indices, the two types may be differentiated readily, for the Bxo figure yields the opposite sign of the true optic character. Dispersion, if moderate to strong, should be noted from either an optic-axis or Bxa figure. Weak dispersion is difficult to recognize.

8. The three indices of refraction should next be measured. Grains that show the maximum birefringence for their thickness are parallel with the optic plane and yield flash figures. On these α and γ may be measured. Grains from which a centered negative bisectrix figure (either Bxa or Bxo) is obtained can be used to measure β and γ. On one yielding a centered positive bisectrix figure, α and β can be measured.

Grains that are normal to an optic axis can be used for measurement of β. Such grains will show the relatively lowest birefringence in the aggregate of grains and may be nearly at extinction throughout a complete stage revolution.

The β index also may be determined by measuring n_2 and n_1 (larger and smaller indices, not equal to α or γ) on a number of grains and noting the

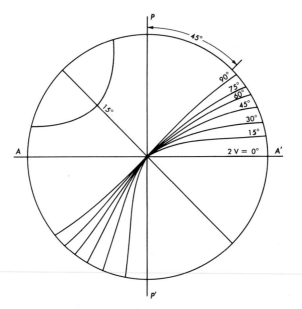

FIGURE 5.1 Variation in curvature of isogyre in optic-axis figure with variation in size of 2V. Mean refractive index is 1.60 (Wright, 1907).

distribution of values. In certain crushed fragments either the lower (n_1) or higher (n_2) index may equal β, but in the same fragment both n_2 and n_1 do not pass beyond β. Thus by measuring n_2 and n_1 in a number of grains, β can be found to lie at that value point at which the n_2 and n_1 ranges meet.

If all three indices are known, a quick inspection will not only reveal whether the mineral is positive or negative but will also give a general approximation of 2V. If the value of β lies nearly halfway between the values of α and γ, 2V will be large, approaching 90°; and as β approaches either α or γ, 2V decreases.

In biaxial minerals α or γ can be determined on those fragments which are so oriented that one of the bisectrices lies in the plane of the microscope stage.[1] Figure 5.2a to e shows the types of interference figures given by fragments with the obtuse bisectrix lying parallel with the plane of the microscope stage. In (a), normal to an acute bisectrix, the obtuse bisectrix is parallel to the plane of transmission of the polarizer. If we now think of the section as rotating to the right about the obtuse bisectrix as an axis of rotation, we successively obtain (b), (c), and (d). Further

[1] Adapted from Slawson and Peck (1936).

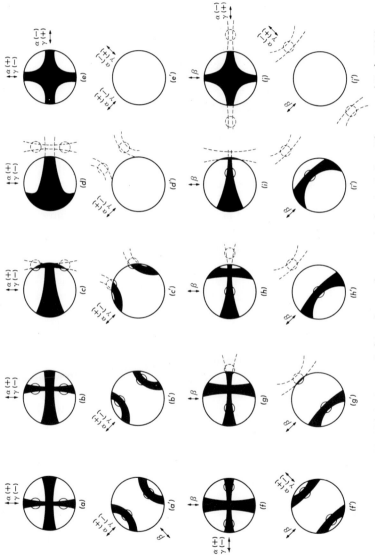

FIGURE 5.2 Symmetrical interference figures and refractive indices (Slawson and Peck, 1936).

rotation brings the section perpendicular to the optic normal as in (e). Rotation to the left would have given similar figures but in reverse position. During rotation, the position of the obtuse bisectrix has remained fixed and the interference figure is always symmetrical to one direction of the field. Consequently, any of these possible positions will permit measurement of α in a positive mineral and γ in a negative mineral. Figure 5.2a' to e' shows the corresponding figures in the 45° position. Here the two brushes are still symmetrical to one direction in the field, when they fall within the field. When the interference figure is perpendicular to the optic normal (Fig. 5.2e), the acute bisectrix also lies in the plane of the microscope stage, and therefore one extinction position will give α and the other γ.

The same argument may be developed for a similar series of interference figures starting with one normal to the obtuse bisectrix, in which case the acute bisectrix would be parallel with the plane of the stage and α would be determined from a negative mineral and γ from a positive mineral. In many cases when the optic angle approaches 90° it is impossible to distinguish the acute and the obtuse bisectrices. This situation is covered by the following generalized conclusion. *If the observed bisectrix is positive* (as determined by the use of a test plate), α *may be determined, and if the bisectrix is negative,* γ *may be determined.* This statement holds whether the mineral is $(+)$ or $(-)$.

In biaxial minerals β may be determined from those fragments so oriented that the optic normal lies in the plane of the microscope stage. Figure 5.2f to j shows the types of figures obtained from such fragments. In (f), perpendicular to the acute bisectrix, the optic normal is parallel to the plane of transmission of the polarizer. If we think of the fragment as rotating to the right about the optic normal as an axis of rotation, we successively obtain (g), (h), and (i). Further rotation brings the section perpendicular to the obtuse bisectrix as in (j). Rotation to the left would give similar figures but in reverse position. During rotation, the position of the optic normal remains fixed, and consequently any of these possible positions will permit β to be determined. When the interference figure is normal to the obtuse bisectrix (Fig. 5.2j), the acute bisectrix also will lie in the plane of the microscope stage, and on rotating the stage through 90° it will be parallel to the plane of transmission of the polarizer. Consequently, in that position γ may be determined in positive and α in negative minerals.

It should be noted that fragments showing interference figures perpendicular to either an optic axis, an acute or obtuse bisectrix, or the optic normal are not the only ones available for determination of the principal indices of refraction. Actually, if any one of the three principal optical sections is vertical, one of the principal indices may be determined in the direction normal to that principal section by rotating the stage so that the normal to the principal section is parallel to the plane of transmission of the polarizer (the principal section being in the E–W position).

Thus the following general rules may be laid down:

a. *From any fragment giving an interference figure possessing one plane of symmetry with reference to the field of vision, one of the principal indices may be determined* (Fig. 5.2b to d and g to i). With the plane of symmetry lying in the E–W position, the index of refraction is determined in the N–S position.

b. *From any fragment giving an interference figure possessing two planes of symmetry with respect to the field of vision, two of the principal indices may be determined* (Fig. 5.2a, e, f, and j). These correspond to the two extinction positions.

Strictly speaking, this holds only for monochromatic light. The moderate dispersion observed in some monoclinic and triclinic minerals offers no serious difficulty to such a generalized statement.

9. In order to describe the optical orientation, some crystallographic direction must be recognizable—by means of cleavage, parting, crystal outline, elongation, crystallographically oriented zoning, or oriented inclusions—to serve as a basis to which the position of optical directions may be referred. A single perfect cleavage commonly may be assumed to be (001); elongations are commonly parallel with c.

In monoclinic minerals one of the three optical directions corresponds with the b crystallographic axis. Not uncommonly $\beta = b$. Minerals with such an orientation will have parallel extinction on (100) and the maximum inclined extinction on (010). This inclined extinction is recorded in terms of an angle, for example, $\gamma \wedge c = 10°$. The value is specific or within a restricted range for a mineral species.

With the above orientation, on (110) or other prism faces, the extinction angle lies between $0°$ and the maximum. Thus in practice the measurement of extinction angles may be done on a statistical basis, and the largest angle of a dozen or more randomly oriented grains approaches the maximum extinction angle of the mineral. If $\beta = b$ as above, the procedure is somewhat facilitated by choosing for measurement sections with the maximum birefringence in proportion to their thickness; for these will more nearly lie parallel with the optic plane.

If the extinction angle is small, it is still possible to speak of the character of the elongation as either length-fast or length-slow. If the extinction angle becomes larger, these terms lose their significance. In some cases a mineral may be elongate parallel with β, and elongate fragments may be either length-slow or length-fast, with β being compared in one case with α' and in another case with γ'.

In monoclinic minerals principal sections show parallel, inclined, or symmetrical extinction. Oblique sections show inclined extinction. In triclinic minerals there is no correspondence between any optical and any crystallographic direction. Each mineral has its own particular orientation, which may be very difficult to determine and usually requires the universal stage. Principal sections of triclinic minerals show inclined extinction.

10. If a mineral is colored and shows pleochroism, the various colors should be referred to the optical directions and an absorption formula established.

Only in rare cases is it necessary to determine all of the optical properties of a mineral in order to identify it. The student should be able to measure all the constants, but in practice this is rarely done. With experience the eye records many optical characteristics automatically, and the more common mineral species come to be recognized on sight or can be distinguished by measuring only one or two particular optical properties.

PART TWO

DESCRIPTIONS OF MINERALS

CHAPTER SIX

INTRODUCTION

Included in this section are descriptions of the mineral species that are of importance as essential or accessory constituents of rocks. In addition, other minerals which are of lesser significance are referred to briefly under species to which they are related. Of the minerals described, some have not generally been included heretofore in similar compilations but have of late assumed importance in petrographic studies.

A general description of the mineral with emphasis on its optical characteristics is given, and specific information is listed on its appearance in thin sections, in crushed fragments, and, if it appears in heavy or light residues, also in detrital grains. The following subheadings have been generally employed for optically complex species:

General. Information relating to a mineral group.

Composition. Formula, isomorphism.

Indices. Indices, variation with composition, indices of cleavage flakes, birefringence, interference colors in thin section.

Color. Color, pleochroism, zoning (refers to mineral in thin section or in crushed pieces, unless otherwise specified).

Form. Crystal system, shape in thin section, cleavage, parting, fracture, shape in crushed and detrital pieces, inclusions, aggregate forms.

Orientation. Position of optical directions with respect to crystallographic directions, 2V, optical character, dispersion, extinction angles, character of elongation, twinning, optical orientation of cleavage particles. Variation of these properties with composition.

Occurrence. Common rock types, associated minerals, alteration.

Diagnostic features. Characteristics useful in distinguishing the mineral from other species it resembles.

However, for opaque and isotropic minerals only the headings Properties, Occurrences, and Diagnostic features are used.

The minerals are presented in the order in which they are most easily and systematically studied, i.e., in increasing complexity of their optical properties. Thus they are arranged in the following groups:

Opaque
Isotropic
Uniaxial
Biaxial

Most natural chemical groups, such as the garnets, feldspars, and zeolites, are not disrupted by this arrangement. As in all classifications, however, a few anomalies appear. For example, magnetite is listed more conveniently under the spinel group rather than separately with the other opaque minerals. The SiO_2 group also is split by an optical classification, quartz appearing under the uniaxials and opal and silica glass under the isotropics. Similarly, members of the feldspathoid group are to be found in several optical categories. Within each optical group the arrangement is by chemical group, following, in general, the new Dana classification. Quartz is regarded as a silicate. Within the biaxial silicates the sequence is (with minor exceptions):

Nesosilicates	single SiO_4 tetrahedra
Sorosilicates	groups of SiO_4 tetrahedra
Inosilicates	chains of SiO_4 tetrahedra
Cyclosilicates	rings of SiO_4 tetrahedra
Phyllosilicates	sheets of SiO_4 tetrahedra
Tectosilicates	three-dimensional frameworks of SiO_4 tetrahedra

A number of substances that do not qualify as minerals under rigid definitions are described: limonite, opal, lechatlierite, chalcedony, iddingsite.

CHAPTER SEVEN

OPAQUE AND ISOTROPIC MINERALS

OPAQUE MINERALS

GRAPHITE

Properties. Composition, carbon (C); hexagonal; black in reflected light; opaque; perfect basal (001) cleavage; brilliant metallic luster from coarse cleavage surfaces. Rounded and platy crystals; tabular cross sections (Fig. 7.1). In thin section edges of coarse flakes appear crinkled or minutely scalloped. In reflected light the soft flakes also may show pronounced, randomly oriented grooves formed by grinding. Thin sections with abundant graphite commonly are characterized by clouds of fine graphitic dust embedded marginally in the mounting medium.

Occurrence. Principally in slates, schists, gneisses, and marble. In schists biotite and quartz are common associates; in marbles calcite, dolomite, phlogopite, and diopside may be found with it. Crystallized during both regional and contact metamorphism. Also present as an uncommon accessory in felsic igneous rocks, particularly pegmatites. Some shales, argillites, sandstones, and limestones contain abundant, minute, opaque particles, shreds, or pellets usually referred to as "carbonaceous material." These organically derived substances are not graphitic.

Diagnostic features. Tabular shape, grooves, crinkly margins, and peripheral dust usually differentiate it from magnetite. Ilmenite also is tabular but commonly is altered to white leucoxene. Graphite greatly resembles molybdenite, which is rare, however, except in mineralized rocks.

FIGURE 7.1   Graphite in graphite schist, Kate Creek, Beaverhead County,
Montana. Polars not crossed, ×28.

PYRITE

Properties. FeS_2; isometric; opaque; pale brass yellow in reflected light. Euhedral crystals common, especially cubes, with rectangular, square, or triangular sections. Anhedral grains, veinlets, and oolites also occur.

Occurrence. In all types of rocks as a common product of hydrothermal mineralization, in or near ore deposits. In some gabbros and peridotites it forms late magmatic anhedra, and in shales it may also be syngenetic. Alters to limonite, goethite, or melanterite. Pseudomorphs of limonite are common.

Diagnostic features. If anhedral, indistinguishable from marcasite in thin section. Chalcopyrite has a deeper yellow color, but the color difference in oblique reflected light in thin section is hard to detect.

MARCASITE

Properties. FeS_2; orthorhombic; opaque; color like pyrite but slightly paler. Concretionary with radial-fibrous structure or anhedral.

Occurrence. In metalliferous veins and their wall rocks (anhedral grains) and in clays associated with coal (concretionary structures). Alters readily to melanterite with by-product sulfuric acid.

Diagnostic features. Difficult to tell from pyrite except in the polished section. FeS_2 in argillaceous rocks not associated with coal is more likely to be pyrite.

PYRRHOTITE

Properties. $Fe_{1-x}S$, with $0.2 > x > 0$; hexagonal; opaque; bronze color in reflected light. Anhedral grains are typical; very uncommon euhedral to subhedral crystals are tabular in cross section, platy parallel with (001). The platy structure may be emphasized by a basal parting.

Occurrence. A high-temperature mineral of ore deposits and enclosing mineralized rocks. An accessory mineral in gabbros, feldspathoidal gabbros, peridotites, and more rarely in marbles of contact metamorphic origin, some schists and amphibolites. Also in stony meteorites, but iron meteorites contain troilite (FeS). Alters to limonite.

Diagnostic features. Bronze color in reflected light separates it from pyrite, marcasite, and chalcopyrite. Troilite is restricted to iron meteorites. Pentlandite, $(Fe,Ni)_9S_8$, is an uncommon mineral found with pyrrhotite in some ore deposits and resembles it closely. A test for nickel or a polished-section determination generally is advisable, if the presence of pentlandite is suspected.

ILMENITE

FIGURE 7.2 Titanian magnetite with exsolved ilmenite that has been altered to leucoxene.

Properties. $FeTiO_3$; hexagonal-rhombohedral; opaque; black like magnetite in reflected light unless altered to leucoxene, then gray white (Fig. 7.2). Elongate cross sections of tabular crystals are common in thin section; skeletal crystals or anhedral grains also appear. Detrital grains are subrounded to irregular.

Occurrences. In ore deposits of close magmatic affiliation, not uncommonly in regular or semiregular intergrowths with magnetite (Fig. 7.2). A widespread accessory mineral of intermediate and especially of mafic and ultramafic igneous rocks; less abundant in felsic rocks. Commonly occurs with magnetite. A main component of black sands and of some heavy detrital suites. If unaltered, it may resemble magnetite closely, and a distinction in thin section is difficult. As detritals, however, the two are easily separated because of the much greater magnetic susceptibility of magnetite.

Leucoxene is an opaque, generally amorphous, chemically variable titanium oxide, usually with small amounts of adsorbed water. It is gray white to gray in reflected light and forms as an alteration of ilmenite and other Ti minerals, such as sphene, rutile, anatase, brookite, and perovskite. It is a useful designation for the alteration products of these minerals, but it does not represent a single mineral species and should be considered analogous to the terms limonite, wad, and bauxite. In addition to forming coatings on ilmenite, leucoxene is also a constituent of high alumina clays, bauxite, and some beach sands. X-ray patterns of leucoxene generally give anatase or rutile lines.

(MAGNETITE)

(see under Spinel Group, pp. 53–55)

ISOTROPIC MINERALS

PERICLASE

Properties. MgO; isometric; colorless; $n = 1.736-1.745$; Fe^{2+} may replace Mg (up to 8.5% FeO); indices increase with increasing Fe^{2+}. High relief in thin section; perfect cubic cleavage. Forms euhedral cubes or octahedra and anhedral granular masses.

Occurrence. Principally as an uncommon high-temperature mineral in dedolomitized marble (pencatite) of contact metamorphic origin. Usually partly or almost entirely altered to brucite (Fig. 7.3) and in some instances to hydromagnesite. Appears typically as cubic remnants in central parts of brucite aggregates. Other associated minerals are calcite, forsterite, chondrodite, serpentine, and spinel. Synthetic periclase, used for electrical insulations, is completely isomorphous with FeO. Periclase also occurs in magnesite bricks and most forsterite refractories.

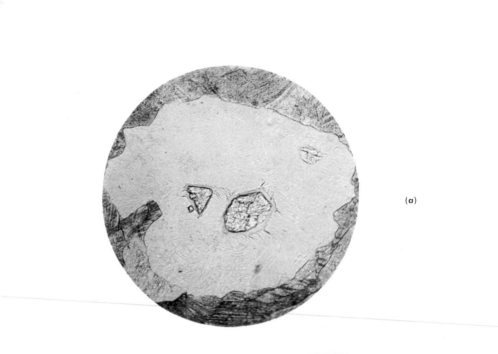

(a)

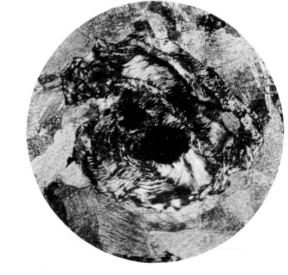

(b)

FIGURE 7.3 Periclase remnant in whorl of brucite, in marble, Organ Mountains, New Mexico, ×70. (a) Polars not crossed; (b) polars crossed (Hunt and Faust, 1937).

Diagnostic features. The combination of high index, isotropism, excellent cubic cleavage, and association are distinctive and serve to separate it even from garnet and associated spinel.

SPINEL GROUP

FIGURE 7.4   Pleonaste in serpentinized peridotite, Winchester, California. Polars not crossed, ×55.

General. The spinel group consist of three series, the Al, Fe^{3+}, and Cr series, each of which has from two to five members. In all, some nine species and many intermediate varieties are recognized, but many are very rare. The more common members are:

MEMBER	FORMULA	COLOR	n
spinel	$MgAl_2O_4$	colorless	1.72–1.74
pleonaste	$(Mg,Fe)Al_2O_4$	green to blue green	1.75–1.79
hercynite	$(Fe,Mg)Al_2O_4$	dark green	1.78–1.80
picotite	$(Fe,Mg)(Al,Cr)_2O_4$	olive brown to brown	2.00
chromite	$FeCr_2O_4$	opaque to deep brown on thin edges and splinters	
magnetite	$Fe^{2+}Fe^{3+}_2O_4$	opaque, metallic black in reflected light	

FIGURE 7.5 Skeletal magnetite crystals in Precambrian diabase, Guffey, Colorado (J. E. Bever). Polars not crossed, ×40.

Degree of transparency and indices vary with composition; in general, the index rises with increasing amounts of Fe^{2+} and Fe^{3+}. Nearly all spinels are isotropic, although some blue-green zinc spinels (gahnite) with anomalous birefringence are known.

Form. A poor octahedral parting is present rarely. In thin section spinels appear euhedral to subhedral, commonly in cubes (Fig. 7.4) or octahedra with characteristic rhomboid cross sections. Detrital grains are sub-rounded, dark-colored octahedra, unevenly pockmarked and with a distinctive conchoidal fracture, or rounded particles. Some magnetite forms skeletal crystals (Fig. 7.5).

Occurrence. Spinel and pleonaste are contact-metamorphic minerals in dolomitic marbles; associates are phlogopite, corundum, forsterite, and chondrodite. They also are found in some aluminous schists and in aluminous xenoliths in igneous rocks, with cordierite, andalusite, biotite, and quartz. Picotite and chromite are confined largely to various types of peridotites and serpentinites as accessory grains and crystals; pleonaste is less common in these rocks. Some gabbros contain accessory anhedral pleonaste, some of which may form a fine-grained, vermicular intergrowth with pale hornblende in a corona around olivine or pyroxene. Uncommonly, pleonaste forms phenocrysts in foidal olivine basalts. Chromite occurs in meteorites and in high-temperature ore deposits of close magmatic affiliation. Aluminous spinels occur in firebricks.

Magnetite, the most common and abundant member of the group, is a widespread accessory in almost all metamorphic and igneous rocks and is also common in many types of ore deposits, particularly those of magmatic (with ilmenite), high-temperature–hydrothermal, pyrometasomatic, and metamorphic origin. In igneous rocks both primary and secondary magnetite occur. Secondary ("dusty") magnetite, which forms minute granules arranged typically in clusters or along fractures and cleavages, results from the alteration of primary iron-bearing minerals, such as the hydration of olivine to serpentine and magnetite. The alteration of biotite to chlorite or the replacement of biotite by muscovite liberates magnetite, and the corrosion of biotite or hornblende phenocrysts in volcanic rocks gives rise to a pyroxene and magnetite. Because magnetite is also a very common heavy detrital mineral (black sands), it usually must be removed before it is practical to study the other constituents. Magnetite alters to hematite or limonite. It is present in most iron oxide slags.

Diagnostic features. In crushed pieces some spinels and garnets are difficult to distinguish from each other. This is also true in heavy residues where red spinel and almandite may be confused; the octahedral shape and lack of inclusions will aid in recognizing spinel. In thin section the outlines may be characteristic. A green to olive-brown, isotropic grain of high relief in an ultramafic rock is generally a spinel, for the usual garnet found in such an environment is pale pink (pyrope). In thin section color can be used in a general way to determine the type of spinel present, and a more accurate determination is possible by measuring the refractive index. A good approximation of the composition may be obtained by measuring the refractive index, the density, and the edge of the unit cell. Chromite resembles magnetie, but the latter is completely opaque. Ilmenite differs from magnetite in its tabular or skeletal outline and its leucoxene coating. Graphite also is tabular. Under high magnification and intense illumination detrital grains of magnetite may show faceting on a very small scale; otherwise, confusion with unaltered ilmenite grains is possible.

LIMONITE

Properties. Optically amorphous $FeO(OH) \cdot nH_2O$ or $HFeO_2 \cdot nH_2O$. Not a true mineral. Water content indefinite; very impure with much admixture of clay minerals, fine-grained quartz, manganese oxides, and other secondary minerals. $n = 2.0\text{-}2.4$. Upon standing in index liquids the index increases. May show very low anomalous birefringence. Shades of red and brown, both in transmitted and reflected light. Usually without shape, commonly interstitial as a cement around other mineral grains. Oolites, coatings, crusts, and pseudomorphs after pyrite, marcasite, siderite, magnetite, pyrrhotite, and many other species also are typical forms. Detrital particles are irregular, rusty, and generally opaque.

Occurrence. A secondary substance formed by weathering, as a precipitate in bogs, lakes, and springs. Appears with goethite in gossans and is an important constituent of residual iron ores. A cement in sandstones. Hematite and manganese oxides are other associates.

Diagnostic features. Difficult to tell from some earthy fine-grained hematite. X-rays have shown that much limonite actually is cryptocrystalline goethite or lepidocrocite. More coarsely crystalline goethite and lepidocrocite are anisotropic and may form by the crystallization of limonite.

OPAL

FIGURE 7.6 Relation between maximum index of refraction and water content in opal. Commonly the index will be less than the maximum possible for a given weight percentage of water.

Properties. Approaches $SiO_2 \cdot nH_2O$; optically amorphous and isotropic; some varieties show low (strain) birefringence; $n = 1.406$–1.460, decreasing with increasing amounts of water (Fig. 7.6); much opal has an index between 1.44 and 1.45 and shows high negative relief in section. Usually colorless in thin section or in crushed pieces, but gray and pale brown varieties also are found; various colored inclusions may be present. Irregular cracks are to be seen in thin section; a conchoidal fracture appears on broken grains. Veinlets, irregular interstitial masses, and vesicle fillings are characteristic forms; the cellular structure of replaced wood may be preserved.

X-ray studies demonstrate that opal consists of highly disordered cristobalite. The disordering stems from the entry of Ca, Mg, Al, and alkalies into the structure; thus opal–cristobalite is a stuffed derivative of silica.

Occurrence. As a secondary mineral in all types of volcanic and intrusive rocks it fills both primary and secondary openings, and veins and replaces rock silicates. Chalcedony is a common associate. In sedimentary rocks it forms concretions and "petrified wood," replaces fossil shells, is the main

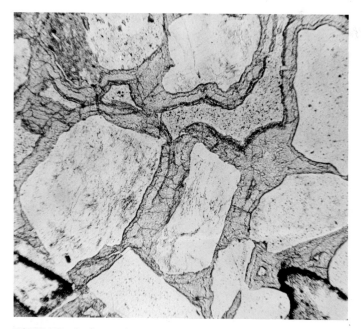

FIGURE 7.7 Opal cementing quartz grains in Tertiary sandstone. Blacktail Range, southwestern Montana. Polars not crossed, ×28.

substance of diatomite and geyserite, and is a cement in some sandstones and arkoses (Fig. 7.7).

Diagnostic features. High negative relief and manner of occurrence. Lechatelierite, with similar optical properties, has a different occurrence. Rhyolitic glass has a higher minimum index (1.480), as have analcite (1.488) and members of the sodalite group (1.483). Some opals (especially the variety hyalite) fluoresce strongly whitish green and are very weakly radioactive, owing to traces of U. Opal bears the same relationship to cristobalite as chalcedony bears to quartz.

LECHATELIERITE

Properties. SiO_2 glass; amorphous; isotropic; $n = 1.457$–1.462. Contains innumerable small vesicles which may be oriented in layers. Index increases with Al, Fe, Mg, Ca, and alkalies which occur as common impurities. Colorless to turbid.

Occurrence. In fulgurites, formed by the action of lightning on sand; in sandstone xenoliths in volcanic rocks; and in crusts adhering to volcanic bombs that fell into sand. Also in impactites, partly melted rocks resulting from meteorite falls. Manufactured for lenses, prisms, and windows in apparatus where the transmission of shortwave ultraviolet radiant energy or an extremely low coefficient of expansion is desired.

Diagnostic features. Opal is similar but occurs in a different geological environment. Other natural glasses have higher indices.

FLUORITE

Properties. Isometric; CaF_2, some Y, Ce, and other rare earths substitute for Ca; $n = 1.434$, usually constant, but increasing with rare-earth content to as high as $n = 1.457$; invariably isotropic. In thin section has high negative relief. Generally colorless but may have nonuniformly distributed patches, spots, or pleochroic haloes or may be zoned in purple or blue. Fine-grained purple fluorite commonly is radioactive, owing to submicroscopically disseminated uraninite. Perfect octahedral cleavage (111). Cleavage particles tend to be triangular in outline. In thin section cleavage traces occur usually as two sets of lines intersecting at 70 and 110° or as three sets crossing at 60 and 120°. In rocks commonly anhedral and interstitial. In detrital residues it appears as triangular plates or as irregular or slightly rounded cleavage pieces.

Occurrence. A persistent mineral that forms under a wide variety of geological conditions: in hot springs deposits, epithermal and mesothermal veins and pipes, greisens, carbonatites, pegmatites, and granites as a late accessory mineral. It is also a constituent of some phosphorites, limestones, and sandstones (as a detrital and rarely as a cement).

Diagnostic features. Low index, shape of cleavage pieces, isotropism, color spots.

HALITE

Properties. NaCl; isometric; almost always istropic; $n = 1.544$, thus low relief in Canada balsam. Colorless but may contain reddish inclusions of iron oxide, also liquid inclusions with bubbles as well as grains of anhydrite and gypsum. Perfect cubic cleavage; crushed fragments are square or rectangular in outline. Euhedral cubes occur in shales, but the texture of rock salt is anhedral. Soluble in water.

Occurrence. As beds of massive rock salt; associated with other evaporate minerals such as sylvite, carnallite, polyhalite, anhydrite, gypsum, and dolomite. As scattered euhedra in some shales. As massive vertical plugs (salt domes) and as the chief precipitate from salt lake brines.

Diagnostic features. Cubic cleavage, isotropism, and index (low relief) are sufficient to identify it. Sylvite (KCl), which is similar in cleavage, solubility, and color, has $n = 1.490$ and therefore shows negative relief in section.

GARNET GROUP

General. The garnet group can be described by means of five common end-member molecules: spessartite ($Mn_3Al_2Si_3O_{12}$), almandite ($Fe_3Al_2Si_3O_{12}$), pyrope ($Mg_3Al_2Si_3O_{12}$), grossularite ($Ca_3Al_2Si_3O_{12}$), and andradite ($Ca_3Fe_2Si_3O_{12}$). The first three are related in that they all contain aluminum; the last two by their content of calcium. Other less common members are uvarovite (Ca–Cr garnet), melanite and schorlomite (Ti-andradite), and rhodolite (intermediate between pyrope and almandite). Spessartite may contain yttrium, with Y^{3+}–Al^{3+} substituting for Mn^{2+}–Si^{4+}. Small amounts of PO_4 occur in some garnets for SiO_4; others contain Zr (kimzeyite). Analyses of many garnets show that their compositions can be expressed adequately for petrographic purposes by three of the end-member molecules. With determinations of both the index and the specific gravity of the garnet the chemical composition may be estimated (Fig. 7.10).

Although garnets are isometric and generally isotropic, some varieties may show weak birefringence. Spessartite may be distinctly anisotropic;

INDICES OF REFRACTION	RANGE OF NATURAL SPECIES	SYNTHETIC END MEMBER (SKINNER, 1956)
spessartite	1.790–1.810	1.800
almandite	1.770–1.820	1.830
pyrope	1.720–1.770	1.714
grossularite	1.735–1.770	1.734
andradite	1.850–1.890	1.887

grossularite not uncommonly is birefringent in low first-order grays and is commonly twinned in "pie-cut" sectors (Fig. 7.8). Garnets of this type are nonisometric above 800°C and may invert near that temperature to an isotropic form or else remain birefringent with inversion twinning.

Color. In thin section spessartite is colorless, in grains salmon pink to pinkish yellow; almandite is colorless to pink in thin section and red in grains; pyrope grains are red to deep red, in thin section pink. Grossularite in thin section is colorless, in grains yellow to brown; andradite is light brown in sections, in grains brown to dark brown. Garnet that is dark brown in thin section is usually melanite. The darker garnets, particularly melanite and grossularite, may show zoning with alternating light and dark bands (Fig. 7.8).

Form. Cleavage is lacking, and crushed pieces are irregular, bounded by

FIGURE 7.8 Zoned and twinned grossularite (Hauswaldt, 1904). Polars crossed, ×20.

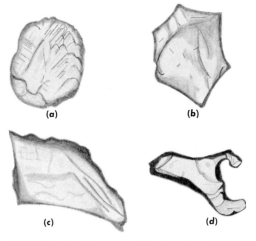

(a)

(b)

(c)

(d)

FIGURE 7.9   Detrital garnet. (a) Lake shore, North Fairhaven, New York. (b) Etched, lake sand, Lake Michigan, Frankfort, Michigan. (c) Lake sand, Lake Erie, Cedar Point, Ohio. (d) Strongly etched, Rimu Dredge, New Zealand (Hutton, 1950).

curved, semiconchoidal fracture surfaces. Detrital grains are angular to rounded (Fig. 7.9). Because of (110) parting, almandite may occur as platy grains. Many almandite grains show strong pitting and grooving and are etched (Fig. 7.9). In thin section garnets appear commonly as euhedra of dodecahedral habit with six-sided cross sections or of trapezohedral habit in eight-sided sections. Masses of irregular rounded grains also occur. Inclusions are common and may be zonally arranged, usually in cores. Among the common included minerals are quartz, feldspar, magnetite, muscovite, biotite, sillimanite, zircon, apatite, rutile, and graphite.

Occurrence. The composition of a garnet varies with the rock in which it is developed. Garnets chiefly of spessartite and varied amounts of almandite are found in pegmatites, a few granites, and metamorphically low-grade schists. High-manganese garnets decompose readily to manganese oxide, which stains surrounding minerals black. They also may be veined and replaced by chlorite or sericite. Garnets from biotite schists and gneisses are largely almandite. Those from amphibole-bearing meta-

FIGURE 7.10   Variation of refractive indices and densities with composition in garnets (Fleischer and Kennedy, 1947). (a) Grossularite, pyrope, almandite. (b) Almandite, grossularite, andradite. (c) Pyrope, grossularite, andradite. (d) Pyrope, spessartite, almandite. (e) Spessartite, grossularite, andradite. (f) Spessartite, andradite, almandite. (g) Grossularite, spessartite, almandite. (h) Andradite, pyrope, almandite.

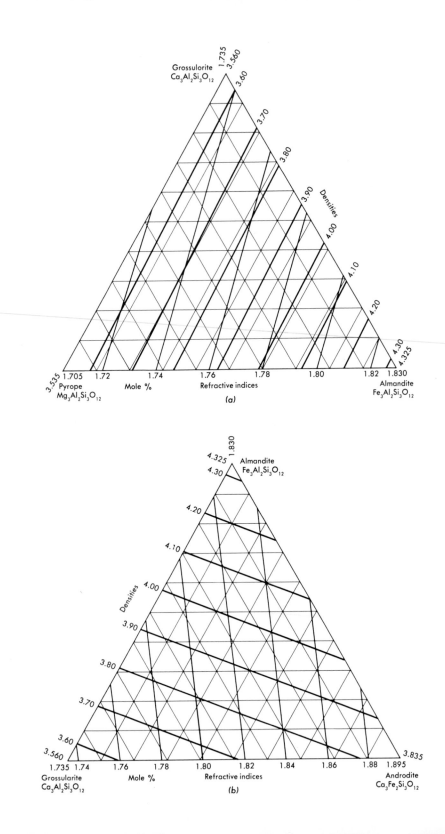

(a)

(b)

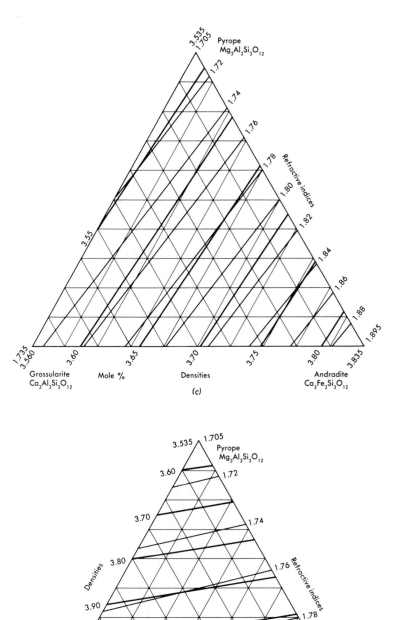

(c)

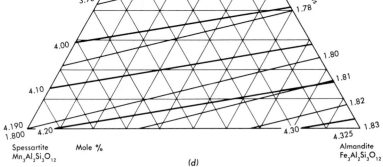

(d)

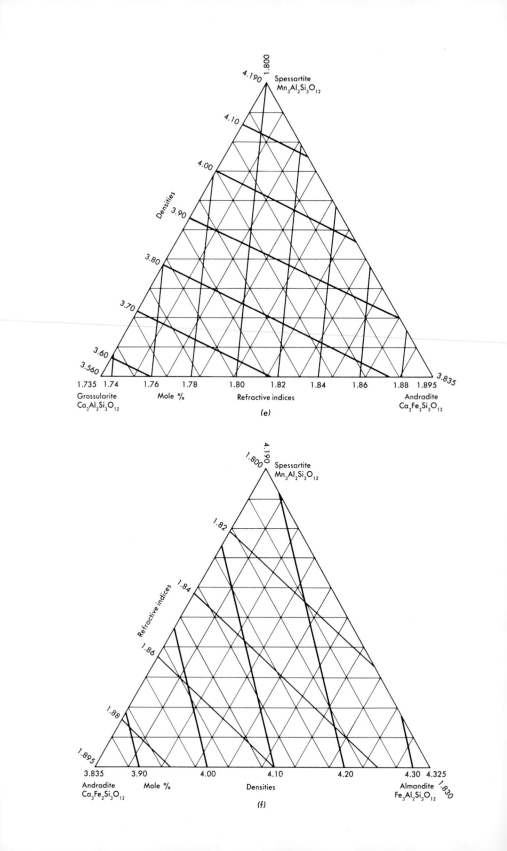

(e)

(f)

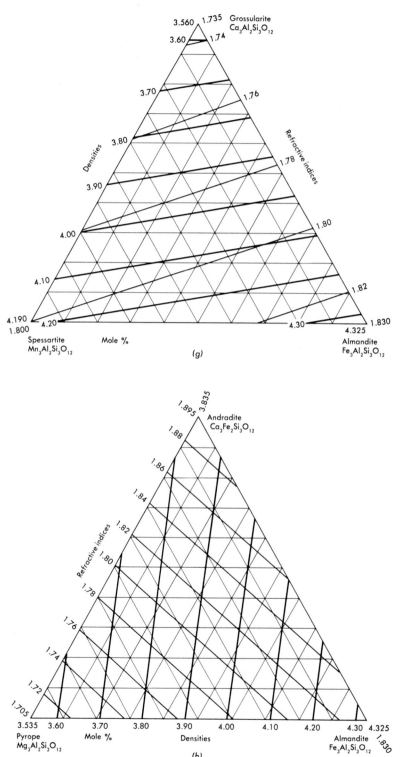

(g)

(h)

morphic rocks fall in the almandite–pyrope range and are similar to those from mafic igneous rocks. In peridotites the garnet is pyrope, but garnets of eclogites fall between pyrope and almandite. The irregular pyrope grain may be enveloped by a kelyphite border consisting of fibrous green amphibole that may be intergrown with feldspar. In contact-metamorphic marbles (tactites) garnet is of the andradite–grossularite group, and may be altered to calcite, epidote, or chlorite. Melanite is characteristic of feldspathoidal syenites and their extrusive equivalents. Uvarovite, which is rare, occurs associated with chromite in altered peridotites.

Diagnostic features. Crushed garnet is difficult to distinguish from spinel. In general, however, garnet indices are higher than those of translucent spinels, and the green tints common in spinel are rare in garnet. In thin section form and color (and the birefringence of some garnets) suffice to separate the two. Garnet is not difficult to identify as such, and to distinguish the various species a knowledge of the parent rock type is very helpful. Within the group a combination of refractive index, density, and cell edge gives a good approximation of the composition.

SODALITE GROUP

General. The sodalite group, which is in the feldspathoid family, contains the species sodalite, hackmanite, nosean, haüyne, and lazurite: silicates of sodium and aluminum with chlorine, sulfur, or sulfate. As feldspathoids they commonly occur together and also with other members of the family, but only in subsilicic igneous rocks in which primary quartz is absent.

Sodalite. $Na_8Al_6Si_6O_{24} \cdot Cl_2$. Hackmanite has the same general composition, with some sulfur replacing chlorine. $n = 1.483–1.487$, increasing with increasing S and with Ca, which may also be present. The colorless or pale blue varieties usually have lower indices than the pink types. Isometric and isotropic; may be slightly birefringent around inclusions. Dodecahedral crystals with hexagonal cross sections are common, also anhedral grains and granular aggregates. Colorless to gray in thin section. Poor (110) cleavage rarely seen.

Nosean. $Na_8Al_6Si_6O_{24} \cdot SO_4$, some Ca may be present. $n = 1.485–1.495$; isometric and isotropic; colorless to neutral gray, locally darker owing to inclusions, rarely blue. Commonly in dodecahedral crystals as phenocrysts (Fig. 7.11) with hexagonal cross sections. Cleavage (110) poor or absent. Minute inclusions are very abundant: glass, gas, or microlites of magnetite, hematite, or ilmenite. They are concentrated typically in layers: in cores, in outer zones, or less commonly in concentric zones alternating with inclusion-free layers. Under high magnification the dark inclusions

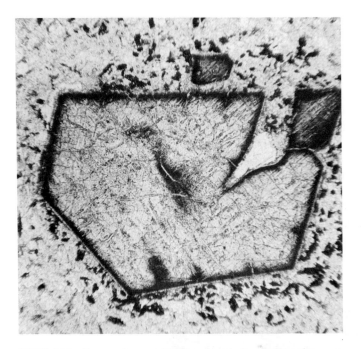

FIGURE 7.11   Nosean phenocryst, embayed, with rim of post-embayment exsolution (?) inclusions, in leucite phonolite, Schildkopf, Eifel Mountains, Rhineland, Germany. Polars not crossed, ×28.

are seen arranged as rows or streaks along several directions (the three-fold crystal axes) and intersect at regular angles. Rarely they parallel (110) cleavages.

Haüyne. $(Na,Ca)_{6-8}Al_6Si_6O_{24} \cdot (SO_4)_{1-2}$; grades into lazurite, $Na_{8-10}Al_6$ $Si_6O_{24} \cdot S_{2\pm}$, by a decrease in Ca and a substitution of S for SO_4. Isometric, isotropic; $n = 1.496$–1.510, increasing with Ca. Color is usually blue, but lilac blue and blue green also occur; may be darker because of abundant inclusions, which are zonally arranged as in nosean. The blue color also may be zonally or irregularly distributed. Occurs commonly as euhedral dodecahedral phenocrysts that show a hexagonal outline.

Occurrence. Sodalite, nosean, and haüyne occur together, and other feld-spathoids (nepheline, leucite) also may be present. They appear typically as phenocrysts with a well-developed outline, locally embayed (Fig. 7.11). Sodalite may also be anhedral. They occur principally in phonolites and related extrusive rocks, in feldspathoidal syenites, and as accessory miner-als in alkali syenites, trachytes, tephrites, and feldspathoidal basalts.

Lazurite is typically of contact-metamorphic origin, forming in limestones as lapis lazuli, a mixture of lazurite, diopside, muscovite, calcite, pyrite, and amphibole. The common alteration products of these minerals are radial natrolite together with minor amounts of other zeolites, diaspore, and gibbsite. The dark inclusions change to limonite, staining the grains red to brown.

Diagnostic features. It is not always possible to distinguish among members of this group by optical properties alone. Sodalite generally is colorless and inclusion-free; nosean is colorless and inclusion-zoned; and haüyne is blue, with or without the inclusions. Sodalite and some nosean have a lower index than haüyne. In some instances, in order to be certain of the species, chemical tests can be made on the powdered mineral on a glass slide. A test for sulfate with HCl and $BaCl_2$ separates nosean and haüyne from sodalite. Haüyne treated with HNO_3 yields white $CaSO_4 \cdot 2H_2O$. Sodalite may be confused with analcite.

LEUCITE

FIGURE 7.12 Multiple lamellar twinning in leucite, Highwood Mountains, Montana. Polars crossed, ×30 (Larsen et al., 1941).

Composition. $KAl(SiO_3)_2$, may have minor Na for K. Synthetic iron leucite is known.

Indices. $\alpha = 1.508–1.511$, $\gamma = 1.509–1.511$, $\gamma - \alpha = 0.001$, or the mineral may be isotropic (smaller crystals). In thin section in large twinned crystals (Fig. 7.12) the interference shades are dark grays and may require the gypsum plate for detection.

Color. Colorless.

Form. Isometric above about 600°C. Pseudoisometric, tetragonal at room temperatures. Phenocrysts are of trapezohedral habit with eight-sided outlines in thin section; matrix crystals are rounded with roughly circular sections; skeletal cruciform microlites also occur. Inclusions are common, usually symmetrically arranged in concentric or radial patterns or combinations thereof (Fig. 7.13). These inclusions are glass or fluid with or without gas, and minerals such as augite, olivine, magnetite, spinel, apatite, melanite and the feldspathoids, haüyne, sodalite, and nepheline; glass, magnetite, and augite are particularly common. (110) cleavage poor and generally not apparent in thin section.

Orientation. Probably biaxial; exact structure in doubt. Optical figures are secured with difficulty and seemingly indicate that it is (+) with a small 2V. Many larger anisotropic crystals consist either of twin sectors or of one basic individual with multiple twin lamellae in two or three directions (Fig. 7.12). The lamellae are broad with fuzzy boundaries.

Occurrence. In subsilicic potassic igneous rocks, in both felsic and mafic types, particularly, though not exclusively, in geologically young extrusive members, including leucite phonolites, leucitites, leucite basalts, leucite

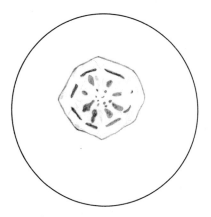

FIGURE 7.13 Leucite phenocryst zoned with both tangential and radial inclusions, from phonolite, Mt. Vesuvius, Italy.

tephrites, leucite basanites, and in similar dike rocks. Not uncommonly replaced by a mixture called pseudoleucite, consisting of orthoclase and nepheline or of orthoclase and a clay alteration of nepheline. In some cases the leucite is rimmed by this material, or the entire crystal is replaced. Other feldspathoids, including sodalite, haüyne, and nepheline, commonly occur with leucite, as does melilite. Leucite has been detected in some firebricks.

Diagnostic features. Resembles sodalite and analcite but has a higher index, distinctive twinning and inclusions.

ANALCITE

Properties. $NaAlSi_2O_6 \cdot H_2O$. Some varieties contain small amounts of CaO and nearly 5% K_2O. The replacement of Si^{4+} by $Na^{1+}Al^{3+}$ also occurs. A member of the zeolite group mineralogically, but paragenetically a feldspathoid. Isometric and isotropic, or very weakly anisotropic (use gypsum plate) with complex twinning. The birefringence is apparently related to a change in structure following loss of water. $n = 1.479–1.493$, with the higher indices for the potassium-bearing varieties. May be anomalously biaxial ($-$), with $2V = 85$–near $0°$. Birefringent analcite may show multilamellar twinning. Colorless. The fair cubic (001) cleavage may appear in thin section as sets of intersecting lines and also influences slightly the shape of crushed grains. Rounded or eight-sided sections characterize phenocrysts of trapezohedral habit; also anhedral and interstitial, as veinlets, cavity fillings, and replacements of plagioclase.

Occurrence. Both as phenocrysts and more abundantly as matrix material in some feldspathoidal basalts and related rocks; as a replacement of sodalite, nepheline, and leucite in subsilic rocks in general; as a hydrothermal mineral in cavities in basalts and diabases with calcite and other zeolites: rarely as a sedimentary (authigenic) mineral of tuffs formed by ash falls into alkaline lakes, of tuffaceous siltstones, of sandstones, of analcimolites (some oolitic, others with laumontite, fluorite, quartz, and calcite), and of pyrite concretions in coal beds. Potash varieties may be altered to pseudoleucite or rimmed by orthoclase. With heulandite it occurs as an alteration of glass shards of tuffs interbedded with graywackes, but replaced at greater depth by laumontite (zeolite metamorphic facies). Also a constituent of boiler scale.

Diagnostic features. Euhedral analcite resembles leucite but has a lower index. Sodalite generally has a slightly lower index than analcite but not sufficiently lower to prevent confusion, especially if both are anhedral. A chemical test for NaCl with HNO_3 on the powdered mineral may be required. The occurrence and cleavage of secondary analcite are usually distinctive to separate it from sodalite. The index of opal is appreciably lower, but interstitial analcite may be mistaken for some rhyolitic glass of low index; however, treatment with HCl causes gelatinization of analcite.

CHAPTER EIGHT

UNIAXIAL MINERALS

CORUNDUM

Composition. Al_2O_3, minor Fe^{3+}, Ti, and Cr.

Indices. $\epsilon = 1.760-1.763$, $\omega = 1.767-1.772$; Cr-bearing varieties have indices near the upper end of the range. Birefringence low ($0.007-0.010$), usually just slightly higher than quartz, but thin sections with abundant corundum may be slightly thicker than normal, thus with slightly higher colors.

Color. Colorless, gray, blue gray, blue, and light red. Blue varieties (sapphire) are colored by Fe or Ti and red (ruby) by Cr. In thin section usually colorless or with pale blue or red areas randomly distributed. Zoned crystals are common; the zoning results in some cases from concentrations of tiny hematite or rutile inclusions; these zones also show pleochroism. In other cases color banding (Fig. 8.1), even alternation of ruby and sapphire bands, is present. Not uncommonly detrital grains show marked blue to blue-green pleochroism.

Form. Hexagonal-rhombohedral. Well-formed and relatively large crystals common. Cross sections are hexagonal; longitudinal sections tabular to prismatic. Smaller crystals may be skeletal. Cleavage absent but basal (0001) parting good to perfect; also rhombohedral (1011) parting less well and less commonly developed. The basal parting may control the shape of broken particles, although detrital grains usually show irregular-fracture outlines. Minute inclusions are hematite, but larger included grains of rutile, garnet, and spinel also appear.

FIGURE 8.1   Zoned corundum porphyroblast in mica schist, Ruby Mountains,
Montana. Polars not crossed, ×33.

Orientation. Uniaxial (−), but basal plates of some varieties may show
an anomalous 2V as large as 58°. Lamellar multiple twinning with (1011)
as the twin plane is common.

Occurrence. In igneous rocks in syenites, monozonites, and nepheline
syenites; in their pegmatitic equivalents; in generally quartz-free plagioclase pegmatites or coarse-grained albitites ("desilicated" pegmatites) that
transect mafic or ultramafic rocks; and rarely in andesite dikes. Found in
regionally metamorphosed shales transformed into corundum schists or
gneisses; in aluminous xenoliths; in contact-metamorphosed shales and
limestones as emery deposits and single crystals of ruby; and in some
metamorphosed quartzites. In emery, magnetite, spinel, and garnet are
typical associates; in "desilicated" pegmatites with margarite, zoisite,
kyanite, chlorite, and tourmaline. Locally abundant in placers; a rather
erratically distributed member of heavy minerals suites. Alters easily
to damourite, margarite, and fine-grained muscovite. Synthetic material is
formed in fused Al_2O_3 bricks, fireclay bricks, laboratory ware, and spark
plugs.

Diagnostic features. High relief plus low-order interference colors as well as parting and multiple twinning. Irregular color blotches aid in recognizing it in sedimentary grains.

HEMATITE

Properties. Fe_2O_3. $\epsilon = 2.94$, $\omega = 3.22$. Hexagonal, uniaxial $(-)$. Usually opaque except along edges or in minute thin crystals, in which the color is deep red (use accessory condenser). Black in reflected light, like magnetite. Birefringent colors are masked. Optical figure and sign are hard to secure. Usually in anhedral grains, flakes, or scales; thin plates or microlites occur as inclusions in many minerals. Larger masses may show parting, (0001) and (1011), caused by multiple twinning.

Occurrence. Rarely in igneous rocks as a primary accessory mineral. Hematite of magnetic ancestry appears as a sublimation of thin scales in volcanic rocks, in hydrothermal veins, and in pyrometasomatic deposits associated with magnetite. Widespread and abundant in certain Precambrian rocks in bands alternating with quartz-rich layers. Grunerite and magnetite are associated with metallic hematite (specularite), and siderite, greenalite, and fine-grained quartz with earthy hematite. Specularite also forms hematite schists. Oolitic hematite-bearing sedimentary rocks contain clay minerals, goethite, and limonite, and some types have chamosite. Hematite is rather common cement, with or without limonite, in sandstones. Minute plates of hematite appear as inclusions (often oriented crystallographically) in muscovite, plagioclase, pyroxene, sylvite, and other species.

Diagnostic features. Earthy hematite resembles goethite and limonite closely but may be distinguished from the latter by its anisotropic character. Goethite is biaxial and shows extreme dispersion, but this characteristic cannot always be observed. The red color of minute flakes, the platy habit, and the oriented texture of hematite-rich rocks aid in separating it from magnetite.

RUTILE

Composition. TiO_2 with variable amounts of Fe^{2+} (maximum ca. 15% FeO), Ta (nearly 40% Ta_2O_5), Nb (ca. 30% Nb_2O_5 max), and Fe^{3+}, Sn, V, and Cr are accessory.

Indices. $\omega = 2.605\text{--}2.616$, $\epsilon = 2.890\text{--}2.903$; for synthetic rutile, which is nearly pure TiO_2, $\omega = 2.605$, $\epsilon = 2.901$. Birefringence very high, but interference colors do not appear because of total reflection and the strong color.

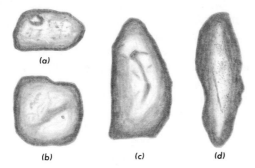

(a)

(b)   (c)   (d)

FIGURE 8.2   Detrital rutile. (a), (b) Beach sand, Florida. (c), (d) Wealden, Rusthall, Kent, England.

Color. In thin section usually red brown, amber, and yellow brown; pleochroism ordinarily not noticeable; $\epsilon > \omega$. Fe, Ta, and Nb cause darker, nearly black color, and thick grains with these elements may be essentially opaque in their maximum absorption position. Cr may induce a greenish hue. Colors best seen with accessory condenser.

Form. Tetragonal. Habit varies; commonly as euhedral, small prismatic crystals or as geniculate ("knee-shaped") twins; less commonly as subhedral grains; as minute slender needles included in other minerals, particularly phlogopite, corundum, and quartz; and as reticulate or network aggregates, owing to twinning. Distinct (110) cleavage; fracture varies from irregular to conchoidal; crushed pieces show minor orientation. Detrital grains are euhedral, knee-shaped, subrounded, or irregular (Fig. 8.2).

Orientation. Uniaxial (+); parallel extinction; some anomalous biaxial twin bands with (110) as the optic plane are observed across (001) sections. Rarely biaxial owing to deformation.

Occurrence. A common and locally abundant constituent of various metamorphic rocks, particularly schists and gneisses and to a lesser extent marbles and quartzites. Associated with kyanite in schists and quartzites; in low-grade argillaceous metamorphics it forms minute felted needles ("clay-slate needles"). To a lesser extent in igneous rocks, especially in hornblende-rich plutonic types and in some anorthosites and pegmatites (both granitic and mafic). Also a constituent of veins, including high-temperature and Alpine types. Very common as a detrital mineral. Alters to leucoxene.

Diagnostic features. The combination of color, shape, and exceedingly high relief is distinctive. The relief of hematite is similar, but its color is deep red and its form flaky. Limonite is isotropic, irregular in outline or pseudomorphous after other minerals. The rare mineral baddeleyite (ZrO_2), in some of its occurrences, resembles rutile but has slightly lower relief ($n = 2.13$–2.30) and is monoclinic, biaxial ($-$), with $2V = 30°$.

Brookite and anatase, which are polymorphous with rutile, are of considerable significance as accessory detrital minerals, although they are encountered only rarely in thin sections.

Brookite	Anatase
1. TiO_2	1. TiO_2
2. $\alpha = 2.583$	2. $\omega = 2.561$
$\beta = 2.584$	$\epsilon = 2.488$
$\gamma = 2.700$	strong birefringence
very strong birefringence	
3. Yellow, orange, light brown, weakly pleochroic, some growth banding.	3. Shades of brown, deep blue and black; thick grains are pleochroic, some growth banding.
4. Orthorhombic; biaxial ($+$); orientation varies with wavelength of light, but $\gamma = b$ for all lengths; dispersion very strong and many grains fail to extinguish in any position.	4. Tetragonal; uniaxial ($-$); perfect (001) and (111) cleavages; basal sections are nonbirefringent and in darker varieties may yield a biaxial figure with a small 2V.

Both minerals have a submetallic luster which is not necessarily conspicuous because of the common leucoxene alteration. Both occur commonly together in various veins, including the Alpine type, associated with adularia, hematite, chlorite, and quartz, and in hydrothermally altered felsic igneous rocks and quartzites. They are important distinctive min-

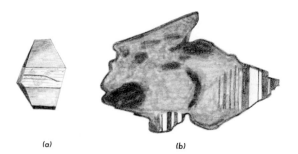

(a)   (b)

FIGURE 8.3 Detrital anatase. (a) Euhedral, yellowish brown, Recent beach sand, Wainui Inlet, New Zealand (Hutton, 1950). (b) Partly altered to leucoxene, pictured with accessory condenser inserted, Cleveland, Yorks, England.

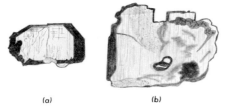

FIGURE 8.4   Brookite. (a) Dartmoor, England. (b) Bokkeveld series, Wuppertal, Cape Province, Africa (Swart, 1950).

erals in detrital suites from sandstones; in these have been considered as authigenic, possibly after ilmenite (Figs. 8.3 and 8.4).

CASSITERITE

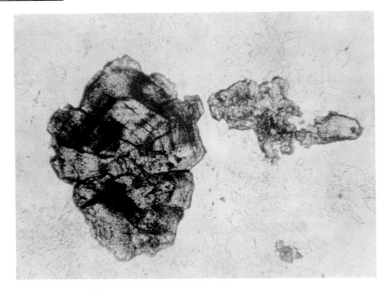

FIGURE 8.5   Zoned cassiterite in greisen, Cornwall, England. Polars not crossed, ×94.

Composition. SnO_2, with Fe^{3+} and smaller amounts of Ti, Ta, and Nb.

Indices. $\omega = 1.992$–1.997, $\epsilon = 2.091$–2.093. The effect of substitution of Fe^{3+} is to increase the indices slightly. Very high relief in thin section. The exceedingly high birefringent colors usually are hidden by the mineral color.

Color. Rarely colorless, usually in shades of yellow, orange red, brown,

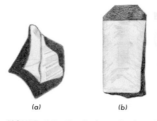

(a)   (b)

FIGURE 8.6   Detrital cassiterite. (a) Red River, Cornwall, England. (b) Dartmoor, England (Brammall, 1928).

or gray. Color zoning or irregular, blotchy color distribution may be present (Fig. 8.5). Pleochroism is normally absent or weak with $\epsilon > \omega$; less usually strong with ω green and ϵ red. Detrital grains are opaque, dark brown, or dark red (Fig. 8.5).

Form. Tetragonal. Euhedral to anhedral. Cross sections tend to be diamond-shaped. Veinlets of subhedral grains are not uncommon. One variety, needle tin, occurs as acicular prisms. Wood tin, which is botryoidal or concretionary, has a radial-fibrous structure of slender needles and may contain intergrown quartz or topaz. Cassiterite from clastic rocks is irregularly rounded or in crystals of prismatic or pyramidal habit (Fig. 8.6).

Orientation. Uniaxial (+). Some types show an anomalous, small 2V. Parallel extinction. The length-slow character may be difficult to determine. Multiple twinning bands parallel with (011) are common.

Occurrence. In hydrothermally altered granites; rarely as a primary accessory constituent in granites. In granitic pegmatites and in greisen with quartz, muscovite, topaz, and tourmaline. In hydrothermal veins or metasomatic deposits of high-temperature origin. Wood tin occurs characteristically in cavities in rhyolites, where it may also be associated with topaz.

Diagnostic features. Anhedral types greatly resemble rutile, and this distinction may be difficult for rounded detrital pieces. The form of cassiterite and its occurrence as a product of mineralization usually separate the two. Sphalerite may have the general color and relief of cassiterite but is isotropic.

BRUCITE

Composition. $Mg(OH)_2$, minor Mn and Fe^{2+}.

Indices. $\epsilon = 1.580-1.600$, $\omega = 1.559-1.590$. Indices increase with Fe^{2+} and probably slightly with Mn. $\omega - \epsilon = 0.010-0.021$, with a maximum of first-order orange in thin section. However, owing to the distinct change in birefringence with wavelength, abnormal deep brown (tobacco brown) may appear.

Color. Colorless.

Form. Hexagonal. Most typically as concentrically arranged, whorly aggregates of scales. Also in subparallel platy masses. Rarely fibrous. Perfect basal (0001) cleavage.

Orientation. Uniaxial (+). Cross sections of plates are length-fast. May be biaxial with a small anomalous 2V probably due to strain. Cleavage flakes yield centered figures.

Occurrence. The common alteration of periclase in calcite-periclase marbles (pencatite). In veins with talc, magnesite, and serpentine that cut peridotites and serpentinites.

Diagnostic features. Resembles some serpentine minerals, but these are length-slow. Talc aggregates show higher birefringent colors, and muscovite or phlogopite basal plates yield biaxial (−) figures. The association with remnants of periclase (Fig. 7.3*a*), the whorls (Fig. 7.3*b*), and the anomalous birefringence are characteristic.

GOETHITE

Properties. $HFeO_2$, usually with small amounts of Mn. $\alpha = 2.26$, $\beta = 2.39$, $\gamma = 2.40$. Orthorhombic. Biaxial (−) with 2V small–30°. Extreme dispersion and, for light of 610–620 mμ, is uniaxial (−). Optic plane parallel with (100) for red light and with (001) for white and yellow light; in each case $\alpha = b$. Yellow, orange, and brownish orange in thin section, with $\gamma > \beta > \alpha$. Massive concretionary, oolitic, fibrous, and earthy; pseudomorphous after pyrite and siderite. Perfect (010) cleavage pieces show parallel extinction and are length-fast for white and yellow light and length-slow for red light. Fibers, which are subparallel with the *c* axis, yield irregular interference figures and have mean refractive indices between 2.12 and 2.30 owing to finely admixed silica and water.

Occurrence. Formed by weathering from pyrite, magnetite, siderite, and glauconite. Much material called limonite actually is goethite. Found in gossans, bog iron ores, laterites, residual brown iron ores, and in minette-type iron ores, associated with limonite, lepidocrocite, hematite, manganese oxides, quartz, and clay minerals. Also a low-temperature hydrothermal mineral in veins with quartz, calcite, and siderite. A constituent of iddingsite.

Diagnostic features. Distinguished from hematite by its biaxial character, dispersion, fibrous habit, and greater translucence. Limonite is isotropic, but much fine-grained or impure goethite may be confused with it. Goethite and lepidocrocite are commonly closely associated, and *x*-ray methods may be required to identify them. Lepidocrocite also is orthorhombic and optically $(-)$, but has $2V = 83°$ and lacks the goethite dispersion.

ALUNITE

Composition. $KAl_3(SO_4)_2(OH)_6$, some Na may replace K.

Indices. $\omega = 1.569–1.578$, $\epsilon = 1.590–1.601$; natroalunite, $NaAl_3(SO_4)_2(OH)_6$, has $\omega = 1.585$, $\epsilon = 1.595$, but is otherwise similar. The interference colors in thin section reach second-order blue. Cleavage pieces show little or no birefringence.

Color. Colorless.

Form. Hexagonal; euhedral to anhedral. Rhombohedra with diamond-shaped or nearly square cross sections occur. Some crystals are tabular, parallel (0001) with rectangular cross sections. Grain size may be highly variable; aggregates are flaky, radial, or plumose.

Orientation. Uniaxial $(+)$. (0001) cleavage distinct, and cleavage pieces present a centered figure. Elongate sections and cleavage traces are length-fast.

Occurrence. A characteristic hydrothermal alteration mineral in silicic and intermediate volcanic rocks, related to metallic mineralization. Associated are quartz, kaolinite, allophane, leucoxene, pyrite, diaspore, and less commonly specular hematite. Some alunite is apparently supergene in gossans and weathered pyrite veins.

Diagnostic features. Alunite resembles brucite in its indices, general birefringence, and cleavage. Brucite occurs in different rocks, shows distinctive aggregate structures, and has anomalous interference colors. Scapolite superficially resembles alunite but is distinguished by somewhat lower birefringence and $(-)$ sign.

JAROSITE

Composition. $KFe_3(SO_4)_2(OH)_6$, may contain minor Na, Pb, or Ag.

Indices. $\omega = 1.815–1.820$, $\epsilon = 1.713–1.715$; $\omega - \epsilon = 0.101–0.105$. Natrojarosite, $NaFe_3(SO_4)_2(OH)_6$, has $\omega = 1.830$, $\epsilon = 1.750$. High relief in thin section. Very strong birefringence, slightly less than that of calcite, which may be partly masked by the pleochroic colors. Cleavage flakes show no birefringence.

Color. May be colorless but usually shows some pleochroism from very pale yellow or pale greenish yellow (ϵ) to deep golden yellow (ω).

Form. Hexagonal. Euhedral grains are cubelike rhombohedral or basal plates. Cross sections are thus squares, diamonds, or rectangles. Anhedral aggregates also occur, typically as clusters of rounded grains.

Orientation. Uniaxial (−); a 2V as large as 10° can be found in homogeneous, untwinned grains. Some basal sections may be composed of six segments each of which has the optic plane parallel with the edge and a small 2V with a (−) sign. Basal cleavage (0001) is distinct.

Occurrence. An oxidation product of weathered sulfide ores and their enclosing rocks, particularly in volcanic types. May be associated with pyrite, barite, kaolinite, allophane, and limonite. Jarosite also forms from hydrothermal solutions. In sediments it appears uncommonly as a detrital mineral and also as an alteration of glauconite and pyrite in siltstones, glauconitic sandstones, diatomaceous shales, and underclays.

APATITE

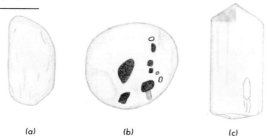

(a)   (b)   (c)

FIGURE 8.7 Detrital apatite. (a), (b) Yorkshire, England. (c) Nutfield fullers earth, England (Newton).

Composition. Fluorapatite is $Ca_5(PO_4)_3F_2$, but F may be partly replaced by Cl (chlorapatite) or OH (hydroxylapatite); CO_3 radical may also be present (carbonate–apatite) or combinations of these, as in carbonate–fluorapatite. Other minor constituents are rare-earth elements, Th, Sr, Mn, Na, Ba, and K for Ca; O for F; and S, As, V, Cr, Al, and Si for P. The most common type is F–OH–apatite; pure F–, OH–, or Cl–apatite is rare. Chlorapatite contains only one Cl ion in substitution for F. Many varieties, especially fluorapatite, collophane, and francolite are weakly radioactive (U). Rare Th types may be strongly so.

Indices of refraction. Ordinary apatite (with F or with F, Cl, and some CO_3 and OH) has $\epsilon = 1.630\text{–}1.645$, $\omega = 1.632\text{–}1.649$. Birefringence = 0.003–0.005, and in thin section the maximum interference color is gray white of the first order. Mangan–fluorapatite with some Cl has $\epsilon = 1.635\text{–}1.654$, $\omega = 1.640\text{–}1.660$, $\omega - \epsilon = 0.005\text{–}0.006$. Carbonate apatite has a mean index of refraction = 1.618–1.631 and birefringence = 0.007–0.017.

Hydroxylapatite has indices $\epsilon = 1.640–1.645$, $\omega = 1.645–1.651$, and birefringence $= 0.004–0.007$. Apatite with 11.6% SrO has $\epsilon = 1.634$, $\omega = 1.638$. The effect of ionic substitutions upon the optical constants of apatite has been summarized by McConnell and Gruner (1940), using as a reference the values for fluorapatite: mean index of refraction $\geq 1.631 \leq 1.634$ and birefringence $\geq 0.003 \leq 0.005$ (Table 8.1).

TABLE 8.1 EFFECT OF IONIC SUBSTITUTION
ON OPTICAL CONSTANTS OF APATITE

ION	MEAN INDEX OF REFRACTION	BIREFRINGENCE
CO_3	decrease	large increase
F	decrease	very slight
Cl	increase	very slight
OH	increase	very slight
Sr	increase	very slight
Mn	increase	very slight
Na	decrease	slight
K	increase	
Mg	decrease	
Rare earths	increase	slight
S, Si		slight

Numerous varietal names have been employed, including francolite = carbonate–fluorapatite; collophane = cryptocrystalline carbonate–fluorapatite; dahllite = carbonate-hydroxylapatite. Basal sections of apatite remain dark, and near-basal sections should be tested with the gypsum plate to reveal their birefringence.

Color. Colorless, but detrital crystals may be pale green or exhibit pleochroic purple centers. Collophane is brown.

Form. Hexagonal. Poor basal (0001) cleavage, which in thin section divides the elongate crystals into nearly equal segments. The orientation of crushed fragments is nearly random. In thin section apatite typically appears as rather small euhedral crystals (Fig. 8.8). Sections tend to be hexagonal in outline (basal) or prismatic (longitudinal). Some crystals have a tabular habit, and the elongate cross sections of these do not show cleavage traces at right angles to the length. Apatite grains of heavy residues are commonly egg-shaped; others are somewhat rounded prismatic crystals that may show corrosion (Fig. 8.7). Inclusions parallel with c may be present and may be zonally concentrated. Overgrowths of carbonate–apatite on fluorapatite are recorded, and these shells may also be zoned concentrically and divided into six biaxial sectors. Much carbonate–

FIGURE 8.8   Apatite in diorite, Lindenfels, Odenwald, Germany.
Polars not crossed, ×94.

hydroxylapatite or carbonate–fluorapatite is spherulitic, oolitic, plumose, radiating, or cryptocrystalline in structure. Commonly in oolites (Fig. 8.9), which are concentrically layered, with some also showing radial fibrous structure, or in ovules, which are structureless, massive. These bodies are generally ellipsoidal or may be flattened and indented, with cores of clastic mineral grains (especially quartz), pyrite cubes, or fossil fragments. Cores of oolites may be massive with zoning marginal. Some show color banding; others show bleaching. In some types zones of collophane and francolite alternate, or collophane alternates with calcite. Oolites may be veined or partly replaced by calcite or chalcedony. Carbonate–fluorapatite also forms larger ovoid masses, including pisolites, nodules, and concretions and is the chief constituent of fossil, bone, teeth, some shells, and fecal pellets.

Occurrence. A very common accessory in all types of igneous rocks from

APATITE   **83**

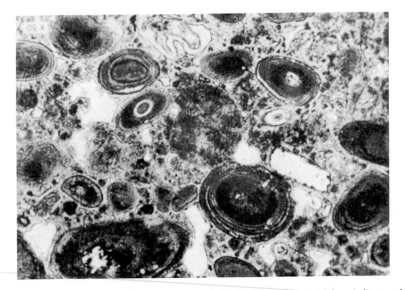

FIGURE 8.9 Oolitic phosphorite, Deer Creek–Wells Canyon Area, Idaho. Oolites and ovules of collophane-francolite in cryptocrystalline matrix of collophane-francolite. Polars not crossed, ×50 (Lowell, 1952).

gabbro to pegmatite and granite, but less abundant in peridotites. Chlorapatite is more common in basic types, fluorapatite in granitic types. Mn–apatite is a late pegmatitic variety; Sr–, Re–, Th– and silicoapatites occur in nepheline syenite pegmatites. Apatite is a widespread accessory to essential mineral of carbonatites. Certain high-temperature veins, phlogopite deposits, and some iron deposits of magmatic affiliation also contain apatite in abundance. It is found in gneisses and schists, marbles of contact-metamorphic origin, and is a widespread constituent of heavy residues. Hydroxylapatite occurs in some talc schists. Phosphatic limestone and phosphorite (Fig. 8.9) contain fibrous carbonate–hydroxyl- or fluorapatite and collophane, which occur as spherulites, oolites, ovules, layers of radiating fibers, fossil replacements, and very fine-grained, concretionary masses. Collophane is isotropic and usually brown in color. Fluorapatite occurs in some types of basic slags.

Diagnostic features. In thin section the form and relief are characteristic. Apatite resembles topaz, but topaz cleavage directions parallel the fast ray. In crushed pieces the low birefringence plus moderate indices and uniaxial (−) character are distinctive. These together with the prismatic or egg shape define the mineral in heavy residues. Greatly resembles beryl, which has lower refractive indices.

FIGURE 8.10   Detrital xenotime, Snowy River Dredge, New Zealand (Hutton, 1950).

Composition. YPO_4, other rare earths (Nd, Pr, Er, Tb) of the yttrium subgroup, especially Er, are also usually present. Some Ca, Ce, U, and Th may replace Y; SiO_4 substitutes to a limited extent for PO_4, as does a small amount of SO_4. Usually at least slightly radioactive.

Indices. $\omega = 1.720–1.724$, $\epsilon = 1.816–1.828$. Indices probably increase with increasing Th. Birefringence very high, 0.095–0.107, with interference colors similar to those of calcite, but may be masked somewhat by mineral color. High relief in thin section.

Color. Colorless to yellow, pale brown. May be weakly pleochroic:
ω = pale rose, buff, pale yellow
ϵ = yellow, gray brown, pale yellow green

Form. Tetragonal. Prismatic crystals resemble those of zircon. Occurs mainly in euhedral crystals of simple development. A parallel intergrowth with zircon has been reported. Inclusions are zircon and magnetite specks. Detrital grains are usually doubly terminated euhedra, slightly abraded or somewhat rounded prisms; some basal plates have been recorded (Fig. 8.10).

Orientation. Uniaxial (+). Interference figures may be difficult to secure because of the small size of the grains. Parallel extinction; length slow. Cleavage (110) is fair, with little tendency for preferred orientation in crushed fragments.

Occurrence. An accessory mineral in some granites, pegmatites, syenites; less commonly in diorite. Pleochroic haloes may surround the grains. Also in quartzose, micaceous, and granitic gneisses as well as in Alpine and several other types of veins. Probably more widely distributed than the record indicates, because of confusion with zircon and monazite. Not uncommon as a heavy detrital mineral in beach and river sands and in sandstones.

Diagnostic features. May be confused with zircon, sphene, monazite, or bastnäsite, $(Ce,La)FCO_3$. Zircon has considerably higher indices and lower birefringence. Sphene also has higher indices and very strong dispersion. Monazite is biaxial with a small 2V, shows a small extinction angle and has a birefringence about equal to that of zircon, i.e., lower than xenotime. Basnäsite also resembles monazite, but its good basal cleavage and effervescence in warm HCl aid in distinguishing it from xenotime. Xenotime shows strong electromagnetism (near that of ilmenite) and may display distinct yellow fluorescence (Tb).

RHOMBOHEDRAL CARBONATE GROUP

GENERAL There are three main series of rock-forming rhombohedral carbonate minerals:

1. Calcite–rhodochrosite
$CaCO_3$–$MnCO_3$
2. Dolomite–ankerite
$CaMg(CO_3)_2$–$Ca(Mg,Fe)(CO_3)_2$
3. Magnesite–siderite
$MgCO_3$–$FeCO_3$

In addition, siderite and rhodochrosite form a series. Calcite may also contain small amounts of Mg, Ba, Sr, Re, Fe, Zn, Pb, and Co. In dolomite there may be very limited contents of Co, Pb, and Zn, and both siderite and magnesite may show minor replacement of Fe and Mg by Ca and Zn. In rhodochrosite small quantities of Zn, Co, and Mg can substitute for Mn. The members of this group are similar in their optical and physical properties and in thin section are distinguished one from another with difficulty.

CALCITE

Composition. $CaCO_3$. Calcite and rhodochrosite ($MnCO_3$) are isomorphous. Small amounts of Mg may be retained in solid solution—as much as 9 mole % $MgCO_3$ (Goldsmith et al., 1955).

Indices. $\epsilon = 1.486$, $\omega = 1.658$ for calcite, increasing with increasing Mn content (Fig. 8.11). Because of the high birefringence, calcite "twinkles" as the microscope stage is rotated; high relief appears when the long diagonal of the cleavage rhomb parallels the vibration direction of the polarizer and low relief when the short diagonal is in this position. The birefringence is extremely high and even in thin section appears as pearl grays or pale pastel tints of higher orders. Crushed fragments appear white under crossed polars. It is not possible to measure ϵ on such grains but only an intermediate, higher value, ϵ', which for pure calcite is 1.566.

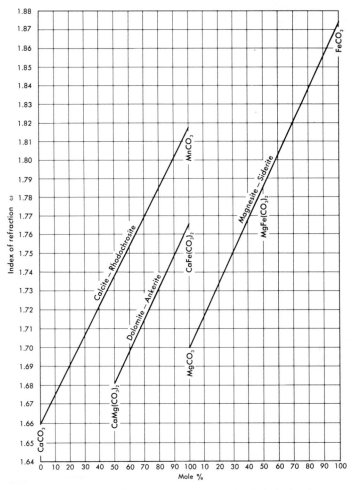

FIGURE 8.11 Variation of ω with composition in rhombohedral carbonates (Ross and Kennedy, 1947).

Color. Colorless to turbid.

Form. Hexagonal-rhombohedral. Perfect rhombohedral cleavage (1011), which in thin section appears as two intersecting lines at oblique angles. If the section is normal to the cleavage traces, the angle is 75°. Crushed pieces are almost invariable cleavage blocks and will lie on cleavage faces. Rarely a parting on (0112) is present.

In thin section calcite usually is anhedral in very fine-grained aggregates. The finer-grained aggregates usually do not show cleavage. Spherulitic, oolitic, and fossiliferous structures are common. Layered and banded forms occur in stalactitic calcite and in travertines. Recrystallized calcite in marbles shows interlocking grains as well as coarse and abundant multiple twinning. In some high-grade metamorphic carbonate rocks and carbonatites calcite contains (exsolved) blebs of dolomite.

Orientation. Uniaxial $(-)$. Basal sections, which are rarely obtained, show a figure with many color rings. Calcite from metamorphic rocks may show $2V = 4-14°$, with the trace of the optic plane parallel with $(10\overline{1}1)$. Grains yielding a 2V do not extinguish completely in any position but show a peculiar mottled-blue interference color. Grains normal to the optic axis may be hard to find, owing to coatings of randomly oriented cleavage dust which prevent extinction. Cleavage pieces yield an eccentric figure. Multiple twinning $(01\overline{1}2)$ is common. If rhombohedral outlines or cleavages appear in thin section, the twin lamellae either parallel the long diagonal of the rhomb or are oblique to it, depending on the orientation of the section. Twin lamellae, which are in nearly reversed optical orientation with respect to adjacent, partly overlapping parts of the crystal, show lower-order interference colors or may not extinguish at all. Calcite twins easily by twin-gliding, and some of the lamellae may form during preparation of the section. Extinction is symmetrical with respect to cleavage traces.

Occurrence. In calc-alkalic igneous rocks most calcite is due either to weathering or hydrothermal alteration, but magmatic calcite occurs in carbonatites and in alkalic rocks. Forms as a low-temperature mineral in basalt cavities, associated with zeolites. Calcite is the chief constituent of calcareous marbles and tactites and also occurs in ophicalcites, calc schists, skarns, and lime-silicate gneisses. It is found in limestones, chalk, calcareous shales, and marls, as cement in sandstones and arkoses, in travertine, calcareous sinters, cave deposits, and limey tuffs. It is a very common gangue mineral of many types of veins and forms under a wide range of geological conditions. It occurs as detrital species in the light fraction.

Diagnostic features. For its distinction from dolomite, see dolomite. Aragonite also has extreme birefringence but is biaxial and lacks rhombohedral cleavage, and all three of its indices are above the index of Canada balsam.

RHODOCHROSITE

Composition. $MnCO_3$, completely isomorphous with calcite and siderite. Limited miscibility with $MgCO_3$.

Indices. $\omega = 1.597$, $\epsilon = 1.816$, decreasing with increasing Ca content

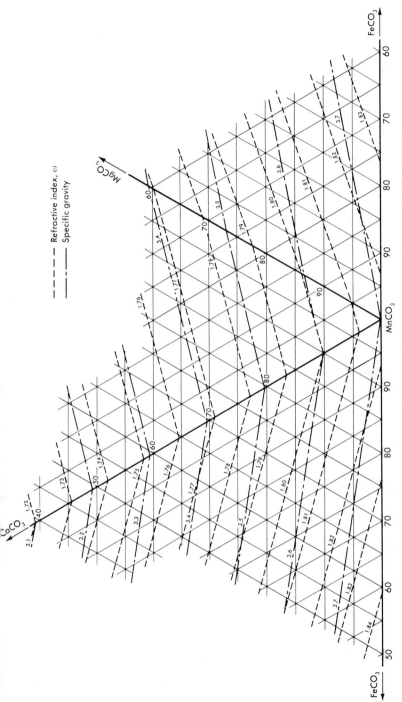

FIGURE 8.12 Relation of ω and specific gravity to composition of rhodochrosite (Wayland, 1942).

(manganocalcite) and increasing with increasing Fe^{2+} content (manganosiderite) (Fig. 8.12). The extreme birefringence results in pearl-gray or white high-order interference colors. ϵ' on cleavage pieces is 1.702, decreasing toward calcite and increasing toward siderite. Birefringence also increases with Fe^{2+}.

Color. Colorless to neutral gray in thin section.

Form. Hexagonal-rhombohedral. In anhedral masses, bands, layers, or euhedral rhombic crystals. Perfect rhombohedral (1011) cleavage. Crushed pieces are cleavage fragments from which values of ϵ' and ω can be obtained. Individual crystals may be zoned, and higher indices appear in more deeply colored parts.

Orientation. Uniaxial $(-)$. Cleavage fragments give an extremely eccentric figure with many color rings.

Occurrence. Chiefly as a mineral of ore deposits, accompanying Pb, Ag, and Cu. To a small extent in sedimentary carbonate rocks and their metamorphosed equivalents. Alters readily to pyrolusite. Rhodonite and quartz are common associates. Rare in carbonatites.

Diagnostic features. The ϵ' index is characteristic for pure rhodochrosite. For varieties rich in Ca or Fe a chemical test for Mn may be necessary.

DOLOMITE AND ANKERITE

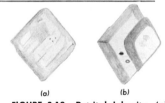

(a)   (b)

FIGURE 8.13 Detrital dolomite. (a) Fassa, Tyrol. (b) Dorset, England.

Composition. Dolomite, $CaMg(CO_3)_2$; ankerite, $Ca(Mg,Fe)(CO_3)_2$. Crystals may be zoned, with variations in Fe^{2+} content.

Indices. $\epsilon = 1.500$, $\omega = 1.680$ for dolomite; increasing with Fe^{2+} content to $\epsilon = 1.526$, $\omega = 1.716$ for ankerite with 22% $FeCO_3$. Ankerite with 38% $FeCO_3$ has $\epsilon = 1.547$ and $\omega = 1.749$. A variety with nearly 5% PbO and 8.7% ZnO has $\epsilon = 1.520$, $\omega = 1.703$. Cleavage prevents measurement of ϵ on fragments. For dolomite ϵ' on cleavage pieces is 1.588. If the values of any two of the three constants ϵ, ϵ', and ω are known, the other may readily be calculated (Loupekine, 1947). Interference colors are high-order white, except on some twin lamellae. Grains twinkle when the stage is rotated; high relief appears when the long diagonal parallels the polarizer vibration direction.

Color. Colorless to turbid gray. Iron-bearing varieties turn brown upon alteration.

Form. Hexagonal-rhombohedral. Perfect rhombohedral (1011) cleavage. Crushed pieces lie on cleavage planes and give very eccentric interference figures. In thin section dolomite is more commonly euhedral than calcite and occurs in rhomboid crystals that may show the two sets of cleavage lines intersecting at oblique angles. In marbles anhedral to subhedral grains, usually of noninterlocking character, appear. Detrital grains can best be recognized by their uniformly simple rhombohedral habit (Fig. 8.13). Casts of dolomite crystals (dolocasts) form an important fraction of insoluble residues from carbonate sedimentary rocks. These rhombohedral imprints occur chiefly in chert but also in shale, glauconite, pyrite, and limonite. Dolomite rhombs also replace nuclei of carbonate oolites.

Orientation. Uniaxial $(-)$. Centered figures are not obtained from broken fragments. Basal sections in thin section tend to remain dark through a complete stage revolution and give a figure with many color rings. It may be difficult to locate such a section owing to the birefringence of fine, abundant, unoriented dolomite dust on the grain surface. Such dust is formed in grinding the thin section and obscures extinction positions in the rhombohedral carbonate minerals. Extinction is symmetrical with cleavage traces. Polysynthetic twinning (0221) is common in dolomite of marbles, although twinning in general is not as abundant as in calcite. The lamellae parallel either the long or short rhomb diagonal.

Occurrence. Dolomite is characteristically a secondary mineral in sediments, and much of it forms in limestones by replacement of calcite. Carbonate rocks in which the MgO content is between 5 and 15% are termed magnesian limestones. With greater MgO content the rocks are called dolomitic limestones. Primary dolomites occur in evaporite sequences. Dolomite marbles may be formed under conditions of dynamothermal or contact metamorphism. Contact metamorphism or metasomatism of impure dolomites produces tactites in which magnesium silicates are common: forsterite, phlogopite, spinel, chondrodite, and dravite. Dolomite and ankerite also occur as vein minerals, and metasomatic dolomitic rocks are formed in association with various types of metallic mineralization. Some carbonatites also are partly or wholly dolomitic or even ankeritic.

Diagnostic features. In thin section it may be difficult to distinguish between calcite and dolomite. The following comparison may be of assistance:

Calcite	Dolomite
1. Anhedral	1. More commonly euhedral (in sedimentary rocks)
2. Colorless	2. Colorless to turbid or stained by iron oxide; zonal structure
3. Twinning common	3. Twinning less common
4. Twin lamellae generally parallel the long diagonal or are oblique but not parallel to the short diagonal	4. Twin lamellae parallel both the short and long diagonals of rhombs

In crushed fragments measurement of ϵ' will suffice to differentiate between the two unless the calcite contains much Mn. The ϵ' indices of manganocalcite overlap those of dolomite. Because of the higher specific gravity dolomite tends to appear in the heavy fraction of detrital suites; calcite remains in the light fraction. Dolomite is insoluble in acetic acid and nearly insoluble in cold dilute HCl if not powdered. Calcite effervesces in both; but minerals effervesce in warm HCl.

A variety of staining techniques is available for color differentiation of the carbonate minerals and some associated species. These may be made on sawed hand specimens which are first etched 2–3 min with dilute HCl (8–10 parts by volume of commercial grade concentrated HCl diluted with H_2O to 100 parts) (see Table 2.4).

MAGNESITE

Composition. $MgCO_3$, completely isomorphous with $FeCO_3$.

Indices. $\omega = 1.700$, $\epsilon = 1.509$, $\epsilon' = 1.602$ for $MgCO_3$. With $Mg:Fe = 1:1$, $\omega = 1.788$, $\epsilon = 1.570$, $\epsilon' = 1.673$. Indices increase with Fe^{2+} content. Birefringence extreme, increasing from 0.191 with Fe^{2+} content. Interference colors are high-order whites and grays. Maximum relief when the long diagonal lies parallel with the polarizer vibration direction; pronounced twinkle. Basal sections show very high relief in all positions of the stage. Cleavage pieces yield values for ω and ϵ'.

Color. Colorless, gray, or turbid. Fe-bearing types show yellow or brown owing to oxidation of Fe^{2+} and incipient alteration to limonite.

Form. Hexagonal-rhombohedral. Perfect rhombohedral cleavage (1011) which does not appear in the fine-grained types. Typically subhedral to anhedral aggregates; some varieties of sedimentary magnesite rock are porcelanoid in texture with exceedingly small grain size. Cleavage pieces are rhombohedra (1011). Detrital grains are typically irregular, without crystal outline. In evaporites (0001) plates are typical.

Orientation. Uniaxial (−). Basal sections present a figure surrounded by numerous color rings. Figures hard to obtain in fine-grained varieties. Extinction symmetrical to cleavage traces. Twinning absent.

Occurrence. As grains, aggregates, and veins in chlorite and talc schists and serpentinites; as large, very fine-grained, porcelanoid replacement masses in metasomatized dolomite; rarely as a gangue constituent of ore veins. In some evaporites and rarely in some carbonatites.

Diagnostic features. Distinguished with difficulty from dolomite, especially if the latter contains some iron; for then the ϵ and ϵ' indices of the two overlap. A chemical test for Ca may be necessary. In thin section magnesite usually lacks crystal outline and twinning, two features characteristic of dolomite. Magnesite and siderite are best separated by the differences in ϵ'.

SIDERITE

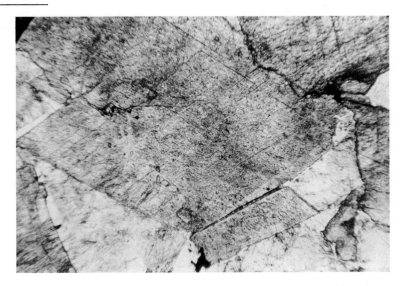

FIGURE 8.14 Euhedral siderite showing cleavage, Siegen, Westphalia, Germany. Polars not crossed, ×55.

Composition. $(Fe,Mg)CO_3–FeCO_3$. Special names such as breunnerite and sideroplessite applied to intermediate members of the siderite–magnesite series are superfluous. Minor Mn and Zn may be present.

Indices. $\omega = 1.788$–1.875, $\epsilon = 1.570$–1.633, $\epsilon' = 1.673$–1.748, $\omega - \epsilon = 0.218$–$0.242$, all increasing with Fe^{2+}. Extreme relief in thin section when long diagonal of rhombs parallels polarizer vibration direction; marked twinkle on stage rotation. High-order whites as interference colors in thin section. High relief for basal sections in all positions. ω and ϵ' can be measured on cleavage fragments.

Color. Gray, neutral; brownish with oxidation of Fe^{2+}.

Form. Oolites, euhedral rhombs (Fig. 8.14), clusters of subhedral grains, interstitial anhedral masses, replacements of fossils. Detrital pieces are spherules or subrounded cleavage particles. Perfect rhombohedral (1011) cleavage, on which crushed material lies.

Orientation. Uniaxial ($-$). Figures show many color rings. Eccentric figures from cleavage pieces. Symmetrical extinction with respect to cleavage traces. Twinning (0112) uncommon, parallel with the long diagonal.

Occurrence. A common constituent of various types of iron-bearing sediments. With chert and greenalite, minnesotaite, and stilpnomelane it forms banded cherty iron carbonate rock; with clay minerals it forms clay ironstone, which may also appear as concretions in other sediments. It also occurs in siderite mudstones, siderite–chamosite mudstones, and sideritic limestones. A relatively common hydrothermal gangue mineral in veins and replacement bodies. Alters readily to limonite or goethite. Forms rare types of carbonatites.

Diagnostic features. Distinguished by its form, high birefringence, and alteration. From magnesite it differs chiefly in higher indices.

QUARTZ

Composition. SiO_2, only traces of other elements (Li, Na, K, C, Fe^{3+}, Al, Ti, Mn).

Indices. $\omega = 1.544$, $\epsilon = 1.553$. In thin section the birefringence shows a maximum of white of the first order. Grains and crushed fragments usually show first- and second-order colors with higher central tints fading into lower-order marginal colors.

Color. Colorless.

Form. Hexagonal-rhombohedral. Rhombohedral cleavage rarely appears, except near the margins of some thin sections or on large single crystals. Nearly random orientation prevails in powdered material. Detrital grains show considerable variation in form, but generally are irregular or subangular (Fig. 8.15); surfaces may be pitted and frosted. Doubly terminated authigenic crystals occur in sediments. Sedimentary grains or crystals

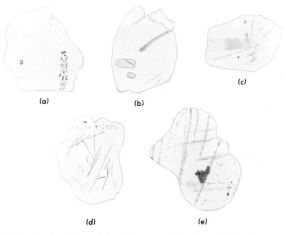

FIGURE 8.15 (a) to (e) Detrital quartz, beach sand, Florida.

may have a detrital nucleus, marked by inclusions or separated from the rim by an iron oxide film, upon which new quartz has been deposited in crystallographic continuity. Phenocrysts of quartz show hexagonal outlines, which may be marred by strong embayment (Fig. 8.16). Quartz of

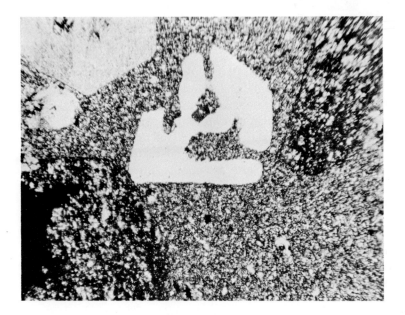

FIGURE 8.16 Deeply embayed quartz phenocryst in granodiorite porphyry, Alpine Gulch stock, Judith Mountains, Montana. Polars crossed, ×50 (S. B. Wallace).

FIGURE 8.17   Undulatory extinction in quartz in amphibole gneiss, Franklin, North Carolina. Polars crossed, ×55.

intrusive igneous rocks is characteristically anhedral. Elongate grains with interlocking margins and undulatory extinction (Fig. 8.17) are characteristic of quartzites and other quartzose rocks. Vein quartz shows a variety of structures—platy, radiating, and feathery.

Inclusions, which are numerous and characteristic, may cause the grains to appear turbid or gray. Some are minute fluid inclusions (CO_2 and H_2O), occurring as bubble trains, which may extend across grain boundaries; others are rutile needles (sagenitic quartz), chlorite, tourmaline, sillimanite, apatite, zircon, and magnetite.

Orientation. Uniaxial (+). Basal sections show no birefringence. Strained quartz, which is characterized by undulatory extinction, may show a small 2V.[1] Owing to rotary polarization, figures from thick grains have centers vaguely defined or absent. Euhedral crystals are length-slow. Twin-

---

[1] Amethystine quartz not uncommonly is biaxial, with 2V as large as 35°. Smoky quartz also may be biaxial, with 2V = 8° or less, and slightly pleochroic, faint brown to faint green.

ning rarely seen in thin section, but narrow deformation lamellae occur in metamorphic quartz.

Occurrence. One of the most widespread of all minerals. In igneous rocks it occurs as an essential mineral in granites, granodiorites, quartz diorites, and their extrusive equivalents and as an accessory constituent of syenites, diorites, and rarely gabbros. In veins and pegmatites it is very abundant. Metamorphic rocks such as schists, gneisses, and quartzites (Fig. 8.18) contain it, as do sandstones, arkoses, shales, and siltstones, as well as novaculite and some cherts. It is a common and abundant detrital mineral and also forms an important part of some insoluble residues. In granitic rocks quartz and potash feldspar are graphically intergrown as graphic granite or micropegmatite (micrographic intergrowth). Grano-

FIGURE 8.18 Anhedral-granular quartz in micaceous quartzite, Petaca, New Mexico. Polars crossed, ×100 (A. A. Corey).

phyre, a plumose intergrowth of quartz in orthoclase, is similar. Vermicular intergrowths of quartz and plagioclase are called myrmekite and occur in silicic intrusive rocks. Quartz in fine-grained form, such as jasperoid, is an important replacement mineral in the wall rocks of various ore deposits.

Diagnostic features. Absence of cleavage, twinning, and alteration. From beryl it is distinguished by lower indices and positive sign. Nepheline has lower indices, is negative, and commonly shows alteration. Cordierite, whose indices may fall near those of quartz, is biaxial with moderate to large 2V, commonly shows some twinning, and is usually somewhat altered to a micaceous aggregate. Untwinned fragments of oligoclase are distinguished from quartz through their better cleavage and biaxial character. Scapolite of low birefringence and low indices resembles quartz but is negative, has cleavage, and is usually in columnar crystals.

All quartz, as it appears in nature, is low-temperature, or α-quartz, for although some quartz forms above the inversion temperature (about 572°C at 1 atm), all of it inverts upon cooling. The following criteria assist in deciding if inversion has taken place: (1) Crystals of inverted quartz are pyramidal in habit with the prism absent or subordinate. (2) The presence of the trigonal trapezohedron and irregular development of the rhombohedral faces indicate crystallization below the inversion point. (3) Twinning that results from the volume change accompanying inversion is very irregular and heterogeneous, whereas growth twins of α-quartz are semiregular in distribution with straight contacts related to crystal outline; this can be detected by etching with HF, generally on polished basal sections. (4) Fractures are more common in quartz that has inverted. The quartz of phenocrysts in rhyolites, of graphic intergrowths, of granites, of the outer shells of zoned pegmatites, and of a few quartz veins was originally β-quartz. That of most veins, of the cores of pegmatites, of geodes, and of authigenic or sedimentary origin crystallized as the α variety.

CHALCEDONY

Composition. SiO_2, with variable amounts of admixed H_2O as submicroscopic bubblelike inclusions, of the order of 0.1 μ in diameter, which can be detected with electron microscope. Most fibrous forms of SiO_2, generally lumped as chalcedony, are quartz, but some are cristobalite (Table 8.2).

Indices. $\omega = 1.531{-}1.544$, $\epsilon = 1.539{-}1.553$. Chalcedony without H_2O has the same refractive indices as ordinary quartz. With increasing H_2O the indices decrease.

Color. Colorless to light brown in thin section. Grains may be white or pale blue.

FIGURE 8.19   Chalcedony, Germany. Polars crossed, ×94.

Form. Hexagonal. Fibrous (Fig. 8.19), commonly radially spherulitic. Also in minute bands with fibers normal to the layers; interstitial, colloform, or vuggy. Pseudomorphs after other minerals and replacements of fossils are common. Spherulites give the typical pseudo-interference figure under crossed polars in parallel light.

Orientation. Uniaxial (+). Fibers that are elongated parallel with *a* are length-fast; if elongated parallel with *c*, they are length-slow (variety called quartzine or lutecite). Fibers of adjoining bands may alternate in the character of their elongation.

Occurrence. Forms either as a product of weathering or of low-grade hydrothermal action, and thus appears as a secondary constituent of a variety of rocks. It is characteristically authigenic material of some sedimentary rocks, including limestones, dolomitic limestones, siderite rocks, and sandstones. In these it forms concretions or bands, replaces fossils, and cements grains. It may be an important part of the insoluble residue

CHALCEDONY   **99** from carbonate rocks; also occurs in diatomite, radiolarite, novaculite, tripoli, chert, and silicified tuffs. Geologically young chert consists of a mixture of chalcedony, opal, fine-grained quartz, and rare cristobalite. With time both chalcedony and opal lose water, and older cherts consist entirely of fine-grained quartz. In some sandstones chert occurs in detrital grains. Many varieties of chert have been named based on color or aggregate structure: agate, onyx, sardonyx, carnelian, and chrysoprase.

Diagnostic features. The fibrous nature in bands or spherulites together with low index of refraction are characteristic. X-rays may be necessary in a few cases to distinguish between fibrous quartz and fibrous cristobalite. Table 8.2 lists the fibrous varieties of SiO_2.

TABLE 8.2 FIBROUS VARIETIES OF SiO_2 (BRAITSCH, 1957)

MINERAL	VARIETAL NAME	FIBER AXIS	CHARACTER OF ELONGATION
quartz	chalcedony	[1120]	length–fast
	chalcedony	[1010]	length–fast
	quartzine	[0001]	length–slow
cristobalite	lussatite	[110]	length–slow
	lussatite	[101]	length–slow
	lussatite	[011]	length–slow
	lussatine	[111]?	length–fast

TRIDYMITE

Composition. All natural tridymite departs from pure SiO_2, usually with variable amounts of Al, Na, K, Ca.

Indices. Synthetic material has $\alpha = 1.469$, $\beta = 1.469$, $\gamma = 1.473$. Indices of the mineral are generally higher, owing to the presence of cations other than Si: $\alpha = 1.471–1.482$, $\beta = 1.472–1.483$, $\gamma = 1.474–1.488$, and $\gamma - \alpha = 0.002–0.004$. The birefringence is so low that it should be confirmed with the gypsum plate.

Color. Colorless.

Form. Orthorhombic. Pure SiO_2 tridymite occurs in three modifications: orthorhombic α-tridymite below $117°C$, hexagonal β_1-tridymite between $117°$ and $163°C$, and hexagonal β_2-tridymite above $163°C$. The presence of cations other than Si has a marked effect on the inversions. The inversion temperatures may be shifted, or may not appear at all, with α-tridymite inverting to β_2-tridymite. Forms tabular euhedral crystals with hexagonal

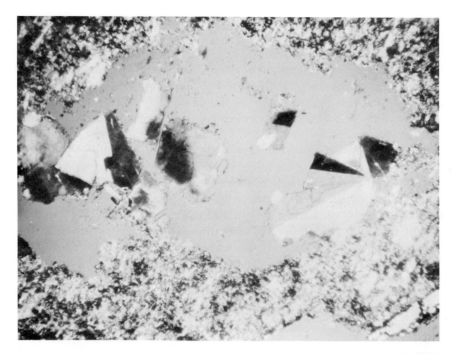

FIGURE 8.20 Twinned tridymite in vesicle in andesite vitrophyre, Cover Mountain, Guffey, Colorado. Polars crossed, ×170 (J. E. Bever).

outline and elongate lathlike cross sections. Laths may be clustered in a network aggregate. Anhedral minute granules also occur. Groups of tabular crystals can be fan-shaped or spherical (Fig. 8.20). More minute, less regular grains form aggregate structures resembling tiles on a roof. Cleavage (1010) and parting (0001), inherited from the hexagonal modification, are both indistinct. Gas inclusions may be present.

Orientation. Biaxial (+). The higher-temperature forms are uniaxial. $\alpha = b$, $\beta = a$, $\gamma = c$. Optic plane = (100). 2V = 36–90°. Synthetic tridymite has 2V = 36°, but in natural tridymite the value of 2V ranges from 40 to 90°, and in many 2V > 75°. The size of 2V may vary in adjoining twins by as much as 4–9°. Elongate rectangular sections are length-fast and show parallel extinction. Twinning is very common and is conspicuous in larger crystals; twin individuals are wedge-shaped pairs or triplets; the twin plane is (1016). In addition, complex lamellar twinning resembling that of leucite may be observed in coarser grains.

Occurrence. Characteristically occurs in extrusive igneous rocks, intermediate to silicic: andesite, dacite, quartz latite, rhyolite, obsidian, and rhyolite tuff, both in gas cavities and in the groundmass. In lithophysae in rhyolite and obsidian it occurs with quartz, sanidine, fayalite or hornblende, cristobalite, specularite, and, less commonly, bixbyite, pseudobrookite, topaz, garnet, and cassiterite. Also occurs uncommonly in gas cavities of some basalts with albite, biotite, and hornblende. In some rhyolites and quartz latites tridymite is the chief SiO_2 mineral, owing to its abundance in the coarser and more vesicular parts of the matrix. Common in stony meteorites, in some up to several percent. In some andesites it has been detected in reaction rims with magnetite around hypersthene. Another less common occurrence is with glass, enstatite, and pseudobrookite in fused quartzose xenoliths in lava. It has been found in sandstone along its fused contact with andesite. Quartz has been found pseudomorphous after tridymite phenocrysts in a quartz latite porphyry and in intrusive granophyre. Cristobalite pseudomorphs after tridymite also are known. Synthetic tridymite occurs in silica brick, associated with quartz, cristobalite, and, less commonly, pseudowollastonite, magnetite, monticellite, and fayalite; also reported from some slags.

Diagnostic features. The lathlike crystals with their wedge-shaped twins are highly characteristic. In order to distinguish some fine-grained tridymite from cristobalite it may be necessary to make either a careful index determination (the indices of cristobalite are usually slightly higher) or to obtain an x-ray powder photograph.

CRISTOBALITE

Composition. Ideally SiO_2. Usually other elements such as Ca, Fe^{2+}, Al, Fe^{3+}, and perhaps Na also are present.

Indices. Two types occur in nature: an isotropic and an anisotropic variety. The isotropic has $n = 1.485–1.487$. For the anisotropic $\epsilon = 1.482–1.484$, $\omega = 1.486–1.488$, $\omega - \epsilon = 0.002–0.006$. For synthetic cristobalite $\epsilon = 1.484$ and $\omega = 1.487$.

Color. Colorless.

Form. The high-temperature form, β-cristobalite, is isometric and inverts, between 275 and 220°C, to tetragonal α-cristobalite. The crystals are commonly of octahedral, less commonly of cubic or cubo-octahedral habit. Etching may appear on the faces of octahedra. Octahedral pseudomorphs of α- after β-cristobalite have the optic axis parallel with an octahedral axis. Fibrous cristobalite occurs intergrown with sanidine needles of spherulites, or occurs in small spheroidal pellets or spherulites with a characteristic radial, wavy fracture pattern (Fig. 8.21). Another type of fibrous cristobalite resembles chalcedony. This is called lussatite (Table 8.2), and some

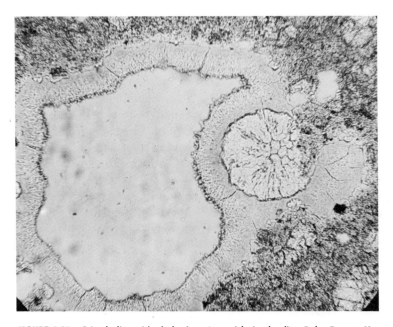

FIGURE 8.21   Cristobalite with chalcedony in vesicle in rhyolite, Ruby Range, Montana. Polars not crossed, ×94.

consists of a mixture of α-cristobalite and about 30% β-cristobalite. Cristobalite forms pseudomorphs after six-sided plates of tridymite. The synthetic material is skeletal or dendritic but may be fibrous. Most opal consists of submicrocrystalline cristobalite.

Orientation. Uniaxial (−) or isometric. Some needles of synthetic cristobalite are described as length-slow, but the natural fibers may be either length-slow or -fast (Table 8.2). It shows parallel extinction, but in elongated aggregates the extinction is not parallel. Twinning may be lamellar like that of albite or in two directions like that of microcline. Octahedra may show twinning on (111).

Occurrence. Chiefly in gas cavities, lithophysae, or along fractures in extrusive and near-surface intrusive igneous rocks, including olivine basalt, basalt, andesite, dacite, trachyte, rhyolite, and obsidian, forming euhedra or spherulitic or spheroidal pellets. It occurs with tridymite and even replaces it pseudomorphously. Octahedral quartz pseudomorphs after cristobalite are recorded. Associated minerals are magnetite, zeolites, opal, and chalcedony. Cristobalite is the common SiO_2 mineral intergrown with potash feldspar in spherulites. In extrusive rocks it may appear as very fine-grained material between spherulites or in the dense, nonporous sub-microscopic parts of the matrix, in which case an x-ray powder photograph may be necessary to identify it. Also detected in meteorites and in sandstone partly fused by basalt. Some bentonites contain considerable cristobalite with montmorillonite. Lussatite has been found in serpentinites and various low-temperature veins. Isotropic cristobalite has been detected through x-rays in some opal. The synthetic form is found in silica and semisilica brick with quartz and tridymite and less commonly with magnetite, fayalite, monticellite, and pseudowollastonite. Both isotropic and anisotropic types have been recorded, even in the same specimen. Some slags contain cristobalite.

Diagnostic features. Shape of pellets and crystals plus curving fracture typify coarser material. The indices are usually slightly above those of tridymite. Fine-grained or fibrous types may require the use of x-rays for identification.

NEPHELINE

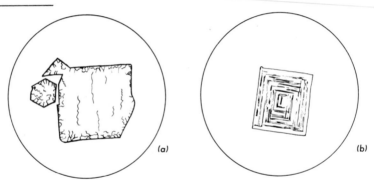

FIGURE 8.22 (a) Nepheline phenocrysts with marginal microfractures in nephelinite, North Kivu, Congo (Sahama, 1953). (b) Nepheline microphenocryst with inclusions along (0001) and (1010) in leucitophyre, Olbrück, Germany (Wülfing and Mügge, 1925).

Composition. $(Na,K)(Si,Al)_2O_4$. Ca may be present in small amounts. The Al:Si ratio is rarely $1:1$; Si predominates slightly. The ratio Na:K may reach $4:3$, but is usually much less. Kaliophilite, in which K predominates greatly over Na, is the rare potash analogue of nepheline. A polymorph of kaliophilite, kalsilite, not isomorphous with nepheline, has the composition $KAlSiO_4$ with minor Na.

Indices. $\epsilon = 1.528-1.544$, $\omega = 1.531-1.549$. The indices of the majority of nephelines fall above 1.54, thus low positive relief in thin section.

Birefringence is weak, maximum interference colors are first-order grays in thin section. Increasing K tends to lower indices slightly, and additions of Ca cause lowering of the birefringence. Kalsilite has $\epsilon = 1.537$, $\omega = 1.542$.

Color. Colorless to gray. Zonal structure may be due to arrangement of inclusions or to alternating layers richer and poorer in SiO_2. Nephelines from young volcanic rocks are commonly comparatively clear, whereas those from plutonic rocks are usually turbid owing to pigment inclusions and/or alteration (Fig. 8.22).

Form. Hexagonal. A poor cleavage (1010) is rarely seen in thin section, where irregular fractures predominate. In intrusive rocks, nepheline is anhedral to subhedral and may be coarse-grained. In extrusive rocks it occurs in smaller, stubby matrix crystals that show rectangular longitudinal sections and square or hexagonal cross sections and may show zoning. Nepheline does not commonly form phenocrysts in extrusive rocks. Kalsilite is anhedral to subhedral, poikilitic, occurring in small stout prisms with prong terminations.

Orientation. Uniaxial ($-$). Basal sections give a figure, but because of the very low birefringence, the cross is hazy and poorly defined and may be undetected unless the gypsum plate is used. Rectangular longitudinal sections have parallel extinction and are length-fast.

Occurrence. Nepheline, because it is unsaturated with respect to SiO_2, does not occur with primary quartz. It is found as an essential constituent in foidal syenites, foidal monzonites, phonolites, and in a few rare mafic foidal rocks and as an accessory constituent in some alkali syenites. Gneissic varieties of foidal syenites also may carry nepheline. Common associates are sodalite, haüyne, nosean, melilite, and melanite. Nepheline alters readily and therefore does not occur as a clastic mineral. Occurs in firebricks attacked by Na-rich slags or gases. Natrolite (hydronephelite) and cancrinite are common alteration products. Other alteration minerals are analcite, sodalite, fine-grained muscovite (gieseckite), and calcite. Graphic intergrowths of nepheline and orthoclase are found in some nepheline syenites. Nepheline is also a product, together with orthoclase, of the alteration of leucite to pseudoleucite. Kalsilite is rare, occurring in mafic, extremely potassic lavas with olivine, melilite, and perovskite and also in blast furnaces. Nepheline and kalsilite may occur together in various ways: (1) an exsolution intergrowth of nepheline in kalsilite (perthite-like structure, (2) nepheline borders on kalsilite, or (3) single euhedra of nepheline with interstitial anhedral kalsilite.

Diagnostic features. The low birefringence, low indices, and ($-$) uniaxial character are distinctive. It resembles quartz superficially, but the latter is free from alteration and is positive. Orthoclase may be confused with nepheline, but has better cleavage, shows Carlsbad twinning, and is biaxial. To distinguish kalsilite from nepheline, x-rays or a staining test for K with HCl and sodium cobaltinitrite is necessary.

CANCRINITE

Composition. $(Na,K,Ca)_{6-8}Al_6Si_6O_{24}(SO_4,CO_3,Cl)_{1-2} \cdot 1\text{-}5H_2O$. Cancrinites high in Ca are also high in CO_3; those rich in Na are sulfatic (vishnevite).

Indices. $\epsilon = 1.491\text{-}1.503$, $\omega = 1.502\text{-}1.528$. Birefringence ranges from 0.005 to 0.029; thus interference colors in thin section vary from low first order to middle second order. With increasing SO_4 content the indices decrease (Fig. 8.23), ω decreasing at a more rapid rate and thus lowering the birefringence.

Color. Colorless.

Form. Hexagonal. Typically in anhedral to subhedral grains that tend to be elongate parallel with c. Excellent $(10\overline{1}0)$ cleavage, on which crushed fragments tend to lie.

Orientation. Uniaxial $(-)$. Extinction parallel with cleavage traces in longitudinal sections. Cleavage traces are length-fast.

Occurrence. Found with other feldspathoids, especially nepheline and sodalite, particularly in nepheline syenite and other foidal intrusive rocks, in which it may be a primary constituent or replace nepheline. Not normally found in phonolites or related extrusives.

Diagnostic features. Superficially resembles muscovite, but has negative relief. The lower indices and distinctive cleavage also distinguish it from

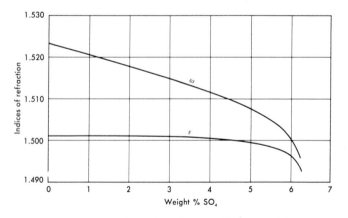

FIGURE 8.23 Variation of ω and ϵ with composition in cancrinite.

members of the scapolite series, which also are uniaxial $(-)$, length-fast, and colorless and show similar variable birefringence.

MELILITE

Composition. $(Ca,Na)_2(Mg,Al)(Si,Al)O_7$. The main variation in composition may be expressed by means of two end members, åkermanite, $Ca_2MgSi_2O_7$, and gehlenite, $Ca_2Al(Al,Si)O_7$. Na probably substitutes for Ca only to the ratio $Ca:Na = 1:1$. Fe^{2+} may also substitute to a small extent for Mg. Synthetic Fe^{3+} melilites are known.

Indices. $\epsilon = 1.626–1.658$, $\omega = 1.632–1.669$. Indices increase with increasing Al, or increasing amounts of the gehlenite end member (Fig. 8.24). The åkermanite end of the series is optically $(+)$; with about 44% of the gehlenite molecule, melilite becomes essentially isotropic. With further increases in Al, melilite becomes optically $(-)$. Thus the birefringence varies from 0.000 to 0.011; in many melilites it ranges from 0.004 to 0.007; first-order gray in thin section. Varieties containing Fe^{2+} show higher birefringent colors. Melilites with low birefringence commonly exhibit abnormal interference colors (Berlin blue).

Color. Colorless to pale yellow, rarely yellow or pale brown. The more strongly colored types are pleochroic: $\omega = $ light yellow, $\epsilon = $ dark yellow.

Form. Tetragonal. Commonly in tabular, euhedral to subhedral crystals with rectangular cross sections. May contain characteristic rodlike inclusions parallel with c and projecting into the crystal from the (001) plane, with the point of the rod at the crystal surface and a thicker knob toward the center (peg structure). The inclusions are isotropic and have a lower index of refraction; they may be glass. Also forms relatively large irregular poikilitic crystals with included leucite.

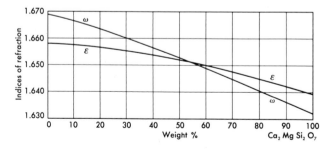

FIGURE 8.24 Variation of ω and ϵ with composition in melilite (Ferguson and Buddington, 1920).

Orientation. Uniaxial $(-)$ or $(+)$. Cross sections are length-slow and show parallel extinction. Cleavage (001) is imperfect.

Occurrence. Relatively rare, being restricted to mafic, generally extrusive igneous rocks very low in Si, relatively high in Al and Ca. In rocks of this type, such as melilite basalts, melilite is associated with nepheline, leucite, olivine, augite, and perovskite. Melilite also occurs in alnöite, an intrusive rock, with biotite, olivine, augite, and monticellite. In uncompahgrite, a rare coarse-grained intrusive rock, melilite is associated with diopside, cancrinite, and perovskite and is altered to cebollite ($H_2Ca_5Al_2Si_3O_{16}$). Melilite (åkermanite) less commonly appears as a high-temperature contact metamorphic mineral in limestones. It also is a constituent of furnace slags and portland cement. Alters to zeolites or carbonates.

Diagnostic features. Resembles vesuvianite in its abnormal birefringent colors but has lower indices and lower relief. Zoisite with abnormal interference colors may also be distinguished by biaxial $(+)$ character. The occurrence of most melilite is distinctive.

APOPHYLLITE

Composition. $Ca_4K(Si_2O_5)_2F_2 \cdot 8H_2O$. Because of the large amount of H_2O and the geological occurrence, it has often been placed in the zeolite group; however, it lacks Al, and the crystal structure is platy. Na may replace K to a limited extent, and some OH substitutes for F.

Indices. $\epsilon = 1.533-1.537$, $\omega = 1.531-1.536$ in those that are $(+)$; $\epsilon = 1.535-1.544$, $\omega = 1.537-1.545$ for those that are $(-)$. Also essentially isotropic in some types, with $n = 1.542$. Birefringence nil to low, 0.000–0.003, with low first-order grays in thin section, or strong abnormal interference colors.

Color. Colorless.

Form. Tetragonal. Anhedral granular to euhedral. Crystals may be equant, platy, or prismatic. Perfect basal (001) cleavage, distinct (110) cleavage. Crushed pieces tend to lie on the base.

Orientation. Uniaxial $(+)$, less commonly $(-)$. May be optically anomalous with biaxial sectors (2V as large as 60°) and abnormal color rings due to strong dispersion of the birefringence. The position of the optic plane varies markedly with the wavelength of light. Longitudinal sections have parallel extinction and usually are length-slow.

Occurrence. A secondary mineral, in amygdaloidal cavities and cracks in basalts, diabases, and other mafic igneous rocks, associated with calcite, pectolite, datolite, prehnite, analcite, and zeolites. Also to a much lesser extent in cavities in granite, gneiss, and contact-metamorphic marbles; in natrolite veins in nepheline syenite. In some metallic mineral deposits with pyrite and chalcopyrite or with native copper. May be replaced by opal or quartz, calcite, or kaolinite.

Diagnostic features. Apophyllite has higher indices than most zeolites. From the uniaxial zeolites, chabazite, and gmelinite it differs in its perfect basal cleavage and abnormal interference colors.

BERYL

Composition. $Be_3Al_2(SiO_3)_6$. Li, Na, K, Cs, Fe^{2+}, as well as some water may be present; trace elements are Cr and V (in emerald) and He.

Indices. $\epsilon = 1.564–1.595$, $\omega = 1.568–1.602$; increasing with increasing alkalies. Birefringence 0.004–0.008, also increasing with the alkali content (Fig. 8.25). Many large beryl crystals are zoned, with the outer zones containing more alkalies and having higher indices. In tapered crystals the end having the larger diameter has higher indices. Maximum interference colors are first-order yellows in thin section.

Color. Colorless.

Form. Hexagonal. Usually euhedral, with six-sided or somewhat rounded cross section and rectangular longitudinal sections. Rarely anhedral, interstitial. With increasing alkali content a tabular habit appears. Inclusions may be liquid (with bubbles), albite, quartz, and muscovite. The included minerals may be arranged zonally or concentrated in the central part of the crystal.

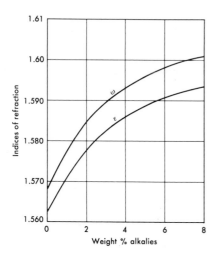

FIGURE 8.25 Variation of ω and ϵ with composition in beryl.

Orientation. Uniaxial $(-)$. Basal sections yield a hazy cross. Rarely biaxial with an anomalous 2V as large as $10°$ in certain parts of zoned crystals. Imperfect (0001) cleavage, rarely seen. Longitudinal sections are length-fast and show parallel extinction. Rectangular cross sections of tabular crystals also show parallel extinction and are length-slow.

Occurrence. Chiefly in granite pegmatites and to a lesser extent in certain of their wall rocks, principally biotite schists and gneisses and altered peridotites in which beryl has been formed metasomatically. Certain high-temperature Sn and W veins, greisens, and skarns also contain it, associated with topaz. Beryl alters to kaolinite or bertrandite.

Diagnostic features. In thin section greatly resembles apatite in outline, sign, and birefringence. However, even alkali-rich beryl has lower indices of refraction. Massive pale-colored beryl may be confused with quartz, from which it differs in sign and higher indices.

ZIRCON

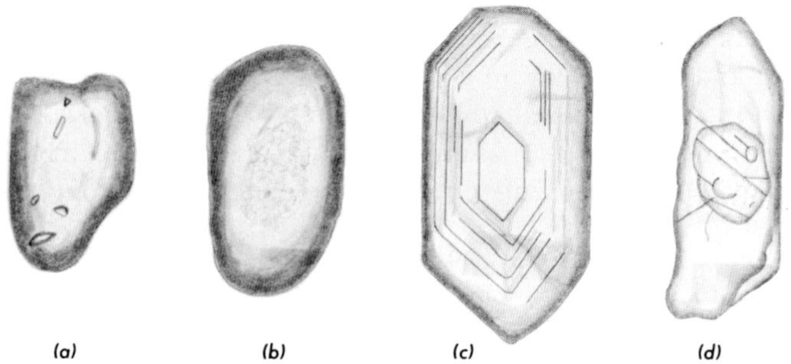

(a)   (b)   (c)   (d)

FIGURE 8.26 Detrital zircon. (a), (b) Beach sand, Florida. (c) Zoned, Burke County, North Carolina. (d) With rounded core, Bokkeveld series, Wuppertal, Cape Province, Africa (Swart, 1950).

Composition. $ZrSiO_4$. Some varieties contain Hf, Fe^{3+}, PO_4, rare earths, Th, and U. Some fine-grained, recrystallized varieties contain water, as $(OH)_4^{4-}$ for $(SiO_4)^{4-}$.

Indices. $\omega = 1.920–1.960$, $\epsilon = 1.967–2.015$. The relationship between optical and compositional variation is not well defined. The birefringence $(0.042–0.065)$ is normally strong, with colors in thin section ranging into third- and fourth-order shades. Grains from heavy residues exhibit pearl-gray tints of higher orders. With increasing alteration the indices and birefringence decrease, and eventually the material becomes isotropic with an index of about 1.82 (metamict). Microcrystalline recrystallized types also have low indices, owing to structural OH.

Color. Most zircon is colorless in thin section. Grains may also be colorless, pale yellow, pink, red, lavender and purple, and, uncommonly, purple brown. Strongly colored grains may show pleochroism with $\omega > \epsilon$. Colorless grains may have a turbid, gray appearance because of a host of minute inclusions (gas, liquid, or dust). Minerals recorded as inclusions are apatite, monazite, xenotime, rutile, hematite, magnetite, ilmenite, cassiterite, biotite, quartz, and tourmaline. Color zoning may also appear in grains, and yellow iron oxide stains coat some grains.

Form. Tetragonal. In igneous and metamorphic rocks euhedral to subhedral, generally as small to minute crystals (Fig. 8.27). Where these grains occur in iron-bearing minerals such as biotite, chlorite, andalusite, tourmaline, cordierite, or hornblende, a strong pleochroic halo (α-particle emanations) surrounds them. In detrital suites zircon exhibits a variety of shapes from rounded to euhedral (Fig. 8.26); angular pieces are rare. The pink, red, and purple varieties (hyacinth) commonly exhibit a more marked rounding than does the colorless type. Other characteristics more typical of the colored zircons are zonal structures and microcracks.

Orientation. Uniaxial (+). Many color rings appear in the figure, which usually can be obtained readily only from the rare larger crystals. Elongate crystals are length-slow, with parallel extinction. Rarely biaxial (+) with 2V about 10°; also isotropic owing to metamictization. Metamict and anisotropic zones may alternate.

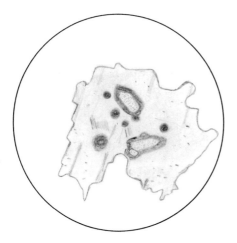

FIGURE 8.27   Zircon inclusions with pleochroic haloes in basal flake of biotite, Pikes Peak granite, Mt. Rosa area, Colorado.

Occurrence. A very common accessory mineral in granitic pegmatite, granite, syenite, and nepheline syenite; less common in granodiorite, quartz diorite, and diorite. Also in mica schists and micaceous and granitic gneisses. In heavy residues from sands and sandstones significant not only quantitatively but also for its diagnostic value through its wide variation in color, shape, and habit. Other useful characteristics of detrital zircon are elongation ratio (length ÷ width or breadth), inclusions, microcracks, overgrowths, radioactivity, and fluorescence. Although it resists alteration strongly, it may become metamict and hydrated to malacon, which is gray white in reflected light under the microscope.

Diagnostic features. Strong birefringence and high relief are characteristic in thin section. Grains and crushed pieces can also be recognized by their high indices and birefringence. Zircon in heavy residues may sometimes be distinguished by means of its crystal form. For its distinction from xenotime see under that mineral. Has relief like that of sphene but with lower birefringence.

VESUVIANITE

Composition. $Ca_{10}(Mg,Fe^{2+},Fe^{3+})_2Al_4Si_9O_{34}(OH)_4$. Highly variable and may also contain generally minor Na, Be, Mn, Zn, Ti, B, and F. Some recent analyses indicate more Al and less OH than the amounts required by the above formula.

Indices. $\epsilon = 1.698$–1.736, $\omega = 1.702$–1.742. Values as low as $\epsilon = 1.655$ and $\omega = 1.682$ have been reported. The relation between indices and composition is imperfectly known. Vesuvianites with very high indices, $\epsilon = 1.730$–1.736 and $\omega = 1.736$–1.742, are marked by a TiO_2 content between 4 and 5%. Both Fe^{3+} for Al and Ti^{4+} for Si cause increases in the indices. The substitution of Fe^{2+} for Mg also brings about a rise in indices; Na for Ca decreases them. Beryllium vesuvianites (BeO = 1–1.5%) have $\epsilon = 1.700$–1.712 and $\omega = 1.712$–1.722. The birefringence ranges from 0.001 to 0.012 and is commonly near 0.005. Thin-section interference colors are usually of first order and may be abnormal Berlin blue and green. The birefringence may also show zonal variation or mottling in randomly distributed areas.

Color. Usually colorless in thin section, although large crystals and grains not uncommonly show color zoning or irregular color distribution. Detrital grains show pleochroism, with $\omega > \epsilon$, brown to gray brown, yellow green to colorless, or red brown to gray.

Form. Tetragonal. Grains may be euhedral to anhedral. Granular or in radial or parallel prismatic clusters; also fibrous. Imperfect (110) cleavage.

Orientation. Uniaxial and usually $(-)$, rarely $(+)$. Some varieties show anomalous biaxial character with $2V = 5\text{–}65°$; these are generally $(-)$, but some are $(+)$. The biaxial and $(+)$ characteristics commonly appear in vesuvianites of high OH content. Crystals with large optic angles may be divided into sectors. Basal sections of such crystals show four sectors with the optic plane of each normal to the outer edge. Parallel extinction and length-fast in elongate sections. Also zoned with uniaxial $(-)$ cores and biaxial $(+)$ margins.

Occurrence. Principally in rocks of contact-metamorphic origin, such as tactite, impure marble, lime silicate gneiss, schists, and hornfels; associated with calcite, diopside, grossularite, wollastonite, sphene, and epidote and less commonly with magnetite, fluorite, and chlorite; also in similarly metamorphosed limestone blocks ejected from volcanoes. Rarely in pegmatites contaminated by limestone assimilation, in nepheline syenites, and veins cutting ultramafic igneous rocks. Rare as a detrital mineral.

Diagnostic features. Because of the high relief and abnormal interference colors, vesuvianite resembles zoisite and clinoziosite. These are biaxial and $(+)$, features that characterize vesuvianite rarely. From melilite it differs in higher relief and usually in occurrence. Andalusite also has lower relief, a very high 2V and normal interference colors.

SCAPOLITE

Composition. $(Ca,Na)_4Al_3(Al,Si)_3Si_6O_{24}(Cl,CO_3)$. The chief variation in composition can be described in terms of two end members: marialite, $Na_4Al_3Si_9O_{24}Cl$, and meionite, $Ca_4Al_6Si_6O_{24}CO_3$. Intermediate varieties are mizzonite, dipyre, and wernerite. K, SO_4, F, and OH may be present in small amounts. Neither marialite nor meionite occurs as a pure compound in nature. Most scapolites have more Ca than Na + K. A zonal structure with the center richer in marialite is rare.

Indices. $\epsilon = 1.522\text{–}1.571$, $\omega = 1.534\text{–}1.607$, increasing with meionite content (Fig. 8.28). K for Na lowers the indices slightly. Both added SO_4 and OH also appear to lower the indices. The birefringence increases toward the meionite end of the series, ranging from 0.008 to 0.038 (owing mainly to CO_3). Thus colors under crossed polars in thin section range from first-order gray to third-order blue.

Color. Colorless.

Form. Tetragonal. Commonly in bladed or prismatic aggregates; granular clusters also occur. Crystals may be poikiloblastic, forming a netlike framework for such species as calcite, diopside, phlogopite, garnet, quartz, actinolite, plagioclase, rutile, muscovite, and tourmaline. Distinct (100)

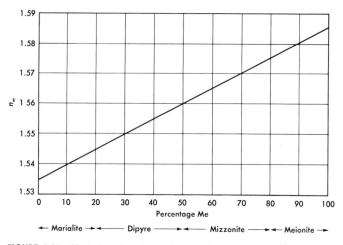

FIGURE 8.28 Variation of median index of refraction in scapolite with percent meionite (Shaw, 1960).

cleavage, imperfect (110) cleavage; in longitudinal sections the traces parallel the outline; in cross sections they intersect at oblique angles.

Orientation. Uniaxial $(-)$, parallel extinction in longitudinal sections, which are elongate parallel with the fast ray. Cleavage pieces yield a flash figure.

Occurrence. A product of contact metasomatism in tactite or skarn and in contact marble or hornfels, associated with grossularite, diopside, sphene, actinolite, and magnetite. Similarly in blocks of metamorphosed limestone ejected from volcanoes. In calcium-rich gneisses, schists, and amphibolite with epidote, hornblende, augite, and sphene. In veins and pegmatites associated with amphibolites and gabbros. A replacement of plagioclase in altered gabbros and similar rocks. Scapolite may be replaced by chlorite, calcite, zeolites, and muscovite.

Diagnostic features. Resembles a number of minerals: quartz is optically $(+)$ and lacks cleavage; cordierite is biaxial; wollastonite and andalusite have higher relief and are biaxial $(-)$. In general only scapolite of low birefringence is confused with these minerals.

TOURMALINE GROUP

GENERAL The tourmaline group is chemically complex, with the general formula $(Na,Ca)(Mg,Fe^{2+},Fe^{3+})_3B_3Al_3(Al,Si_2O_9)_3(O,OH,F)_4$. In addition, Mn, Li, Al, Ti, and Cr may also appear for Mg or Fe^{2+}; Na and Ca are apparently completely isomorphous. The main variations in composition can be expressed adequately in terms of four end members: schorl, Na–Fe^{2+}–tourmaline; dravite, Na–Mg–tourmaline; uvite, Ca–Mg–tourmaline; and elbaite, Na–Li–tourmaline.

SCHORL

Composition. $NaFe^{2+}_3B_3(Al,Si_2O_9)_3(OH)_4$. Some Ca may replace Na; variable amounts of Mg and Fe^{3+} appear, and Cr and Ti may be present.
Indices.

		GRAY, GRAY–BROWN, BLACK SCHORL	BLUE, BLUE–GREEN, GREEN SCHORL
maximum	ϵ	1.627–1.675	1.622–1.658
range	ω	1.649–1.698	1.640–1.677
	$\omega - \epsilon$	0.013–0.040	0.012–0.029
most	ϵ	1.633–1.640	1.630–1.638
common	ω	1.660–1.672	1.653–1.662
range	$\omega - \epsilon$	0.020–0.029	0.018–0.026
average	ϵ	1.638	1.636
	ω	1.663	1.658
	$\omega - \epsilon$	0.025	0.022
No. of determinations		250	30

Both birefringence and indices increase with increasing Fe^{2+} over Mg (Fig. 8.29). Additions of Cr and Ti also cause an increase in indices and in birefringence; for example, with $Cr:Mg:Fe^{2+} = 7:6:2$, $\epsilon = 1.641$, $\omega = 1.687$, $\omega - \epsilon = 0.046$. The maximum interference colors vary from upper first to middle second order in thin section but may be difficult to see in the more strongly colored varieties.

Color. Schorl in thin section or crushed pieces shows great color variety, usually shades of gray, gray brown, buff, blue, blue green, green, olive, purple, or black. In general, colors deepen and pleochroism becomes more pronounced with increased total Fe, markedly with Fe^{3+}. Darker-colored schorl also has higher indices; with ω less than 1.662, light pleochroic colors appear. The blue and blue-green tints have been ascribed to the presence of minor Cu. Cr–tourmalines are deep green in color. ω always shows greater absorption than ϵ, and therefore the maximum absorption occurs when the c axis (and the long direction) is normal to the polarizer vibration direction. The pleochroic formula shows considerable variety:

ε	ω
colorless	gray
light gray	deep gray
buff	olive buff
light blue	deep blue
pale green	dark green
green	blue
smoky gray	deep green
pale gray	black
buff	black

Basal sections are nonpleochroic. Commonly the colors are zonally arranged, usually concentrically with respect to the c axis (Fig. 8.30). Less commonly, different color zones appear along the c direction. Irregularly distributed color variations also occur.

Form. Hexagonal. Commonly euhedral in thin section with trigonal,

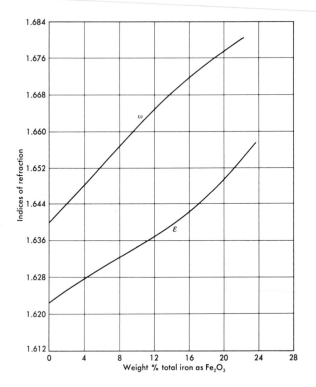

FIGURE 8.29 Variation of ω and ε with composition in schorl (Ward, 1931).

FIGURE 8.30 Zoned tourmaline, quartzose pegmatite, southern Maine. Polars not crossed, ×28.

hexagonal, or rounded outlines in cross section. Stubby to acicular or even fibrous in radial groups ("suns"), which give a pseudo-interference figure under crossed polars with parallel light. Graphic intergrowths with quartz are known. Tourmaline needles appear as inclusions in other minerals, especially quartz, feldspar, and mica. Cleavage is lacking, but fractures normal to c may be well developed and are "healed" by veinlets of quartz or quartz and albite. Inclusions of quartz, albite, muscovite, and iron oxides are present in some crystals and may be concentrated in zones, particularly in the core. Detrital grains, although showing a variety of shapes, commonly are irregular to rounded (Fig. 8.31). Detrital prismatic pieces may show striations parallel with c, but terminated crystals are rare in sediments (Fig. 8.32).

Orientation. Uniaxial ($-$). Longitudinal sections are length-fast and have parallel extinction. Basal sections show a figure with a few color rings. These characteristics are difficult to determine in dark-colored varieties.

Occurrence. Schorl is a magmatic accessory mineral in granites and peg-

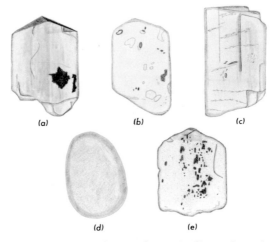

FIGURE 8.31   Detrital tourmaline. (a), (b) Beach sand, Florida. (c), (d) Cornwall, England. (e) With core rich in opaque inclusions, Bokkeveld series, Wuppertal, Cape Province, Africa (Swart, 1950).

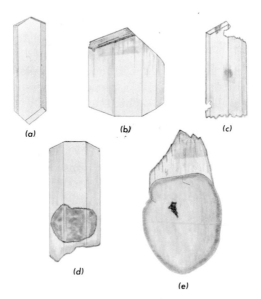

FIGURE 8.32   Authigenic tourmaline. (a), (b) Euhedral. (c) Corroded. (d) With rounded core. (a)–(d) From various sediments from France (Topkaya, 1950). (e) Overgrowth on detrital nucleus, Devonian sandstone, Monroe, Michigan.

UNIAXIAL MINERALS

matites. It also occurs as a hydrothermal mineral in these rocks, as well as in high-temperature veins. Granite that has been strongly replaced by tourmaline is called luxullianite. Greisen also may contain tourmaline, associated with topaz and cassiterite. Some schists, gneisses, and phyllites carry tourmaline, which may be of metasomatic or of recrystallized detrital origin. Schorl is a very abundant and widespread mineral in heavy residues. Authigenic tourmaline has been reported as colorless cappings (elbaite) on only the negative ends of rounded darker-colored detrital grains (Fig. 8.32).

Diagnostic features. Only two other common minerals, biotite and hornblende, have comparable characteristic pleochroism, and both have good cleavage. Schorl is further separated from them by having its greater absorption at right angles to the polarizer direction.

DRAVITE

Composition. $NaMg_3B_3Al_3(AlSi_2O_9)_3(OH)_4$. Some Fe^{2+} usually replaces Mg, with a gradation into schorl. In some brown tourmalines Ca may replace Na almost entirely (calcium tourmaline or uvite).

Indices. $\epsilon = 1.612\text{–}1.630$, $\omega = 1.631\text{–}1.655$, $\omega - \epsilon = 0.019\text{–}0.026$. Indices increase with Fe^{2+} for Mg and to a lesser extent with Ca for Na. The thin-section birefringent colors of upper first order to lower second order are somewhat modified by pleochroic tints.

Color. Shades of brown, buff, and yellow; or colorless. $\omega > \epsilon$:

ϵ	ω
pale yellow	dark yellow brown
colorless	light yellow

Color zoning may be present.

Form. Hexagonal. Commonly euhedral, prismatic but also in coarse subhedral, nearly equidimensional grains. Cleavage absent but fractures at right angles to c may be present.

Orientation. Uniaxial $(-)$. Prismatic sections are length-fast with parallel extinction.

Occurrence. Dravite is characteristic of metasomatically altered magnesian or dolomitic limestones in contact-metamorphic aureoles. Less common in adinoles. With it may occur apatite, tremolite, scapolite, chondrodite, serpentine, and spinel. Some brown tourmalines have also been found in mica schists.

Diagnostic features. Superficially resembles chondrodite, with which it may occur; it may be necessary to obtain a figure to note the biaxial $(+)$ character and large 2V of chondrodite.

ELBAITE

Composition. $Na(Al,Fe^{2+},Mn,Li)_3B_3Al_3(AlSi_2O_9)_3(OH,F)_4$. Elbaite or alkali tourmaline is generally high in Al and low in Mg. Fe^{2+} varies but is usually low, but the Mn content may become high. Ca substitutes for Na only to a very limited extent. Fe^{3+} is low to absent. There is probably a complete gradation between elbaite and schorl.

Indices. $\epsilon = 1.615–1.632$, $\omega = 1.635–1.650$, $\omega - \epsilon = 0.015–0.024$; increasing with Fe^{2+}. Interference colors are of upper first order and lower second order in thin section.

Color. Usually colorless or very pale colored in thin section or crushed pieces. A number of varietal names are given to different macroscopically colored types: achroite for colorless, rubellite for red or pink, indicolite for blue. Many crystals are color-zoned, commonly with a pink core and a green shell ("watermelon" tourmalines); others also show color variation along the c direction. Mn has been suggested as the cause of the pink color, Fe^{2+} of the green, and Cu of the blue.

Form. Hexagonal, euhedral with trigonal or hexagonal cross sections. In parallel or radiating groups or even needlelike clusters.

Orientation. Uniaxial $(-)$. Parallel extinction and length-slow character in longitudinal sections.

Occurrence. In granite pegmatites characterized by hydrothermal replacement. Typical associates are lepidolite, cleavelandite, cookeite, spodumene, and alkali beryl. Rarely authigenic in sediments.

Diagnostic features. Has the lowest indices and weakest colors of all the tourmalines. From apatite it differs in stronger birefringence.

CHAPTER NINE

BIAXIAL MINERALS

Nonsilicates

DIASPORE

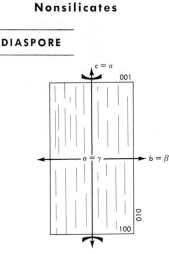

FIGURE 9.1   Orientation of diaspore, section parallel with (100).

Composition. $HAlO_2$, with minor Fe^{3+} and Mn substituting for Al; commonly SiO_2 is present as an impurity.

Indices. $\alpha = 1.700–1.702$, $\beta = 1.720–1.722$, $\gamma = 1.747–1.752$, $\gamma - \alpha = 0.048–0.052$. Slightly higher values may be recorded for ferrian varieties. Maximum interference colors in thin section are of middle third order.

Color. Colorless. Very rarely pale blue or pale pink (manganian). Larger broken pieces may show pleochroism, with $\alpha < \beta < \gamma$.

	α	β	
	colorless	colorless	pale blue
	colorless	pale red	reddish brown

Form. Orthorhombic. Thin tabular crystals platy parallel with (010) and elongate parallel with c are common. Also in scales in interstitial, subhedral to euhedral grains. Perfect (010) cleavage, imperfect (110) cleavage. Broken pieces tend to lie on (010) cleavage faces on which α and γ can be measured. Rare detrital grains also are tabular with ragged ends.

Orientation. Biaxial (+). $\alpha = c$, $\beta = b$, $\gamma = a$. Optic plane is (010) (Fig. 9.1). $2V = 80–88°$, $r < v$ weak. Parallel extinction in principal sections and length-fast in elongate sections. Cleavage pieces yield a flash figure. Bxa figures are secured from sections showing well-developed cleavage. May show twinning on (010).

Occurrence. Found in some altered extrusive igneous rocks of aluminous composition, associated with alunite, kaolinite, and pyrophyllite. Also recorded as a mineral of aluminous xenoliths. A common constituent of emery or of chlorite schists that contain emery, with magnetite, corundum, margarite, chlorite, chloritoid, and spinel. In aluminous quartzites and quartz-mica schists with kyanite or andalusite, rutile, corundum, alunite, pyrophyllite, and lazulite. Less common in marbles with corundum. In sedimentary rocks widespread as a constituent of bauxite, aluminous clays, some laterites, and rarely in aluminous sandstones or shales with zunyite. Bauxite is properly used only as a rock name; for it consists of a mixture of diaspore, gibbsite, and boehmite, usually with some quartz and hydrous iron oxide. Diaspore is a rare detrital, usually found with corundum.

Diagnostic features. The occurrence of diaspore in bauxites and emery is distinctive. In association with corundum it might be mistaken for sillimanite or andalusite. The former is length-slow, and both have lower birefringence.

GIBBSITE

Composition. $Al(OH)_3$. Fe^{3+}, Si, and PO_4 are common admixed impurities.

Indices. $\alpha = 1.568–1.570$, $\beta = 1.568–1.570$, $\gamma = 1.586–1.587$. $\gamma - \alpha = 0.016–0.019$. Maximum interference colors in thin section are in the upper first order or lower second order.

Color. Colorless to neutral gray.

Form. Monoclinic. Crystals are tabular (001) with hexagonal format.

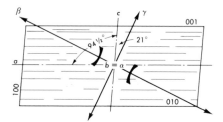

FIGURE 9.2 Orientation of gibbsite, section parallel with (010).

Also concretionary with radial structure or finely granular. Perfect (001) cleavage, which may not be visible in fine-grained material.

Orientation. Biaxial (+). $\alpha = b$, $\beta \wedge a = 16°$, $\gamma \wedge c = 21°$. Optic plane is normal to (010)˙ (Fig. 9.2). 2V = 0°–small. With increasing temperature the orientation changes markedly: $\alpha \wedge a = -40°$, $\beta = b$, $\gamma \wedge c = 45°$, the optic plane becomes (010), and 2V becomes as large as 40°. Dispersion either $r > v$ or $r < v$ strong. Fibers show inclined extinction and can be either length-slow or -fast. Interference figures may be difficult to obtain because of the fine grain. Twinning is common, either on (001) or about [130]. The twinning commonly is multiple, and combinations of several twin types may occur.

Occurrence. The main occurrence is in bauxite, residual clays, and laterites; in these it is associated with diaspore, boehmite, quartz, and limonite. It occurs in chert breccia, replacing quartz. Also a low-temperature hydrothermal mineral in cavities in altered igneous or metamorphic rocks. It may replace corundum, and pseudomorphs after feldspar are known.

Diagnostic features. Much like brucite in indices; birefringence and (+) optical sign but different in inclined extinction; lack of anomalous interference colors and occurrence. Muscovite is (−), has a smaller extinction angle, and shows somewhat higher birefringence. Kaolinite has about the same relief but much lower interference colors. Fibrous gibbsite resembles chalcedony but has higher relief and birefringence. Fine-grained sedimentary carbonate apatite may possibly be confused with gibbsite but has lower birefringence and higher indices.

BOEHMITE

Composition. AlO(OH); generally impure; minor Fe^{3+} may substitute for Al.

Indices. $\alpha = 1.646$, $\beta = 1.653$, $\gamma = 1.661$. Determinations on the pure mineral are few, and a range in the mean index of refraction from 1.624

to 1.645 has been reported for fine-grained material of probably varying degrees of purity.

Color. Colorless.

Form. Orthorhombic. Minute crystals that are tabular parallel with (001) have lenticular sections. Typically in very fine-grained masses or oolitic. Perfect (001) cleavage is rarely seen because of the fine grain.

Orientation. Biaxial $(+)$. $\alpha = a$, $\beta = b$, $\gamma = c$. Optic plane $= (010)$. $2V =$ about $80°$. The orientation requires further study. It also has been regarded as $(-)$ with $\alpha = a$, $\beta = c$, $\gamma = b$, thus with (001) as the optic plane. Basal cleavage pieces give a centered figure that can be secured only on the exceptional coarse-grained material. Elongate sections are length-fast and have parallel extinction.

Occurrence. Chiefly in bauxite, in which it may be the main constituent and is associated with diaspore, gibbsite, quartz, and limonite. Less commonly as an alteration of nepheline with gibbsite and natrolite.

Diagnostic features. Because of restricted occurrence, boehmite can be readily recognized. It has somewhat higher relief than gibbsite and lower relief and lower birefringence than diaspore.

PEROVSKITE

Composition. $(Ca,Na,Ce)(Ti,Nb)O_3$. Commonly essentially $CaTiO_3$. Niobian perovskite may also contain Fe^{2+} (dysanalyte) or Na- and Ce-earths (loparite). Cerian perovskite (knopite) also usually contains some Na and Fe^{3+}. Loparite is usually radioactive owing to Th (as much as 11% Th).

Indices. $n = 2.30$–2.37. Small crystals are isotropic, especially those of the niobian and cerian varieties; larger crystals are anisotropic and complexly twinned with very weak to weak birefringence. Relief in thin section is extreme.

Color. Commonly yellow, orange brown, or light brownish red in thin section. Also gray, black (niobian types), and, rarely, green (dysanalyte). A weak pleochroism may be seen with $\gamma > \alpha$. The darker-colored types may transmit light only in very thin fragments or around the edges, and in thin section the color of the mineral is best observed by inserting the accessory condenser. Minute crystals may otherwise appear opaque owing to total reflection. In a few larger crystals a zonal arrangement of colors is present with brownish-red core and yellowish-red margin.

Form. Small crystals are isometric, but larger anisotropic ones are either orthorhombic or monoclinic. Neither optical nor x-ray data have given an unequivocal determination. Crystals are cubic in habit with square cross sections, or, less commonly, cubooctahedral or skeletal. The cleavage, (100), is distinct but can be seen only in large individuals.

Orientation. Biaxial ($+$). Orthorhombic interpretation: $\alpha = a$, $\beta = c$, $\gamma = b$, optic plane is (001): Monoclinic interpretation: $\alpha \wedge a =$ about 46°, $\beta = b$, $\gamma \wedge c = 45°$, optic plane is (010). 2V = nearly 90°, $r > v$. Twinning is invariably present in the larger crystals and is commonly multiple on (111).

Occurrence. An accessory mineral of subsilicic igneous rocks including intrusive melilite, melilite–nepheline, and nepheline–pyroxene rocks and peridotites and extrusive melilite basalts, nepheline–melilite basalts, olivine–melilite basalts, and leucite basalts. Also as an accessory mineral in some foidal syenitic pegmatites. Also formed in alkalic contact-metamorphic aureoles in carbonate rocks or in marble blocks ejected from alkalic volcanoes and in some carbonatites. To a lesser extent in chlorite schist, talc schist, and pyroxene gneiss. May alter to leucoxene, and has been noted as an alteration of sphene.

Diagnostic features. Larger crystals are distinctive because of the combination of habit, extreme relief, and twinning. Smaller isotropic crystals may resemble melanite or picotite, both of which have much lower relief and different crystal habit. The dark-colored, nearly opaque varieties might be confused with ilmenite in thin section, especially if alteration has taken place. In this case the form and radioactivity are distinctive.

CARNALLITE

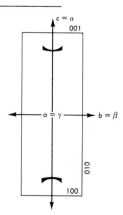

FIGURE 9.3 Orientation of carnallite, section parallel with (100).

Composition. $KMgCl_3 \cdot 6H_2O$. Minor Br may replace Cl. Minor Fe^{2+} for Mg. Soluble in drilling brine of NaCl and $MgCl_2$, so that it may be dissolved from cores. Deliquesces readily and dissolves in the adsorbed moisture. Thin sections should not be ground in water.

Indices. $\alpha = 1.465–1.467$, $\beta = 1.472–1.475$, $\gamma = 1.494–1.497$, $\gamma - \alpha = 0.028–0.031$. In thin section the maximum interference colors lie between upper second and lower third orders.

Color. Colorless.

Form. Orthorhombic. Forms irregular blebs and elongated granular streaks and seams; also in rectangular to irregular aggregates. Crushed material is unoriented. Oriented hematite plates form inclusions.

Orientation. Biaxial $(+)$. $\alpha = c$, $\beta = b$, $\gamma = a$. Optic plane is (010) (Fig. 9.3). $2V = 70°$, $r < v$ weak. May show twinning lamellae.

Occurrence. In saline deposits and evaporites, especially those rich in potassium. Common associates are sylvite, halite, polyhalite, kainite $(KMgSO_4Cl \cdot 3H_2O)$, anhydrite, langbeinite $[K_2Mg_2(SO_4)_3]$, and kieserite $(MgSO_4 \cdot H_2O)$. May be replaced by halite.

Diagnostic features. The only common massive mineral of moderate birefringence in evaporites that has all indices less than 1.50. Epsomite has about the same birefringence but is never massive and has all of its indices below α of carnallite.

ANHYDRITE

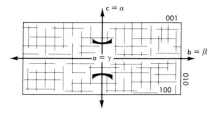

FIGURE 9.4 Orientation of anhydrite, section parallel with (100).

Composition. $CaSO_4$; very minor Sr or Ba may be present.

Indices. $\alpha = 1.569–1.573$, $\beta = 1.572–1.579$, $\gamma = 1.613–1.618$, $\gamma - \alpha = 0.044–0.047$. In thin section the interference colors are as high as third-order green. Grains show higher-order tints. Twinkling, although not as pronounced as in calcite, is noticeable.

Color. Colorless in thin section. Grains may be colorless, pink, or blue;

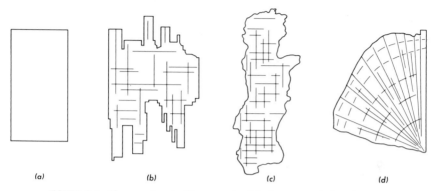

(a)   (b)   (c)   (d)

FIGURE 9.5 Shapes of anhydrite crystals. (a) Common tabular shape. (b) Steplike margins. (c) Corroded relict. (d) Fan-shaped radial group. Carlsbad area, New Mexico and Texas (Schaller and Henderson, 1932).

some show alternating color zones. Others have cores that are gray to bluish in color.

Form. Orthorhombic. Three directions of cleavage: (100), (010), and (001). Crushed pieces tend to be cubic in shape. In thin section anhedral to subhedral aggregates showing two sets of cleavage traces are typical. In some rocks elongated crystals and fibrous types occur (Fig. 9.5). Detrital grains show a wide variety of shapes (Fig. 9.6), including the cubic cleavage pieces, "axehead" type crystals, very irregular jagged grains, and rounded "pebbly" forms (Fig. 9.7).

Orientation. Biaxial (+). $\alpha = c$, $\beta = b$, $\gamma = a$. Optic plane is (010) (Fig. 9.4). $2V = 36$–$45°$, $r < v$. Cleavage pieces on (010) give a flash figure; those on (001) a Bxo figure; and those lying on (100) yield a Bxa figure. Multiple twinning in two sets of lamellae intersecting at nearly 90° occurs in some aggregates. Twins cut the two cleavage traces on the side pinacoid at almost 45°. Extinction is parallel with cleavage traces.

Occurrence. A constituent of evaporite beds with such minerals as halite, gypsum, sylvite, carnallite, and polyhalite. Found in limestones and dolomites, from which it may constitute an important part of the insoluble residue. Associated minerals include dolomite, calcite, magnesite, and celestite. Also a constituent of the caprock of salt domes. It occurs uncommonly as a vein mineral and in contact magnetite deposits. Detrital grains are common locally. A constituent of locomotive boiler scale. Alters readily to gypsum, with which it may be intimately mixed.

Diagnostic features. The pseudocubic cleavage, high birefringence, and

FIGURE 9.6   Radial anhydrite in anhydrite rock, Sun City, Kansas. Polars crossed, ×18 (D. J. McGregor).

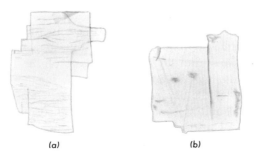

(a)   (b)

FIGURE 9.7   (a), (b) Detrital anhydrite, Persia.

BIAXIAL MINERALS   128

twinkle are characteristic. Gypsum has lower relief and lower birefringence and lacks the three perfect cleavage directions. Grains in heavy residues may be confused with barite, which has higher indices but much lower interference colors.

GYPSUM

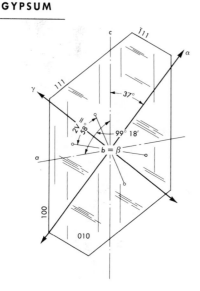

FIGURE 9.8   Orientation of gypsum, section parallel with (010).

Composition. $CaSO_4 \cdot 2H_2O$.

Indices. $\alpha = 1.520{-}1.521$, $\beta = 1.522{-}1.523$, $\gamma = 1.529{-}1.530$, $\gamma - \alpha = 0.009{-}0.010$. .Maximum thin-section interference colors are like those of quartz, i.e., grays and whites of the first order.

Color. Colorless in thin section.

Form. Monoclinic. Fine-grained to coarse-grained and subhedral to anhedral in rock gypsum (Fig. 9.9). Crystals may show elongation parallel with c and be acicular to fibrous or be platy flattened parallel with (010). Bent and twisted crystals are not uncommon. Perfect (010) cleavage, imperfect (100) and (111) cleavages. Crushed material tends to lie on (010). Detrital gypsum may appear as cleavage plates, single fibers, mats of irregular fibers, spherulites, or euhedral crystals.

Orientation. Biaxial (+). $\alpha \wedge c = -37°$, $\beta = b$, $\gamma \wedge c = 46°$. Optic plane is (010) (Fig. 9.8). $2V = 58°$, $r > v$, with strong inclined dispersion. (010) cleavage pieces show a flash figure; rare (110) cleavage fragments display

FIGURE 9.9   Gypsum in rock gypsum, elongate with c and showing
lamellar twinning, Göttingen, Germany. Polars crossed, ×55.

an eccentric Bxo figure; and (100) cleavage surfaces (also rare) yield a
nearly centered optic-axis figure. Twinning on (100) is common and may
be multiple.

Occurrence. In evaporites and as massive rock, usually associated closely
with anhydrite, from which it may form by hydration. Calcite, dolomite,
and halite are other typical associates. In the caprock of salt domes with
anhydrite, rhombohedral carbonates, aragonite, and native sulfur. Also in
limestone and dolomitic limestone, together with anhydrite, barite, celestite and fluorite, and in some shales. Some ore deposits also contain it. The
grinding of thin sections containing gypsum may generate sufficient heat
to dehydrate the mineral partially to $CaSO_4 \cdot \frac{1}{2}H_2O$, as in the manufacture
of plaster of paris. If the section is moistened, rehydration takes place.

Diagnostic features. Anhydrite has higher indices, stronger birefringence,
and pseudocubic cleavage. The hemihydrate has $\alpha = 1.559$, $\gamma = 1.583$, and

$\gamma - \alpha = 0.024$. Thus it has higher birefringence and indices than gypsum and lower indices and lower birefringence than anhydrite.

BARITE

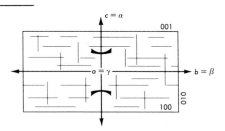

FIGURE 9.10 Orientation of barite, section parallel with (100).

Composition. $BaSO_4$. Small amounts of Ca and Pb replace Ba, and minor Ra may also substitute. There is probably a complete series to celestite, with Sr for Ba.

Indices. $\alpha = 1.634$–1.637, $\beta = 1.636$–1.638, $\gamma = 1.646$–1.648, $\gamma - \alpha = 0.010$–0.013. The birefringence is low, with a maximum in thin section of orange of the first order. The interference tints may be irregularly distributed to give a mottled appearance. Front pinacoid sections have very low birefringence. The indices increase with Pb and decrease slightly with Sr.

Color. Colorless generally, but pale blue and yellow grains appear in heavy residues. May be dark or turbid because of inclusions.

Form. Orthorhombic. Perfect (001) and (210) cleavages, imperfect (010) cleavage. Crushed pieces tend to lie on either of the good cleavages. Cleavage traces appear in thin section generally as two sets of lines at right angles on (100) or on (010), and as a single set on (001). On other sections the traces join at oblique angles. Fractured, irregular pieces on cleavage flakes are characteristic of heavy residues; rounding is rare in grains (Fig. 9.11). Inclusions of clay minerals and carbonaceous material may be present, and abundant sand grains also occur as inclusions in larger crystals. Usually in granular anhedral aggregates or in masses of subhedral crystals, either bladed, elongated parallel with b, or tabular parallel with (001). Oolitic, pisolitic, concretionary, and radial-nodular structures are common. Much barite is interstitial, and some replaces fossils.

Orientation. Biaxial (+). $\alpha = c$, $\beta = b$, $\gamma = a$. Optic plane is (010) (Fig. 9.10). $2V = 36$–$40°$, $r < v$ weak. (100) sections give a Bxa figure; (010) sections yield a flash figure; basal cleavage plates give a Bxo figure on

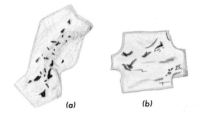

(a)   (b)

FIGURE 9.11   (a), (b) Detrital barite, Trias, Cheshire, England.

which α and γ can be measured. Cleavage pieces that lie on (110) show an off-center Bxa figure. Multiple twinning with (110) as the twin plane may be present.

Occurrence. Barite occurs in various types of metalliferous veins as a gangue mineral, and in other veins as the chief constituent. It is also found in vesicles in rhyolites and basalts and in carbonatites. In sandstones it forms cement. Also occurs as concretions, oolites, and nodular masses in limestone and some sedimentary iron ores. Detrital barite is not common except locally.

Diagnostic features. The combination of cleavages, indices, and birefringence is distinctive in thin section and in crushed fragments. Shape, vitreous luster, and cloudy appearance are important in grains. Celestite resembles barite closely but has lower indices and a slightly larger 2V. A flame test may be necessary to distinguish between the two.

CELESTITE

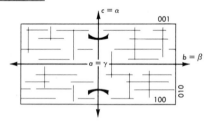

FIGURE 9.12   Orientation of celestite, section parallel with (100).

Composition. $SrSO_4$, with minor Ca; there probably is a complete series to barite.

Indices. $\alpha = 1.621–1.622$, $\beta = 1.623–1.624$, $\gamma = 1.630–1.632$, $\gamma - \alpha = 0.008–0.009$. Maximum interference colors in thin section are in the first order, similar to those of quartz.

Color. Colorless in thin section and in crushed grains.

Form. Orthorhombic. Either elongate parallel with b (rarely a) or tabular parallel with (001). Also anhedral, fine-grained. Perfect (001) cleavage, less perfect (110) cleavage, imperfect (010) cleavage. Detrital grains are very irregular in shape. Crushed grains tend to lie on (001) or (110). May be fibrous, nodular, or interstitial.

Orientation. Biaxial (+). $\alpha = c$, $\beta = b$, $\gamma = a$. Optic plane is (010) (Fig. 9.12). $2V = 50\text{–}51°$, $r < v$. Tabular, fibrous, or bladed crystals are length-slow. (001) cleavage pieces yield a Bxo figure; (010) cleavage surfaces give a flash figure; (110) cleavages an eccentric figure.

Occurrence. Chiefly in sedimentary rocks, including limestone, dolomitic limestone, dolomite–anhydrite rocks, chalk, and rock salt. Associated minerals are calcite, dolomite, anhydrite, barite, gypsum, and halite. It may be a cementing mineral in sandstones. Also a constituent of veins, with strontianite, pyrite, quartz, calcite, and fluorite. An accessory of anhydrite caprock of salt domes.

Diagnostic features. Difficult to tell from barite optically. A flame test for Sr may be necessary.

GLAUBERITE

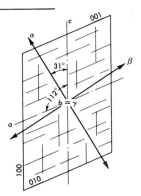

FIGURE 9.13 Orientation of glauberite, section parallel with (010).

Composition. $Na_2Ca(SO_4)_2$. In water becomes translucent, dissolves partly, and leaves a residue of calcium sulfate. Thus thin sections should not be ground in water.

Indices. $\alpha = 1.507\text{–}1.515$, $\beta = 1.527\text{–}1.535$, $\gamma = 1.529\text{–}1.536$, $\gamma - \alpha = 0.021\text{–}0.023$. In thin section the maximum interference colors are in the middle of the second order.

Color. Colorless.

Form. Monoclinic. Usually massive and anhedral. Crystals projecting into cavities tend to be basal (001) plates or prismatic. Perfect (001) cleavage, poor (110) cleavages.

Orientation. Biaxial ($-$). $\alpha \wedge c = 31\text{--}34°$, $\beta \wedge a = -9\text{---}12°$, $\gamma = b$. Optic plane normal to (010) (Fig. 9.13). $2V = 0\text{--}7°$, $r > v$ strong. The optical orientation changes markedly with temperature.

Occurrence. In marine evaporites and other saline deposits with anhydrite, halite, polyhalite, magnesite, thenardite (Na_2SO_4), bischofite ($MgCl_2 \cdot 6H_2O$), kieserite ($MgSO_4 \cdot H_2O$), and clay minerals. It may replace magnesite. Rare as a fumarole deposit.

Diagnostic features. Polyhalite has a larger 2V, and its best cleavage fragments show an eccentric optic-axis figure. Anhydrite has parallel extinction and stronger birefringence.

POLYHALITE

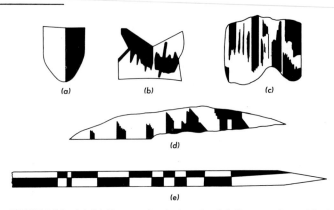

FIGURE 9.14 (a)–(e) Shape and twinning of polyhalite crystals, Carlsbad area, New Mexico and Texas (Schaller and Henderson, 1932).

Composition. $K_2Ca_2Mg(SO_4)_4 \cdot 2H_2O$. Partly water-soluble with deposition of gypsum.

Indices. $\alpha = 1.546\text{--}1.548$, $\beta = 1.558\text{--}1.562$, $\gamma = 1.567$, $\gamma - \alpha = 0.019\text{--}0.021$. In thin section the interference colors reach a maximum of second-order blue.

Color. Colorless, or pale red owing to hematite inclusions.

Form. Triclinic. A variety of forms: fibers in coalescing spherulites, aggregates of minute subparallel wisps, granular masses, (010) tablets,

TABLE 9.1   OTHER EVAPORITE MINERALS *

NAME	CRYSTAL SYSTEM	FORMULA	REFRACTIVE INDICES α β γ		SIGN, 2V	CLEAVAGE	EXT. ANGLE TWINNING	SOLUBILITY	CONTINENTAL OR MARINE DEPOSITS
sylvite	I	KCl	$n = 1.490$			(100)		H₂O sol.	M
langbeinite	I	$K_2Mg_2(SO_4)_3$	$n = 1.535$					H₂O sol.	M
thenardite	O	Na_2SO_4	1.464 1.474 1.485		(+) 83°	(001) dist.		H₂O sol.	C
epsomite	O	$MgSO_4 \cdot 7H_2O$	1.433 1.455 1.461		(−) 51°	(010) perf., (011) dist.		H₂O sol.	M
bischofite	M	$MgCl_2 \cdot 6H_2O$	1.492 1.506 1.519		(+) 79°		$\beta \wedge c = 10°$	H₂O sol.	M
kainite	M	$KMgCl(SO_4) \cdot 3H_2O$	1.494 1.505 1.516		(−) 85°	(100) (110) dist.	$\alpha \wedge c = 8°$	H₂O sol.	M
kieserite	M	$MgSO_4 \cdot H_2O$	1.523 1.535 1.586		(+) 57°	(110) (111) perf.	$\gamma \wedge c = 76°$	H₂O sl. sol.	M
nahcolite	M	$HNaCO_3$	1.375 1.505 1.583		(−) 75°	(101) perf. (111) dist.	$\alpha \wedge c = 28°$ (101) twin.	H₂O sol.	C
trona	M	$HNa_3(CO_3)_2 \cdot 2H_2O$	1.412 1.492 1.540		(−) 72°	(100) perf.	$\beta \wedge c = 17°$	H₂O sol.	C
colemanite	M	$Ca_2B_6O_{11} \cdot 5H_2O$	1.586 1.592 1.614		(+) 56°	(010) perf. (001) dist.	$\gamma \wedge c = 83°$		C
ulexite	T	$NaCaB_5O_9 \cdot 8H_2O$	1.496 1.505 1.519		(+) 73°	not noted in fibers	$\gamma \wedge c = 22°$		C

* See also halite, calcite, dolomite, carnallite, anhydrite, gypsum, glauberite, polyhalite, aragonite, borax, and kernite

euhedral rectangular or euhedral double wedge-shaped crystals, corroded crystals, and thin veinlets along halite cleavages (Fig. 9.14). Fibers are elongated parallel with *b*. Good (100) cleavage and a (010) parting; crushed material tends to be oriented on either of these faces.

Orientation. Biaxial (−). Optical orientation not known completely. One optic axis is eccentric to the (100) cleavage, the other is nearly parallel with *b*. 2V = 60–62°. Multiple or paired twinning with (010) as the twin plane is common, also in various combinations with (001) twinning to produce irregular grating or crosshatch type of twinning (Fig. 9.14). On twinned (100) cleavage faces the extinction angle between the composition plane and $\gamma' = 28°$.

Occurrence. In evaporites with anhydrite, halite, magnesite, and kieserite. It may replace halite and anhydrite. Readily altered by water to gypsum.

Diagnostic features. Gypsum has lower double refraction and lower indices. Glauberite is monoclinic with a smaller 2V.

MONAZITE

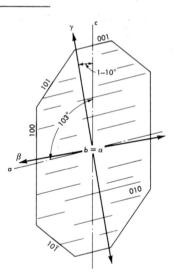

FIGURE 9.15   Orientation of monazite, section parallel with (010).

Composition. $(Ce,La)PO_4$. Nd-, Pr-, Sm-, and Y-earths and Th also are usually present, less commonly some Ca, U, and Fe^{3+}. Si may replace P in Th-rich varieties by $Th^{4+}+Si^{4+}-Ce^{3+}+P^{5+}$.

Indices. $\alpha = 1.785–1.800,\quad \beta = 1.787–1.801,\quad \gamma = 1.837–1.849,\quad \gamma - \alpha =$

0.045–0.055. In general the indices rise with increasing Th and Si. Relief very high, and interference colors in thin section show a maximum of high third order to low fourth order. Cleavage pieces (001) have low birefringence and yield indices close to α and β.

Color. Colorless to gray in thin section. Detrital grains may be pale yellow, yellow brown, or dark red brown. Large grains may show pleochroism, $\beta > \alpha = \gamma$:

$\alpha =$ light yellow
$\beta =$ dark yellow
$\gamma =$ greenish yellow

Form. Monoclinic. Occurs as small euhedral crystals which may be tablets flattened parallel with (100) or elongated parallel with b. Longitudinal sections are four- or six-sided parallelograms. (001) cleavage distinct. Detrital pieces may be (100) tablets with rounded corners, tending to be egg-shaped, (001) cleavage pieces, or euhedral crystals of varying complexity (Fig. 9.16). Inclusions of zircon, rutile, and opaque dust appear in some grains.

Orientation. Biaxial (+). $\alpha = b$, $\beta \wedge a = 4\text{--}13°$, $\gamma \wedge c = 1\text{--}10°$. The extinction angle, $\gamma \wedge c$, apparently increases slightly with Th. Optic plane is normal to (010) (Fig. 9.15). $2V = 6\text{--}19°$. Cleavage pieces yield a slightly eccentric Bxa figure. Dispersion either $r < v$ or $r > v$ weak to very weak. Tablets (100) elongate parallel with c are length-slow; sections of crystals elongate parallel with b are length-fast.

Occurrence. An accessory mineral in granite, granodiorite, aplite, granite pegmatite, syenite, syenite pegmatite, carbonatite, and some quartz veins. Also in biotite gneiss, granulite, mica schist, and sillimanite gneiss. May be an important detrital mineral of beach and river placers and some sandstones. May show superficial alteration of brown stains.

Diagnostic features. Resembles xenotime and bastnäsite. For the distinction between it and xenotime see under xenotime. Bastnäsite is uniaxial and has lower indices: $\omega = 1.716\text{--}1.722$, $\epsilon = 1.816\text{--}1.824$, and higher bi-

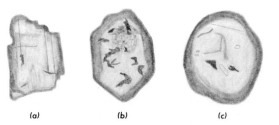

(a) (b) (c)

FIGURE 9.16 Detrital monazite. (a), (b) Beach sand, Florida. (c) Brazil.

refringence: $\epsilon - \omega = 0.100-0.102$. Sphene has higher birefringence than monazite and much stronger dispersion and shows different crystal outlines.

LAZULITE

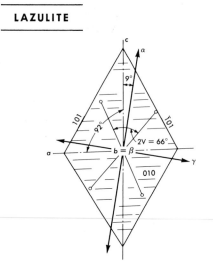

FIGURE 9.17   Orientation of lazulite, section parallel with (010).

Composition. $(Mg,Fe^{2+})Al_2(PO_4)_2(OH)_2$. The series can be expressed in terms of two end members, lazulite, $MgAl_2(PO_4)_2(OH)_2$, and scorzalite, $FeAl_2(PO_4)_2(OH)_2$.

Indices. $\alpha = 1.604-1.637$, $\beta = 1.631-1.667$, $\gamma = 1.642-1.677$, $\gamma - \alpha = 0.034-0.040$. Both indices and birefringence increase with increasing Fe^{2+} (Fig. 9.18).

Color. Pleochroic from colorless to blue in thin section and crushed pieces:

α = colorless, colorless
β = light blue, blue green
γ = light blue, blue green

$\alpha < \beta < \gamma$. There is no definite correlation between color and composition, but lazulite from quartz veins and pegmatite dikes shows darker blue colors than that from quartzite.

Form. Monoclinic. Usually in anhedral masses; rarely in bipyramidal crystals of diamond-shaped outline in thin section. Cleavages (110) and (101) good. Quartz blebs may be poikilitically included. Some specimens contain liquid inclusions with bubbles.

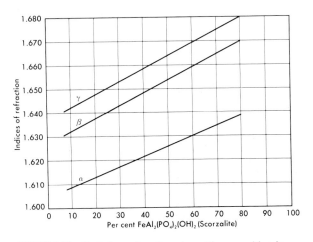

FIGURE 9.18 Variations of α, β, and γ with composition in lazulite-scorzalite series (Pecora and Fahey, 1950).

Orientation. Biaxial $(-)$. $\alpha \wedge c = -9$--$-10°$, $\beta = b$, $\gamma \wedge a = 11$--$12°$. Optic plane is (010) (Fig. 9.17). $2V = 61$--$70°$, decreasing with increasing Fe^{2+}; $r < v$ weak. Cleavage pieces give off-centered interference figures. The long diagonal of euhedral sections is nearly parallel with the fast ray. Multiple twinning is common, with c as the twin axis.

Occurrence. Occurs in quartzite and quartz schist, quartz veins and associated replacement lodes ("aluminous quartzites"), and in granite pegmatites. Lazulite-bearing rocks are usually aluminous and may contain such associated species as sericite, kyanite, dumortierite, andalusite, corundum, pyrophyllite, and rutile.

Diagnostic features. Characterized by strong pleochroism combined with high birefringence. Blue tourmaline is uniaxial and shows different crystal outlines; dumortierite has higher indices and lower birefringence and shows parallel extinction.

ARAGONITE

Composition. $CaCO_3$. Dimorphous with calcite. May contain important amounts of Sr, Pb, and Zn.

Indices. Normally $\alpha = 1.529$--1.530, $\beta = 1.680$--1.682, $\gamma = 1.685$--1.686, $\gamma - \alpha = 0.155$--0.156. With increasing Pb the indices rise to $\alpha = 1.542$, $\beta = 1.695$, $\gamma = 1.699$, $\gamma - \alpha = 0.157$. With additions of Sr the indices may decrease to $\alpha = 1.527$, $\beta = 1.670$, $\gamma = 1.676$, $\gamma - \alpha = 0.149$. In thin section

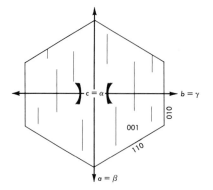

FIGURE 9.19 Orientation of aragonite, section parallel with (001).

the maximum interference colors are whites of the high orders. The high birefringence gives rise to a twinkle effect upon rotation of the stage, as in calcite. Basal sections display first-order colors; for $\gamma - \beta$ is small.

Color. Colorless.

Form. Orthorhombic. Columnar, acicular, or fibrous. Thicker crystals show a six-sided outline in cross section. Radial aggregates or randomly oriented masses of needles are common. Also stalactitic, massive-granular, and in thin parallel or concentric bands with calcite. Elongation is parallel with c. Imperfect (010) cleavage.

Orientation. Biaxial ($-$). $\alpha = c$, $\beta = a$, $\gamma = b$. Optic plane is (100) (Fig. 9.19). $2V = 18–19°$, $r < v$ weak. With more Pb $2V$ increases to 23°. Twinning with (110) as the twin plane is common and may give rise to pseudohexagonal groups. Needles and fibers are length-fast. Cleavage faces yield a Bxo figure.

Occurrence. Common as a low-temperature hydrothermal mineral in cavities in andesites, basalts, and serpentinites, usually with zeolites. Also widespread in tufa, calcareous sinter, onyx, dripstone, cave oolites, pearls, and shells; less common in some varieties of gypsum caprock of salt domes and other evaporites. Inverts readily to calcite and may be replaced by dolomite. Also occurs in boiler scale and, rarely, as a detrital species.

Diagnostic features. Resembles calcite, but is biaxial and lacks the rhombohedral cleavage. Crushed calcite pieces consistently give an eccentric uniaxial figure; aragonite cleavage pieces tend to give a centered Bxo figure. A staining test may be necessary in some instances to distinguish the two (Table 2.4).

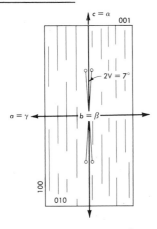

Composition. $SrCO_3$. May contain some Ca.

Indices. $\alpha = 1.516–1.520$, $\beta = 1.664–1.667$, $\gamma = 1.666–1.668$, $\gamma - \alpha = 0.147–0.150$. Interference colors are high-order whites except in basal sections, which are nearly dark under crossed polars, since $\gamma - \beta$ is very small.

Color. Colorless.

Form. Orthorhombic. Acicular, also anhedral-granular. (110) cleavage distinct.

Orientation. Biaxial $(-)$. $\alpha = c$, $\beta = b$, $\gamma = a$. Optic plane is (010) (Fig. 9.20). $2V = 7°$, $r < v$ weak. Lamellar twinning on (110) is very common. Pseudohexagonal groups also are formed by this twinning.

Occurrence. In limestone, dolomite, magnesite rock, and clay as nodules, seams, and in cavities. Minerals found with it are celestite, which it may replace, gypsum, quartz, and dolomite. In salt dome caprock. Also as a gangue mineral with barite and celestite in veins of lead and silver minerals. Rare as a detrital mineral.

Diagnostic features. Aragonite has higher indices, and its optic plane parallels (100). Witherite has slightly higher indices and a larger 2V.

BORAX

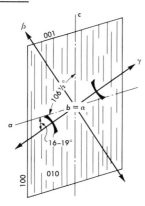

FIGURE 9.21   Orientation of borax, section parallel with (010).

Composition. $Na_2B_4O_7 \cdot 10H_2O$. Water-soluble. Dehydrates to tincalconite, $Na_2B_4O_7 \cdot 5H_2O$.

Indices. $\alpha = 1.447$, $\beta = 1.469-1.470$, $\gamma = 1.472$, $\gamma - \alpha = 0.025$.

Color. Colorless.

Form. Monoclinic. Crystals are prismatic. Granular-anhedral aggregates also occur. Perfect (100) and good (110) cleavages.

Orientation. Biaxial $(-)$. $\alpha = b$, $\beta \wedge c = 33-36°$, $\gamma \wedge a = -16--19°$. Optic plane normal to (010) (Fig. 9.21). $2V = 39-40°$, $r > v$ distinct. Strong crossed dispersion, and sections normal to an optic axis display abnormal blue and brown interference colors and do not extinguish. Twinning on (100) may be present.

Occurrence. In saline lake deposits together with other borates and halite. An efflorescence on soils in arid regions.

Diagnostic features. Low indices plus strong crossed dispersion. Tincalconite is uniaxial $(+)$ with $\omega = 1.461$, $\epsilon = 1.474$.

KERNITE

Composition. $Na_2B_4O_7 \cdot 4H_2O$.

Indices. $\alpha = 1.454$, $\beta = 1.472$, $\gamma = 1.488$, $\gamma - \alpha = 0.034$. The maximum thin-section interference colors are in the upper second order.

Color. Colorless.

Form. Monoclinic. Anhedral to euhedral. In granular, irregular masses, cross-fiber veins, rounded individuals, and single crystals; (001) and

BIAXIAL MINERALS

142

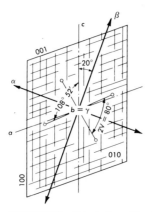

FIGURE 9.22 Orientation of kernite, section parallel with (010).

(100) cleavages perfect and govern the shape and orientation of broken material; γ can be measured on either of these cleavages. (201) cleavage fair; (011) parting.

Orientation. Biaxial $(-)$. $\alpha \wedge a = 38°$, $\beta \wedge c = -20°$, $\gamma = b$. Optic plane is normal to (010) (Fig. 9.22). $2V = 80°$, $r > v$ distinct. (100) cleavage pieces show an eccentric Bxa figure; (001) pieces, an eccentric optic-axis figure.

Occurrence. As veins, aggregates, and crystals in clay. Locally the rock becomes massive kernite, and clay is subordinate. Also as crystals in borax. Other associated species are tincalconite, probertite, and calcite. Kernite is slowly soluble in cold water, so that thin sections must be prepared in an anhydrous medium.

Diagnostic features. Good cleavages, water solubility, and association.

Ortho- and Ring Silicates

OLIVINE GROUP

GENERAL The olivine group includes the forsterite–fayalite series, the fayalite–tephroite series, monticellite, as well as some rarer types. Within the forsterite–fayalite series various varietal names (chrysolite, hyalosiderite, hortonolite, etc.) have been used in different ways to express chemical variation based on molecular percentage of Fe_2SiO_4. Also in the fayalite–tephroite series a number of names have been developed for intermediate members (knebelite, etc.). These names are generally unnecessary.

Forsterite and ferroan forsterite suffice for olivines of composition Fo_{100}–Fo_{50}, and magnesian fayalite and fayalite are enough for the range Fo_{49}–Fo_0. An analogous subdivision and nomenclature are applicable to the fayalite–tephroite series. There have been suggestions that the optical properties of the olivines do not depend exclusively on their composition, but are, in part, a function of their thermal history; for some olivine phenocrysts from extrusive rocks are reported to have smaller 2V's than chemically identical olivines from intrusive rocks.

FORSTERITE

Composition. Mg_2SiO_4 to $(Mg,Fe^{2+})_2SiO_4$ or Fo_{100}–Fo_{50}. Mn can be present in small amounts, and a few rare types apparently are intermediate between forsterite and tephroite.

Indices. $\alpha = 1.636$–1.730, $\beta = 1.650$–1.759, $\gamma = 1.669$–1.772. $\gamma - \alpha = 0.033$–0.042. In thin section maximum interference colors range from upper second to lower third order. Indices and birefringence increase with increasing Fe^{2+} (Fig. 9.23).

Color. Colorless. Rare detrital pieces are yellow to green.

Form. Orthorhombic. Generally polygonal-anhedral in intrusive rocks, and euhedral as phenocrysts in extrusive rocks (Fig. 9.24). Euhedral sections commonly are somewhat elongated parallel with c, six-sided and outlined by traces of side pinacoid (010) or prisms (110) and steep domes (021). In metamorphic rocks it is anhedral, commonly rounded. In meteorites occurs as equant subhedral grains, as poikilitic inclusions in enstatite, and as rounded aggregates (chondrules) of various types. Inclusions are volcanic matrix material, grains of apatite, picotite, and magnetite, and oriented plates of ilmenite (?). Skeletal crystals, shaped like irregular x's, occur in some lavas. Cleavages are usually imperfect and do not affect shape of crushed material, but good (010) and poorer (100) cleavages may be present. Phenocrysts may be rounded or corroded. Sinuous fractures are common, and detrital fragments are rounded or chipped, with conchoidal fracture surfaces.

Orientation. Biaxial. $\alpha = b$, $\beta = c$, $\gamma = a$, optic plane is (001) (Fig. 9.25). Fo_{100}–Fo_{88} (+), with 2V increasing from 87–90°; Fo_{88}–FO_{50} (−), with 2V decreasing from 90 to 74°. Dispersion weak, $r > v$ for the (−) range, $r < v$ for the (+) range. Parallel extinction in principal sections. Sections showing cleavage traces are length-slow. Twinning on (011), (012), or (031) not common. Twin individuals are broad and poorly defined. Peridotitic olivines may display a banded structure (deformation lamellae). Zoned crystals are not uncommon and normally possess a magnesian core

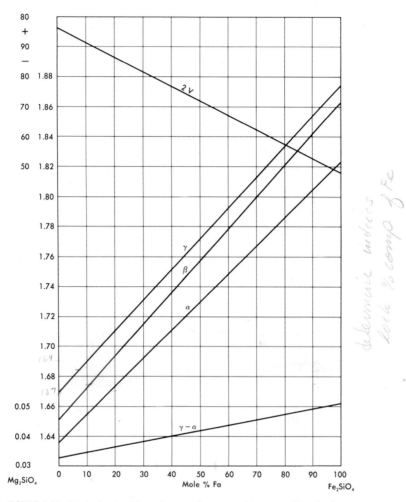

FIGURE 9.23 Variation in 2V, α, β, γ, and $\gamma - \alpha$ with composition in the olivine series (Poldervaart, 1950).

and margins enriched in iron. Reverse relations are very rare. Zoning can be detected by variations in 2V and birefringence.

Occurrence. In ultramafic and mafic igneous rocks, both intrusive and extrusive types, such as stony meteorites, dunite and other peridotites, troctolite, gabbro, norite, basalt, diabase, essexite, and teschenite. Less common in intermediate types. In impure dolomitic marbles of contact or regional origin, associated with garnet, phlogopite, and calcite. Rare as detrital mineral because of its ease of alteration. In extrusive rocks it

FIGURE 9.24   Olivine phenocryst in melilite-bearing basalt,
Hochkohl, Württemberg, Germany. Polars crossed, ×94.

alters typically to iddingsite or to chlorophaeite or to interlayered smectite–chlorite; in intrusive and metamorphic rocks mainly to antigorite, chrysotile, or both, with separation of by-product magnetite. Also replaced by chlorites, talc, carbonates, nontronite, and other poorly defined hydrous magnesium silicates. Commonly encompassed by reaction rims of augite, pigeonite, hornblende, biotite, and magnetite. Although forsterite and quartz are not in equilibrium, forsterite phenocrysts are recorded from quartz basalts or from lavas with either cristobalite or tridymite in gas cavities. In rocks that contain both phenocryst and matrix olivine, the latter contains more iron. In a general way the iron content of olivines also increases with SiO_2 content of the rocks in which they occur. Forsterite is rarely found as a detrital mineral, except in some Recent or Pleistocene shore sands. It is the chief constituent of forsterite refractory bricks and occurs in the matrix of some other bricks.

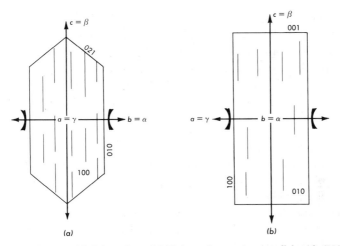

(a)   (b)

FIGURE 9.25 (a) Orientation of (+) forsterite, section parallel with (100). (b) Orientation of (−) forsterite and fayalite, section parallel with (010). The orientation of monticellite is similar.

Diagnostic features. Has lower indices and larger 2V than fayalite. Differs from pyroxenes by absence of cleavage and extinction angles. Partial alteration to serpentine or iddingsite is also highly characteristic.

FAYALITE

Composition. $(Fe,Mg)_2SiO_4$ to Fe_2SiO_4 or Fo_{50}–Fo_0. Mn may be present in considerable amounts, and apparently minor Fe^{3+} and Zn as well. Tephroite is Mn_2SiO_4.

Indices. $\alpha = 1.731$–1.824, $\beta = 1.760$–1.864, $\gamma = 1.773$–1.875, $\gamma - \alpha = 0.042$–0.051. In thin section the maximum interference colors are of the third order. Both birefringence and indices increase with Fe^{2+} (Fig. 9.23). In the tephroite–fayalite series indices also increase with Fe^{2+} over Mn. Tephroite has $\alpha = 1.759$, $\beta = 1.786$, $\gamma = 1.797$.

Color. Colorless to pale yellow. $\beta > \alpha > \gamma$. α, γ = greenish yellow, pale yellow, pale amber; β = orange yellow.

Form. Orthorhombic. Usually anhedral with irregular to polygonal outlines. In cavities it may form euhedral tabular crystals. Cleavages normally absent to very imperfect, but irregular cracks common. Minute oriented plates of iron minerals may be present.

Orientation. Biaxial (−). $\alpha = b$, $\beta = c$, $\gamma = a$. Optic plane is (001) (Fig. 9.25b). $2V = 74$–$47°, r > v$ weak. 2V decreases with Fe^{2+}. Parallel extinc-

tion in principal sections. Twinning in broad lamellae, as in forsterite. Tephroite is $(-)$ with $2V = 65°$, and in this series 2V decreases with Fe^{2+}.
Occurrence. Rare in igneous rocks. Found in a few alkalic granites, syenites, nepheline syenites, granophyres, as well as in rhyolites, trachytes, phonolites, and dacites. Fayalite is in equilibrium with quartz or nepheline. In rhyolites and obsidians it occurs in lithophysae, with quartz, tridymite, cristobalite, sanidine, tourmaline, biotite, garnet, and magnetite. Occurs as a high-temperature metamorphic mineral in eulysites, commonly associated with hedenbergite, hypersthene, grunerite, anthophyllite, garnet, quartz, and magnetite. Characteristic of slags. Tephroite occurs in manganiferous contact ore deposits. Fayalite alters to antigorite, hematite, and limonite. May be replaced by grunerite.

Diagnostic features. Has higher indices and lower 2V than forsterite. Higher birefringence, parallel extinction, and poor cleavage distinguish it from pale-colored pyroxenes.

MONTICELLITE

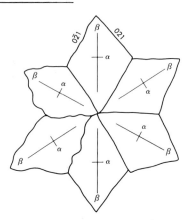

FIGURE 9.26 Orientation of monticellite twinned on (031), monticellite peridotite, Haystack Butte, Montana (Larsen et al., 1941).

Composition. $CaMgSiO_4$. Fe^{2+} and Mn replace Mg. $CaMnSiO_4$ occurs as the rare mineral, glaucochroite. In most monticellites the FeO content is usually below 10%. There is a continuous series to kirschsteinite, $CaFeSiO_4$.

Indices. $\alpha = 1.639–1.663$, $\beta = 1.645–1.674$, $\gamma = 1.653–1.680$, $\gamma - \alpha = 0.014–0.017$. Maximum interference tints for thin sections are upper first

TABLE 9.2 OPTICAL PROPERTIES OF HIGH-TEMPERATURE CALCIUM SILICATES

NAME	COMPOSITION	SYSTEM	α	β	γ	2V	ORIENTATION	CLEAVAGE, ETC.
cuspidine (custerite)	$Ca_4Si_2O_7(F,OH)_2$	mono.	1.586–1.592	1.595–1.596	1.598–1.606	63° (+) $r > v$ marked	$\gamma = b$, $\gamma \wedge c = 6°$; optic plane $\perp$ (010)	(001) good, (110) distinct, lamellar twinning
larnite	β-Ca_2SiO_4	mono.	1.707	1.715	1.730	large (+)	$\gamma = b$, $\alpha \wedge c = 13$–$14°$; optic plane $\perp$ (010)	(100) good $\perp$ to an imperfect cleavage; tabular ∥ (010); fine (100) multiple twinning
merwinite	$Ca_3MgSi_2O_8$	mono.	1.706–1.708	1.711–1.712	1.718–1.724	67° (+) $r > v$	$\gamma = b$, $\alpha \wedge c = 36°$; optic plane $\perp$ (010)	(010) perfect; multiple (110) twinning
rankinite	$Ca_3Si_2O_7$	mono.	1.640–1.641	1.644	1.650	64° (+)	$\beta = b$, $\alpha \wedge c = 15°$; optic plane is (010)	none
scawtite	$Ca_4Si_3O_8(CO_3)_2$	mono.	1.597–1.603	1.606–1.609	1.618–1.621	74–78° (+)	$\beta = b$, $\gamma \wedge a = 29°$; optic plane is (010)	(001) perfect, (010) poor
spurrite	$Ca_5Si_2O_8CO_3$	mono.	1.640–1.641	–1.674––1.676	1.679–1.681	38–40° (–)	$\alpha = b$, $\beta \wedge c = -33°$; optic plane $\perp$ (010)	(100) distinct; fine multiple twinning in two directions at 57°
tilleyite	$Ca_5Si_2O_7(CO_3)_2$	mono.	1.612–1.617	1.632–1.635	1.652–1.654	85–90° (+) $r < v$	$\beta = b$, $\alpha \wedge c = 12$–$18°$; optic plane is (010)	(100) perfect, (101)? good

order. Indices and birefringence increase with Fe^{2+} and Mn. The indices increase slightly more for Fe^{2+} than Mn; the birefringence increases slightly less for Fe^{2+} than Mn. For the rare part of the series, ferroan monticellite–kirschsteinite, $\alpha = 1.674–1.693$, $\beta = 1.694–1.734$, $\gamma = 1.706–1.735$, $\gamma - \alpha = 0.032–0.047$. For glaucochroite, $\alpha = 1.679–1.686$, $\beta = 1.716–1.723$, $\gamma = 1.729–1.736$, $\gamma - \alpha = 0.045–0.051$, decreasing with Mg and increasing slightly with Fe^{2+}.

Color. Colorless.

Form. Orthorhombic. Usually anhedral-granular, but may form good crystals of prismatic habit. Poor (010) cleavage. Six-pointed euhedral trillings divided into six segments are known (Fig. 9.26). Overgrowths on olivine also occur. Augite may be included.

Orientation. Biaxial. $\alpha = b$, $\beta = c$, $\gamma = a$, optic plane is (001) (Fig. 9.26). Usually ($-$) with $2V = 88–65°$, decreasing with increasing Fe^{2+} and Mn; $r > v$, distinct. 2V for pure synthetic $CaMgSiO_4$ is essentially $90°$; for $CaFeSiO_4$, $2V = 48–50°$, ($-$); for glaucochroite $2V = 61°$, ($-$). Both have similar orientations. Parallel extinction. Sections showing cleavage traces are length-slow. Trillings twinned on (031) have been observed (Fig. 9.26).

Occurrence. Primarily as a high-temperature contact metamorphic mineral in impure marbles, with vesuvianite, gehlenite, diopside, wollastonite, and the rare minerals tilleyite, spurrite, merwinite, cuspidine, larnite, wilkeite, crestmoreite, and xanthophyllite (Table 9.2). Also a constituent of ultramafic alkalic rocks such as alnöite, monticellite peridotite, nepheline basalt, olivine lamprophyre and melilite–diopside rocks. Kirschsteinite occurs in melilite nephelinites. Also a constituent of furnace slags and some magnesite bricks. Alters to serpentine and augite and is replaced by vesuvianite. Glaucochroite occurs in Zn–Mn contact ores.

Diagnostic features. Monticellite resembles forsterite but has much lower birefringence and has a ($-$) sign, whereas some forsterite is ($+$). The combination of indices plus 2V serves to distinguish olivines of the forsterite–fayalite series from those of the monticellite–kirschsteinite series.

IDDINGSITE

Composition. Formerly regarded as a hydrated, hydrous Mg–Fe^{3+}-silicate of variable and rather uncertain composition, it is now known, by means of x-rays, that the material is a mixture (Sun, 1957; Wilshire, 1958; Brown and Stephen, 1959). Goethite or hematite is an invariable constituent; the Mg-silicate has been variously described as an amorphous substance, a mixed-layer smectite–chlorite, or "a layer lattice silicate."

Form. As rims around olivine or as complete pseudomorphs after olivine

FIGURE 9.27 Iddingsite rim on embayed olivine phenocryst in basalt, Ruby Mountains, Montana. Polars not crossed, ×94.

(Fig. 9.27). Although the material is polycrystalline and consists of two components, grains show optical properties of single crystals. This homogeneity results because the very small grains of both components are strictly oriented throughout a single olivine pseudomorph. This parallelism results from the inheritance of the oxygen framework of the replaced olivine, completely by goethite or hematite and partly by the Mg silicate. Thus several cleavages have been reported for the "mineral."

Optical properties. Shades of reddish brown, golden brown, brownish red, and ruby red. Direction of the fast ray shows greatest absorption.

	GENERAL RANGE	LOW	HIGH
n_1	1.67–1.73	1.61	1.79
n_2	1.72–1.77	1.66	1.86

$n_2 - n_1 = 0.04$–0.05. In thin section maximum interference colors are in the third order but are masked by the mineral color.

Occurrence. Principally as an alteration product of olivine in extrusive rocks, formed as a deuteric mineral by the escape of gases. Overgrowths of augite may surround the rims of iddingsite on olivine, or zones of fresh olivine may enclose corroded olivine phenocrysts, partly converted to iddingsite. Also reported in the matrix of such rocks. An uncommon detrital mineral.

Diagnostic features. The combination of color, relief, and mode of occurrence is distinctive.

HUMITE GROUP

GENERAL   The humite group includes the species:

Norbergite, $Mg_3SiO_4(F,OH)_2$
Chondrodite, $Mg_5(SiO_4)_2(F,OH)_2$
Humite, $Mg_7(SiO_4)_3(F,OH)_2$
Clinohumite, $Mg_9(SiO_4)_4(F,OH)_2$

Norbergite and humite are orthorhombic; chondrodite and clinohumite are monoclinic. The simplified formulae above show that successive members of the group increase their composition by addition of Mg_2SiO_4 units. In a general way this increase is accompanied by increase in the indices of refraction, norbergite having the lowest range of indices and clinohumite the highest. But there is considerable overlap between the species. Norbergite, the rarest of the group, approximates $Mg_3SiO_4(F,OH)$ closest, with minor Fe^{2+},Fe^{3+} and considerable predominance of F over OH. It has $\alpha = 1.561$–1.563, $\beta = 1.566$–1.573, $\gamma = 1.587$–1.590, $\gamma - \alpha = 0.027$, $2V = 44$–$50°$, $(+)$. New data on the group by Sahama (1953) include reorientation of the crystallographic axes in a manner consistent with those of forsterite; thus he lists extinction angles on (010) as $c \wedge \alpha$ instead of $a \wedge \alpha$.

CHONDRODITE

Composition. $(Mg,Fe)_5(SiO_4)_2(F,OH)_2$. Fe^{2+} may be present in considerable amounts, with a maximum of 10.54% FeO reported. Some Mn, Al, Fe^{3+}, and Ti may be present. The OH:F ratio varies widely, both fluor- and hydroxyl-rich types are known.

Indices. $\alpha = 1.592$–1.643, $\beta = 1.602$–1.655, $\gamma = 1.619$–1.675, $\gamma - \alpha = 0.025$–0.037. The indices increase with Fe^{2+}, Mn, Fe^{3+}, and Ti for Mg

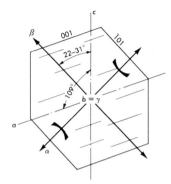

FIGURE 9.28   Orientation of chondrodite, section parallel with (010).

and with OH for F (Fig. 9.31). Maximum birefringent colors in thin section vary from lower to upper second order.

Color. Shades of yellow and brown. $\alpha > \beta > \gamma$. $\alpha =$ yellowish brown,
deep chrome yellow, reddish brown, brownish yellow. $\beta =$ pale yellowish
brown, yellow, pale reddish brown, yellowish green. $\gamma =$ colorless, light
yellow, pale brown, pale green.

Form. Monoclinic, anhedral to euhedral. Rounded grains to (010)
tablets. In complex parallel intergrowths composed of clinohumite, chon-

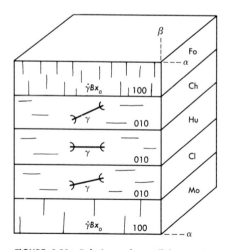

FIGURE 9.29   Relations of parallel growths of the humite-group minerals (chondrodite, humite, clinohumite), forsterite, and monticellite in the skarns of the contact zone
of the Beinn an Dubhaich granite, Great Britain (Tilley, 1951).

CHONDRODITE   **153**

drodite, monticellite, and forsterite (Fig. 9.29). Enclosed by monticellite. (001) cleavage poor.

Orientation. Biaxial normally $(+)$, very rarely apparently $(-)$. $\alpha \wedge a = -22$––$31°$, $\beta \wedge c = 22$–$31°$, $\gamma = b$, optic plane normal to (010) (Fig. 9.28). $2V = 64$–$90°$, $r > v$ moderate to weak. Multiple or simple twinning (001) common, and the mineral may be twinned with another member of the group. Twinning on (105) and (305) also occurs. The extinction direction closest to the trace of the twin plane is the fast ray.

Occurrence. In contact-dolomitic marbles and associated veins with calcite, magnetite, pleonaste, fluorite, humite, clinohumite, monticellite, fosterite, clinochlore, tremolite, diopside, dravite, and vesuvianite. Less commonly in pencatites with periclase and brucite. Alters to serpentine. Exceedingly rare as a detrital species.

Diagnostic features. Humite and norbergite are orthorhombic; clinohumite has a smaller extinction angle. May be mistaken for dravite, which is uniaxial. Staurolite has higher refractive indices and is orthorhombic.

HUMITE

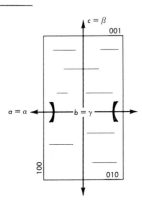

FIGURE 9.30 Orientation of humite, section parallel with (010).

Composition. $(Mg,Fe)_7(SiO_4)_3(F,OH)_2$, nearly 11% FeO has been reported. Fe^{3+}, Ti, Al, and Mn may be present as minor elements. Considerable F–OH variation.

Indices. $\alpha = 1.607$–1.643, $\beta = 1.619$–1.653, $\gamma = 1.639$–1.675. $\gamma - \alpha = 0.028$–0.036. Maximum interference colors are of the second order in thin section. Indices increase with Fe^{2+}, Mn, Fe^{3+}, and Ti for Mg and with OH for F (Fig. 9.31).

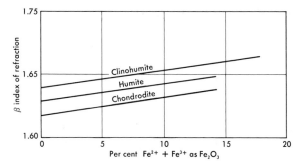

FIGURE 9.31 Variation of β with composition in humite group (Thiele, 1940).

Color. Colorless to pale yellow in thin section. $\alpha > \gamma > \beta$. α = pale golden yellow, deep golden yellow. β = pale yellow, colorless. γ = golden yellow, pale yellowish.

Form. Orthorhombic. Rounded or ellipsoidal anhedral grains and subhedral (100) or (001) plates, (001) cleavage poor.

Orientation. Biaxial (+). $\alpha = a$, $\beta = c$, $\gamma = b$, optic plane is (001). 2V = 68–81°, $r > v$ weak (Fig. 9.30).

Occurrence. Apparently not as common as clinohumite or chondrodite, more common than norbergite. A contact-metamorphic mineral in dolomitic marbles, with such species as calcite, clinohumite, forsterite, fluorite, and magnetite.

Diagnostic features. Colorless types resemble forsterite, but the birefringence is normally somewhat less and 2V somewhat larger. Clinohumite shows inclined extinction. Staurolite has higher refractive indices and has γ = golden yellow.

CLINOHUMITE

Composition. $(Mg,Fe)_9(SiO_4)_4(F,OH)_2$. As much as 15% FeO may be present and also important smaller amounts of Mn, Fe^{3+}, and Ti (titanoclinohumite) and minor Zn and Pb. The F:OH ratio varies considerably.

Indices. α = 1.623–1.702, β = 1.636–1.709, γ = 1.651–1.728. $\gamma - \alpha$ = 0.028–0.045. Maximum birefringent colors in thin section are second and third order. Indices increase with Fe^{2+} and Mn for Mg and with increasing Ti and Fe^{3+}; with OH for F (Fig. 9.31). Abnormal interference tints in titanian types.

Color. Colorless to yellow. $\alpha > \gamma > \beta$. α = brownish yellow, pale golden yellow. β = pale yellow, greenish yellow, colorless. γ = pale yellow, colorless.

Form. Monoclinic. Rounded to irregularly bounded anhedra. In lamellar intergrowths of chondrodite, monticellite, and forsterite (Fig. 9.30). May be euhedral, tabular. Enclosed as minute rounded grains in monticellite. Poor (001) cleavage. Lamellar twinning on (001).

Orientation. Biaxial (+). $\alpha \wedge a = -7-\!-15°$, $\beta \wedge c = 7-15°$, $\gamma = b$. Optic plane normal to (010). $2V = 52-90°$, $r > v$ weak. 2V apparently decreases with Fe^{2+}, Fe^{3+}, and Ti. The dispersion increases markedly with Ti.

Occurrence. In contact-dolomitic marbles and related veins in which humite, forsterite, calcite, fluorite, magnetite, spinel, phlogopite, tremolite, clinochlore, and xanthophyllite are associates. Also in altered peridotites and gabbros with such species as enstatite, actinolite, anthophyllite, forsterite, phlogopite, prehnite, epidote, clinochlore, magnetite, and spinel. May replace forsterite.

Diagnostic features. Differs from humite and olivine in showing inclined extinction and from chondrodite by its lower extinction angle.

AXINITE

Composition. $(Ca,Fe^{2+},Mn,Mg)_3Al_2BSi_4O_{15}OH$. Small to moderate amounts of Fe^{3+} and minor Ti may replace Al, and some Na substitutes for Ca.

Indices. $\alpha = 1.672-1.693$, $\beta = 1.677-1.701$, $\gamma = 1.681-1.704$, $\gamma - \alpha = 0.009-0.013$. In general axinite high in Fe^{2+} or Fe^{3+} has the higher refractive indices, whereas that with high Mn has lower indices. Maximum interference colors in thin section vary from first-order yellow-gray to orange.

Color. Colorless to lavender in thin section. Grains or fragments may be pink, red brown, or purple. Pleochroism moderate: α = colorless, yellow, light brown, yellow brown, pale green, olive green. β = colorless, deep yellow, deep blue, pale violet, violet. γ = colorless, pale green, olive yellow, brown, dark violet. The three directions of minimum, intermediate, and maximum absorption do not coincide with the three vibration directions.

Form. Triclinic. Euhedral to anhedral. Well-developed or subhedral crystals are usually wedge-shaped or bladed and obliquely tabular with (010) or (011) and show acute-angled cross sections. Also lamellar, granular, rarely columnar elongated parallel with c, and also in rosette-like clusters. (100) cleavage is distinct; (001), (110), and (011) cleavages are poor. Detrital pieces are wedge-shaped, rounded, or irregular, showing

minute conchoidal fractures. May contain inclusions such as actinolite.
Orientation. Biaxial ($-$). Positions of the three optical directions vary with composition. In some manganese-rich axinite α is nearly normal to (111), whereas in others α more closely approaches the normal to (011). 2V = 69–87°, $r < v$ strong. Zoning, including variation in 2V, occurs in some crystals. Extinction is inclined to the crystal outlines and to cleavage traces.

Occurrence. In contact-metamorphosed limestones or tactites. Here the associated mineral assemblages include zoisite, epidote, vesuvianite, diopside–hedenbergite, grossularite, danburite, tourmaline, orthoclase, pyrrhotite, sphene, quartz, and calcite. As a deuteric mineral in miarolitic cavities or vesicles in granite, monzonite, diorite, albitophyre, basalt, and diabase with a variety of associates such as epidote, actinolite, hornblende, chlorite, prehnite, datolite, danburite, tourmaline, pectolite, albite, orthoclase, apatite, and calcite. Also in veins that occur in various ore deposits with quartz, calcite, clinozoisite or epidote, fluorite, and pyrite. Axinite may alter to chlorite and is veined by calcite. Intergrowths with prehnite are reported. Rare as a detrital species. Found in granitic pegmatites as a rare accessory.

Diagnostic features. The combination of crystal outline, low birefringence, high relief, and inclined extinction is characteristic. The association of other boron minerals such as tourmaline, datolite, and danburite may point to the possible presence of axinite.

TOPAZ

Composition. $Al_2SiO_4(F,OH)_2$. The $F:OH$ ratio varies considerably.

Indices. $\alpha = 1.606$–1.629, $\beta = 1.609$–1.631, $\gamma = 1.616$–1.638. $\gamma - \alpha = 0.011$–0.009. With increasing OH the indices rise and birefringence decreases slightly (Fig. 9.32). In thin section maximum interference tints are first-order pale yellow. Detrital grains show first-order red to second-order blue colors.

Color. Colorless in thin section. Thick detrital pieces may show a yellow (α and β) to pink (γ) pleochroism.

Form. Orthorhombic. Perfect (001) cleavage. Usually in anhedral grains or aggregates or in vesicular masses. Columnar groups are found in some deposits, as are stubby euhedral crystals. Grains from heavy residues are subangular and commonly platy (001), or may be nearly square in outline (Fig. 9.33). Crushed material tends to lie on cleavage faces. Liquid inclusions are common.

Orientation. Biaxial ($+$). $\alpha = a$, $\beta = b$, $\gamma = c$. Optic plane is (010) (Fig.

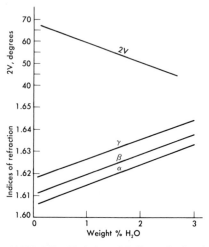

FIGURE 9.32 Variation of indices of refraction and 2V with composition in topaz.

9.34). Basal cleavage plates yield a Bxa figure, with 2V = 48–68° and noticeable dispersion, $r > v$. 2V decreases with increasing substitution of OH for F. Parallel extinction in vertical sections. Eight-sided basal sections show symmetrical extinction. Cleavage traces are length-fast. Some basal sections show a division into sectors around a core, which may have slightly different indices than the marginal units.

Occurrence. In greisen with quartz, lithium muscovite, tourmaline, apatite, and cassiterite and locally with beryl, wolframite, and fluorite. In granitic pegmatites; also in high-temperature ore deposits and veins with Sn, W, and Au mineralization and in the wall rocks of these and other ore deposits, locally associated with pyrite, diaspore, pyrophyllite, and siderite. Of metasomatic origin in such rocks as (1) quartzite that contains diaspore, corundum, fluorite, chloritoid, pyrophyllite, and rutile; (2) fine-grained topaz rock with accessory rutile, magnetite, sericite, and corundum;

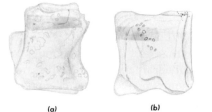

(a)   (b)

FIGURE 9.33 (a), (b) Detrital topaz. (a) Nigeria. (b) Okarito, New Zealand (Hutton, 1950).

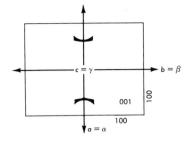

FIGURE 9.34 Orientation of topaz, section parallel with (001).

(3) muscovite schist with disseminated accessory topaz; and (4) musco-vite–topaz–kyanite–quartz rock with some rutile and tourmaline. In lithophysae of rhyolites and obsidians together with the wood-tin type of cassiterite, specular hematite, quartz, sanidine, tridymite, and less commonly the rare minerals bixbyite and pseudobrookite. Intergrown with cassiterite as long fibers in some wood-tin nodules. May be alluvial with cassiterite and columbite. Not abundant in heavy residues except locally. Alters to sericite or kaolinite.

Diagnostic features. The moderate relief coupled with low birefringence and (+) biaxial character is distinctive. Quartz resembles it only superficially and has lower indices. Altered topaz is in some cases hard to differentiate from sericitized andalusite in detrital grains. In apatite the cleavage traces are length-slow.

SILLIMANITE GROUP

GENERAL The sillimanite group includes andalusite, sillimanite, and kyanite. The first two are orthorhombic; kyanite is triclinic. Their composition is Al_2SiO_5. All dissociate at different temperatures to mullite ($Al_6Si_2O_{13}$) and silica glass. Sillimanite is stable to the highest temperature. Mullite, rarely found in nature, is described here because of its close genetic relations to the sillimanite minerals.

ANDALUSITE

Composition. Al_2SiO_5. Variable amounts of Fe^{3+}, Mn^{3+}, and Ti may be present. Manganian andalusite has been called viridine.

Indices. $\alpha = 1.629–1.640$, $\beta = 1.633–1.644$, $\gamma = 1.638–1.650$. Manganian andalusite, with about 10% Mn_2O_3 and some Fe^{3+} and Ti, has $\alpha = 1.662$,

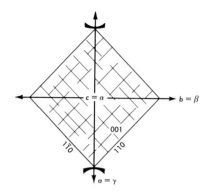

FIGURE 9.35 Orientation of andalusite, section parallel with (001).

$\beta = 1.671$, $\gamma = 1.691$. $\gamma - \alpha = 0.009\text{–}0.011$ for ordinary andalusite, 0.029 for the manganian variety. The indices increase with increasing Fe^{3+} + Mn^{3+} + Ti. In thin section the maximum interference colors are of middle first order.

Color. Some andalusite is colorless in thin section, but usually it is pleochroic:

α = pink, light red, yellow
β = colorless, pale yellow, green
γ = colorless, pale yellow, greenish yellow

The color shows considerable variation, even in single grains. In some, zones are present; in other examples, irregularly distributed patches of darker color occur, owing to local concentrations of Fe^{3+}. Detrital grains vary from colorless to blood red.

Form. Orthorhombic. (110) cleavage is good; in thin section two sets of cleavage traces at about 90° appear in cross sections. Longitudinal sections show one cleavage direction. Coarse columnar aggregates or euhedral crystals with nearly square cross sections predominate. Cleavage pieces

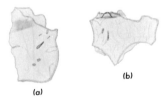

(b)

(a)

FIGURE 9.36 (a), (b) Detrital andalusite, Cornwall, England.

tend to lie on (110). Detrital grains are usually irregular to subangular; some are prismatic (Fig. 9.35). Inclusions of graphite and carbonaceous material are not uncommon (Fig. 9.37) and in the variety chiastolite are symmetrically arranged and appear as a black cross in basal section.

Orientation. Biaxial $(-)$. $\alpha = c$, $\beta = b$, $\gamma = a$. Optic plane is (010) (Fig. 9.35). $2V = 71\text{--}86°$, $r < v$, weak to strong. Basal sections show symmetrical extinction; grains and section elongate parallel with c are length-fast with parallel extinction. Cleavage pieces yield an eccentric Bxo figure. The size of 2V decreases with increasing $Fe^{3+} + Mn^{3+} + Ti$. Varieties rich in Fe^{3+} and Ti display strong dispersion.

Occurrence. Typically in contact-metamorphic slates, hornfelses, phyllites, and low-grade schists. Associated minerals are cordierite and biotite. Also in some regional metamorphic schists (Fig. 9.37) with biotite, garnet, staurolite, and sillimanite and much less commonly in granitic pegmatites. Also in metasomatic "hydrothermal quartzites" and massive andalusite

FIGURE 9.37 Andalusite porphyroblast with zonally distributed inclusions in micaceous schist, Gefrees, Fichtelgebirge, Bavaria. Polars not crossed, ×94.

ANDALUSITE

rocks. Rather widespread as a detrital mineral (Fig. 9.36). Has been found replaced by sillimanite and kyanite and alters readily to sericite.

Diagnostic features. In detrital grains the pleochroism distinguishes it from hypersthene, which has a = pink and c = green. Sillimanite is length-slow. Topaz, which also resembles pale-colored andalusite in heavy grains, is (+) and has a smaller 2V.

SILLIMANITE

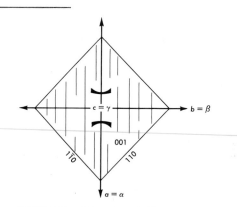

FIGURE 9.38   Orientation of sillimanite, section parallel with (001).

Composition. Al_2SiO_5. Minor Fe^{3+} may be present.

Indices. α = 1.653–1.661, β = 1.654–1.670, γ = 1.669–1.684. The slight increase in the indices is probably due to increasing Fe^{3+}. $\gamma - \alpha$ = 0.020–0.023, with maximum interference colors in thin section in upper first and and lower second orders.

Color. Colorless to neutral gray in thin section. Detrital grains are color-less, yellow, buff, or light brown and uncommonly pleochroic with:

α = pale brown, yellow
β = brown, gray green
γ = brown, blue

Form. Orthorhombic. Usually forms slender prisms or needles, minutely fibrous, subparallel intergrowths, fibrous radial aggregates (Fig. 9.39), or felted masses with curving, interlacing fibers (fibrolite). The (010) cleav-age shows in thin section as the long diagonal in rhombic cross sections of larger euhedra. Fractures across elongated prisms also are common. Grains from heavy residues tend to be slender, flattened prisms with splin-tered ends (Fig. 9.40). Spinel and biotite occur as inclusions.

Orientation. Biaxial $(+)$. $\alpha = a$, $\beta = b$, $\gamma = c$. Optic plane is (010) (Fig. 9.38). $2V = 20\text{--}30°$, $r > v$ strong. Bxa figures can be obtained from basal sections of uncommon, large crystals. Fibers and prismatic sections are length-slow. Cleavage pieces yield a flash figure. Fibrous radial groups show a pseudo-interference figure without convergent light under crossed nicols.

Occurrence. Typically a metamorphic mineral. Occurs in contact hornfelses and slates but also is common in gneisses, schists, and granulites, with garnet, cordierite, biotite, graphite, corundum, staurolite, andalusite,

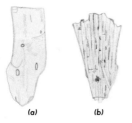

(a)   (b)

FIGURE 9.40   Detrital sillimanite. (a) Ceylon. (b) An aggregate of fibers, stream sand, southwestern Montana.

kyanite, and dumortierite. Replaces biotite. May occur as pods or lenses that consist of nearly pure sillimanite. Sillimanite forms pseudomorphs after kyanite and andalusite and is replaced by sericite. Much less common in some pegmatites and quartz veins. A widespread detrital species.

Diagnostic features. The fibrous nature, length-slow character, rhombic cross section with diagonal cleavage, and parallel extinction are sufficient to separate it from kyanite and andalusite. Mullite resembles sillimanite closely, and careful measurement of a powder x-ray photograph may be required to distinguish between them.

MULLITE

Composition. Variable from Al_4SiO_8 to $Al_6Si_2O_{13}$. Small amounts of Ti and Fe^{3+} may be present.

Indices. $\alpha = 1.642$–1.653, $\beta = 1.644$–1.655, $\gamma = 1.654$–1.679, $\gamma - \alpha = 0.012$–0.031. The indices and the birefringence increase with $Fe^{3+} + Ti$. The maximum interference colors in thin section range from middle first to middle second order.

Color. Colorless in thin section. Coarse crushed material rich in Fe^{3+} and Ti may show weak pleochroism:

$\alpha = $ colorless
$\beta = $ colorless
$\gamma = $ rose

Form. Orthorhombic. Elongated prismatic crystals, needles, or fibers are typical. Cross sections are nearly square in outline; (010) cleavage is distinct.

Orientation. Biaxial (+). $\alpha = a, \beta = b, \gamma = c$. Optic plane is (010). 2V = 20–50°, $r > v$. Cross sections show symmetrical extinction. Longitudinal sections and fibers are length-slow. A synthetic dimorph of mullite, which is apparently of rare occurrence in slags and firebricks, has the properties: $\alpha = c, \alpha = 1.600, \gamma = 1.610$.

Occurrence. Rare as a natural mineral. Occurs at a few localities as a constituent of argillaceous xenoliths. Formed by calcining andalusite, kyanite, sillimanite, dumortierite, and topaz. A constituent of firebricks, refractory porcelains, and slags.

Diagnostic features. The best method of distinguishing between mullite and sillimanite is by means of their x-ray powder diffraction patterns. Mullite is very uncommon in nature, however, and the danger of confusing the two in ordinary rocks is therefore slight.

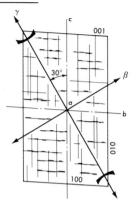

FIGURE 9.41 Orientation of kyanite, section parallel with (100).

Composition. Al_2SiO_5. Minor Fe^{3+}, Cr, and Ti occur in some types.

Indices. $\alpha = 1.712$–1.718, $\beta = 1.720$–1.725, $\gamma = 1.727$–1.734, $\gamma - \alpha = 0.012$–0.015. Maximum interference colors in thin section lie in the upper first order and lower second order range.

Color. Colorless in thin section. Grains are neutral gray or pale blue.

Form. Triclinic. Bladed, prismatic, or rarely fibrous. Narrow elongate sections are parallel with (010), broad elongate sections with (100) (Fig. 9.43). Perfect (100) cleavage and imperfect (010) cleavage. Conspicuous

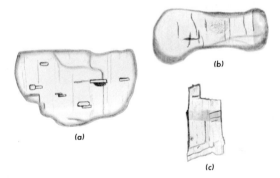

FIGURE 9.42 Detrital kyanite. (a), (b) Beach sand, Florida. (b) "Dogbone" type of grain. (c) River sand, southwestern Montana.

cross fractures or parting planes parallel with (001), 85° to the elongation. Detrital grains are prismatic and slightly rounded, some with ragged terminations (Fig. 9.42). Cleavage pieces tend to lie on (100). Inclusions of carbonaceous material, sericite, and quartz may be present.

Orientation. Biaxial ($-$). Optic plane nearly normal to (100) (Fig. 9.41); the trace of the plane on (100) makes an angle of 27–32° with c. On (010) $\gamma \wedge c = 5$–8°, and on (001) $\gamma \wedge c = 0$–2½°. $2V = 82°$, $r > v$ weak. (100) cleavage pieces, with the maximum extinction angle, yield a nearly centered Bxa figure; (010) cleavage pieces show an eccentric flash figure. Paired twins are common with (100) as the twin plane, and multiple twins with (001) as the twin plane also are known. In the former type the twin

FIGURE 9.43 Kyanite porphyroblast in kyanite schist, Burnsville, North Carolina. Polars not crossed, ×28.

BIAXIAL MINERALS

individuals may extinguish in the same positions, but show opposing character of elongation.

Occurrence. A metamorphic mineral of schists (Fig. 9.43), gneisses, and granulites. Common associates are garnet, biotite, muscovite, staurolite, sillimanite, and rutile. Less commonly kyanite is found in chloritoid-muscovite schists. It may be replaced by sillimanite and is commonly partly altered to fine-grained muscovite. Kyanite is a common accessory detrital in sandstones and sands (Fig. 9.42). Widespread but minor in many types of quartzose veins and some pegmatites.

Diagnostic features. Distinguished from sillimanite by the extinction angle and indices. In grains the shape, cross fractures, and birefringence are distinctive.

STAUROLITE

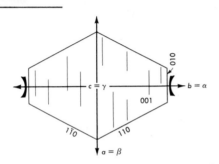

FIGURE 9.44   Orientation of staurolite, section parallel with (001).

Composition. $Fe^{2+}_2Al_9Si_4O_{22}(O,OH)_2$. Mg substitutes to a small extent for Fe^{2+}, and minor Fe^{3+} replaces Al. Mn, Ni, Zn, and Co are rare replacements of Fe^{2+}.

Indices. $\alpha = 1.736–1.747$, $\beta = 1.740–1.754$, $\gamma = 1.745–1.762$. Mg for Fe^{2+} decreases the indices; Fe^{3+} for Al increases them (Fig. 9.47). $\gamma - \alpha = 0.009–0.015$, with maximum interference colors in thin section falling between middle and upper first order. The birefringence also increases slightly with Fe^{3+}. The interference tints may be somewhat altered by the color of the mineral.

Color. In thin section pleochroic with:

$\alpha =$ colorless, pale yellow
$\beta =$ colorles, pale yellow, yellowish brown
$\gamma =$ light yellow, orange yellow, reddish brown

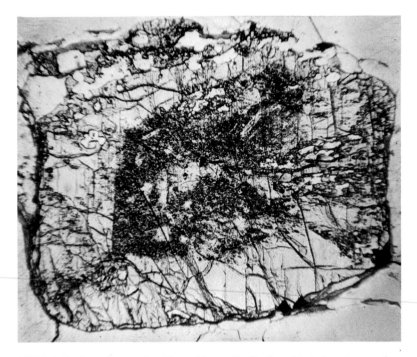

FIGURE 9.45 Staurolite porphyroblast with zonally distributed inclusions in staurolite-kyanite schist, Cherry Creek, Montana. Polars not crossed, ×28.

$\gamma > \beta > \alpha$. Cobaltian staurolite is pleochroic in shades of blue and violet. Color zoning may be present. Detrital grains display yellow to brown colors.

Form. Orthorhombic. Poor (010) cleavage. Crushed pieces are generally unoriented. Commonly euhedral or subhedral with six-sided basal sections in which (110) faces predominate over (010). A short prismatic habit is common. Crystals may be relatively large (porphyroblasts) and usually contain numerous inclusions (Fig. 9.45). A poikiloblastic ("sieve") structure may be strongly developed, so that the crystal is skeletal or

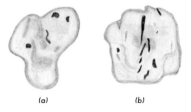

(a)   (b)

FIGURE 9.46 (a), (b) Detrital staurolite, beach sands, Florida.

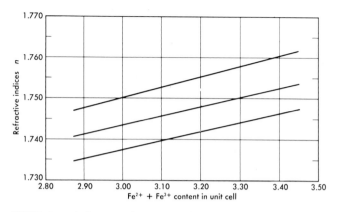

FIGURE 9.47 Refractive indices of staurolite as a function of the Fe^{2+} and Fe^{3+} content in the unit cell (Juurinen, 1956).

netlike. Quartz is the most common inclusion, but garnet, graphite, biotite, muscovite, tourmaline, magnetite, rutile, and zircon also appear. Zircon grains develop a pleochroic halo. Detrital grains tend to be somewhat platy with irregular to ragged edges (Fig. 9.46). Euhedral detrital grains are encountered rarely, and etched terminations on them have been recorded.

Orientation. Biaxial ($+$). $\alpha = b$, $\beta = a$, $\gamma = c$. Optic plane is (100) (Fig. 9.44). $2V = 80$–$89°$; dispersion $r > v$ weak to strong. Rarely ($-$), with $2V = 88°$. Elongate sections have maximum absorption parallel with c, are length-slow, and show parallel extinction. Cruciform twins on (031) and (231) are common but are rarely seen in thin section or as grains. Contact twins on (031) may appear in thin section.

Occurrence. In rocks of regional metamorphic origin such as phyllites, schists, and gneisses. Associated minerals are quartz, biotite, chlorite, muscovite, kyanite, andalusite, sillimanite, plagioclase, tourmaline, garnet, rutile, zircon, graphite, and magnetite. Rarely associated with anthophyllite. A widespread and abundant detrital mineral in sandstones and alluvial deposits. Alters to sericite, chlorite, or limonite.

Diagnostic features. The pleochroism and sieve structure are distinctive. Brown tourmaline, which it resembles, is uniaxial and length-fast and has its maximum absorption at right angles to c.

FIGURE 9.48 Dumortierite in bladed, semiradial aggregates, in pegmatite, Dehessa, California. Polars crossed, ×55.

Composition. $Al_8BSi_3O_{19}(OH)$. Fe^{3+} may substitute for Al (a maximum of about 2.5% Fe_2O_3 has been reported), and Ti usually is present either as Ti^{3+} or Ti^{4+}.

Indices. $\alpha = 1.659\text{--}1.678$, $\beta = 1.684\text{--}1.691$, $\gamma = 1.686\text{--}1.692$, $\gamma - \alpha = 0.011\text{--}0.027$. Maximum interference colors range from middle first to middle second order in thin section. Indices and birefringence rise with increasing Fe^{3+} and Ti.

Color. Strongly pleochroic from colorless to shades of blue, greenish blue, pink, or lilac:

[1] A tectosilicate but grouped here, paragenetically, with the sillimanite group.

α = blue, violet, greenish blue, deep rose
β = colorless, red violet, pale lilac, yellow
γ = colorless, pale blue, pale green

Absorption $\alpha > \beta \geqq \gamma$, thus maximum absorption occurs when the fiber length is parallel with the polarizer vibration direction. Single fibers may show a gradation in color in the c direction and some crystals show different pleochroism in their centers than in their margins. It has been suggested that the blue color is due to Ti^{3+} replacing Al and the pink to Ti^{4+} replacing Si.

Form. Orthorhombic. Usually fibrous, acicular or bladed, elongated parallel with c. Felted aggregates may be randomly oriented, subparallel to parallel, or in radial arrangement (Fig. 9.48). Twinned coarser crystals show a six-sided cross section. Cleavages, (100) distinct and (110) poor, are seen only in larger crystals. Detrital pieces are broken prismatic grains with marked striations parallel with the length.

Orientation. Biaxial $(-)$. $\alpha = c, \beta = b, \gamma = a$. Optic plane is (010) (Fig. 9.49). $2V = 20\text{--}52°$, $r < v$ or $r > v$, strong. Fibers show parallel extinction and are length-fast. Twins on (110) form trillings, with optic planes approximately 60° to one another.

Occurrence. In granite pegmatites, aplites, and quartz veins with other aluminous minerals such as cordierite, andalusite, kyanite, and sillimanite. In quartzite with kyanite, muscovite, and rutile. In various rocks that have been hydrothermally altered, such as granite, granodiorite, trachyte, sericite schist, and granite gneiss, associated with quartz, sericite, anda-

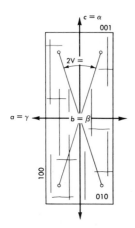

FIGURE 9.49 Orientation of dumortierite, section parallel with (010).

lusite, sillimanite, pyrophyllite, tourmaline, rutile, and pyrite. A relatively rare species in the heavy mineral suites of sandstones and siltstones. Alters to fine-grained muscovite.

Diagnostic features. Blue dumortierite resembles some varieties of tourmaline but which have the greater absorption at right angles to the polarizer direction. Minute dumortierite fibers, in which pleochroism is pale or absent, resemble sillimanite, but that mineral is length-slow. Piedmontite, which also occurs in needles and has vivid red pleochroic colors, is ($+$) with a larger 2V and shows a small extinction angle.

SAPPHIRINE

Composition. $(Mg,Fe^{2+})_2Al_4SiO_{10}$. Fe^{2+} substitutes for Mg to Mg:Fe $=$ $3:1$. Minor Fe^{3+} may replace Al; 2Al may replace a Mg–Si pair.

Indices. $\alpha = 1.701–1.729$, $\beta = 1.703–1.732$, $\gamma = 1.705–1.734$, $\gamma - \alpha =$ $0.004–0.006$. Maximum interference colors in thin section are first-order grays and whites. Sections normal to a may show grayish purple anomalous interference colors.

Color. In thin section pale blue or gray and weakly pleochroic.

$\alpha =$ colorless, greenish blue, buff, pale brown, yellow

$\beta =$ greenish blue, blue, dark blue

$\gamma =$ deep blue, yellow green, green

$\alpha < \beta < \gamma$. The color increases with increasing Fe.

Form. Monoclinic. Anhedral granular or subhedral, tabular parallel with (010) and somewhat elongated parallel with c. No distinct cleavages. Rims of sapphirine around feldspar adjacent to ilmenite are recorded. Inclusions of spinel (hercynite and pleonaste) may be present; magnetite, in parallel rods or unoriented, also appears as inclusions.

Orientation. Biaxial ($-$). $\alpha \wedge a = 15–6°$, $\beta = b$, $\gamma \wedge c = 6–15°$. Optic plane is (010). $2V = 51–69°$; $r < v$ with distinct inclined dispersion. Elongated sections are length-slow. Wide multiple twin bands are present in material from some localities, with (010) as the twin plane.

Occurrence. As a high-temperature mineral of metamorphic origin, it is found in rocks low in SiO_2 and high in MgO and Al_2O_3. In schists, gneisses, granulites, charnockites, and some hornfelses associated with anthophyllite, cordierite, biotite, hornblende, hypersthene, sillimanite, magnetite, pleonaste, anorthite, corundum, and diaspore. Also in ilmenite-rich rock containing rutile. Occurs as a contact-metamorphic mineral in emery, aluminous xenoliths, and altered wall rocks of granitic to gabbroic intrusives. The rare mineral kornerupine, $(Mg,Fe,Al)_4(Al,B)_6(SiO_4)_4$, $(O,OH)_{5-6}$, may be associated in some of the above occurrences. Sapphirine has been reported as a detrital mineral. It alters readily by hydration and may break down to a mixture of corundum and biotite. It may replace spinel.

Diagnostic features. Differs from cordierite in higher indices, from corundum in its biaxial character. Kyanite has good cleavages. Kornerupine is orthorhombic with lower indices and higher birefringence.

SPHENE

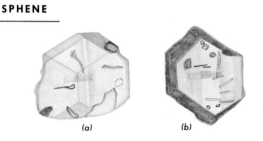

(a)   (b)

FIGURE 9.50 (a), (b) Detrital sphene, Surrey, England.

Composition. $CaTiSiO_5$. Some varieties contain considerable amounts of other elements, such as Na and rare-earth elements substituting for Ca; Fe^{3+}, Al, and Nb for Ti; and F and OH for O. Minor Fe^{2+}, Cr, V, Mn, Mg, and Sr may also be present. Another variety contains relatively large amounts of Sn, and some types are slightly radioactive owing to Th.

Indices. $\alpha = 1.840-1.950$, $\beta = 1.870-2.034$, $\gamma = 1.943-2.110$, $\gamma - \alpha = 0.100-0.192$. Substitution of Al for Ti and F for O lowers the indices. Interference colors in thin section are high-order white but somewhat masked by total reflection and the mineral color. Because of very strong dispersion, sections normal to an optic axis display abnormal interference tints and incomplete extinction. Zoned crystals have marginal zones with higher indices.

Color. May be colorless in thin section but usually is pale gray brown without pleochroism. Detrital grains are yellow brown, less commonly nearly black, green, or orange. Deeply colored grains may show distinct pleochroism with:

$\alpha = $ greenish yellow, colorless, yellow brown
$\beta = $ pink, greenish yellow, yellow brown, yellow green
$\gamma = $ pink, red orange, pale yellow, grass green, red brown

FIGURE 9.51 Euhedral sphene in nepheline syenite, Serrade, Monchique, Portugal. Polars not crossed, ×55.

In sphene with more than 0.5% of combined iron oxides the color values can be correlated with extent of oxidation of the iron.

Form. Monoclinic. Euhedra are common and have a rhombic cross section (Fig. 9.51). Fair (110) cleavage and a prominent (221) parting may be developed. Anhedral sphene occurs as clusters of small rounded pellets, and secondary sphene commonly forms minute elongated lenses or thin granules along the cleavages of such minerals as altered biotite. Detrital sphene is usually rounded or irregular, somewhat modified in shape by the (110) cleavage. Euhedral diamond- or spindle-shaped grains also are known (Fig. 9.50). Inclusions, such as carbonaceous material, feldspar, quartz, or chlorite, are rare except in large crystals.

Orientation. Biaxial (+). $\alpha \wedge a = -6 - -21°$, $\beta = b$, $\gamma \wedge c = 36-51°$. Optic plane is (010) (Fig. 9.52). $2V = 20-56°$, $r > v$ extreme. With decreasing Ti, 2V increases. In zoned crystals 2V increases outward. Rhombic

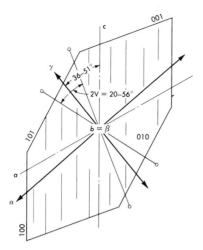

FIGURE 9.52 Orientation of sphene, section parallel with (010).

cross sections show symmetrical extinction, with the long diagonal as the fast ray. In wedge-shaped longitudinal sections the fast ray approaches parallelism with the longest direction across the crystal. Twinned cross sections are bisected along the long diagonal by the (100) twin plane. Multiple (221) twinning also occurs.

Occurrence. A widely distributed accessory mineral in (1) intrusive igneous rocks, particularly syenites, feldspathoidal syenites, monzonites, granodiorites, granitic pegmatites, and less commonly hornblende granites and gabbros; (2) extrusive igneous rocks, especially phonolites; (3) metamorphic rocks including marble, skarn, granitic gneiss, slate, mica schist, chlorite schist, talc schist, and amphibolite; and (4) in sediments as a detrital constituent of sands and sandstones and as a reportedly authigenic mineral in fullers earth. Sphene alters to leucoxene. Secondary sphene can be found as a by-product in the alteration of biotite to chlorite or of titaniferous augite to hornblende.

Diagnostic features. Characteristic is the combination of form, high relief, very strong birefringence, and extreme dispersion. Monazite has lower birefringence. Sedimentary grains of sphene may be confused with grains of xenotime, cassiterite, or rutile, which are all, however, normally uniaxial.

LAWSONITE

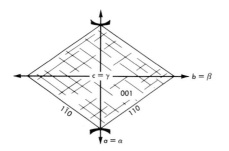

FIGURE 9.53 Orientation of lawsonite, section parallel with (001).

Composition. $CaAl_2Si_2O_7(OH)_2 \cdot H_2O$. A minor amount of Fe^{2+} may be present.

Indices. $\alpha = 1.665$, $\beta = 1.672$–1.676, $\gamma = 1.684$–1.686, $\gamma - \alpha = 0.019$–0.021. Maximum interference colors in thin section vary from uppermost first to lower second order.

Color. Colorless in thin section and crushed fragments. Thick detrital grains may show pleochroism: $\alpha = $ blue, $\beta = $ yellow, $\gamma = $ colorless.

Form. Orthorhombic. Commonly in euhedral crystals that may be prismatic or platy (001); also as needles. Cross sections are usually rhombic in outline. Good (001) and (010) cleavages, poor (110) cleavage, which tend to control the orientation of the powdered mineral. Detrital grains may be prismatic or irregular.

Orientation. Biaxial (+). $\alpha = a$, $\beta = b$, $\gamma = c$. Optic plane is (010) (Fig. 9.53). $2V = 84$–$85°$, $r > v$ strong. Cleavage fragments show a Bxa figure on (001) or a flash figure on (010). In rhombic cross sections the long diagonal is parallel with the slower ray, and the extinction is symmetrical. Longitudinal sections are length-slow and show parallel extinction. Multiple twinning on (110) is common in one or two directions. The thin twin individuals may show curvature.

Occurrence. In metamorphosed gabbros and diabases with glaucophane or aegirine and albite. Relatively common in glaucophane schists, in which the associated mineral assemblage may include garnet, chlorite, albite, pumpellyite, sphene, rutile, and allanite. Less common in some marbles, chlorite schists, and in some types of veins with calcite. May be replaced by pumpellyite.

Diagnostic features. Clinozoisite and zoisite display abnormal interference colors. Andalusite has lower birefringence; prehnite has higher birefringence. Scapolite is uniaxial.

CORDIERITE GROUP

GENERAL Minerals previously described as a single species, cordierite, are now known to represent several species closely related in structure and composition (Miyashiro et al., 1955; Miyashiro, 1956, 1957a). In addition, two other modifications have been synthesized.

Indialite (Miyashiro and Iiyama, 1954) is dimorphous with cordierite. Indialite itself probably can exist in two forms: "high" indialite, synthesized above 830°C, and "low" indialite, synthesized below 830°C. Cordierite also appears to occur in two modifications, a high and a low type, which are apparently not dimorphs but differ in composition. Actually there is a structural series between indialite (hexagonal) and cordierite (pseudo-hexagonal) resembling that between sanidine and microcline. Upon heating to high temperatures cordierite is changed into indialite continuously through all gradations, probably by means of an order-disorder type of transformation. The structure of cordierite may be derived by slight distortion of the indialite structure. By means of x-rays the amount of distortion can be determined. Cordierite that shows the maximum degree of distortion (i.e., the structurally "true" cordierites) have been referred to as "perdistortional cordierites," whereas those types structurally intermediate between indialite and perdistortional cordierite have been termed "subdistortional cordierites." The distortion probably results from some order-disorder change of the Si and Al ions in the rings of the cordierite–indialite structure, Si_5AlO_{18}.

The members of the cordierite group, both natural and synthetic, are listed in Table 9.3.

TABLE 9.3   POLYMORPHS OF $Mg_2Al_5Si_5O_{18}$ AND OSUMILITE (MIYASHIRO ET AL., 1955)

NAME	SYSTEM	COMPOSITION	SIGN, 2V	NATURAL OR SYNTHETIC
high indialite	hexagonal	$\alpha-(Mg,Fe^{2+})_2Al_4Si_5O_{18}$	uniaxial $(-)$	N, S $>$ 830°C
low indialite	hexagonal	$\beta-Mg_2Al_4Si_5O_{18}$	uniaxial? $(-)$	S $<$ 830°C
high cordierite (anhydrous)	pseudohexagonal, orthorhombic	$(Mg,Fe^{2+})_2Al_4Si_5O_{18}$	$(+)$ and $(-)$ $2V_\gamma = 80$–110°	N, S
low cordierite (hydrous)	pseudohexagonal, orthorhombic	$(Mg,Fe^{2+})_2Al_4Si_5O_{18}\cdot nH_2O$	$(+)$ and $(-)$ $2V_\gamma = 75$–140°	N, S
$\mu-Mg_2Al_4Si_5O_{18}$	?	$\mu-Mg_2Al_4Si_5O_{18}$	?	S
osumilite	hexagonal	$(K,Na,Ca)(Mg,Fe^{2+})_2(Al,Fe^{3+}, Fe^{2+})_3(Si,Al)_{12}O_{30}\cdot H_2O$	uniaxial $(+)$	N

CORDIERITE

CORDIERITE

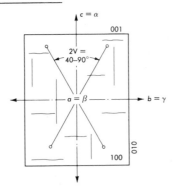

FIGURE 9.54 Orientation of cordierite, section parallel with (100).

Composition. $(Mg,Fe^{2+})_2Al_4Si_5O_{18}$. Usually $Mg > Fe^{2+}$, but some cordierite is Fe^{2+}-rich with a maximum of about 15% FeO reported (87% $Fe_2Al_4Si_5O_{18}$). Small amounts of Fe^{3+}, Mn, Ca, Na, and K are present in many cordierites.

Small amounts of water (up to nearly 3% H_2O) may be present in some cordierites, and apparently such hydrous cordierites (Schreyer and Yoder, 1960) correspond to the "low (temperature) cordierites" of Miyashiro et al. (1955). The water is very likely almost entirely structurally non-essential, probably "stuffed," as H_2O molecules, in the pseudohexagonal channels.

Indices. "High (temperature)," anhydrous cordierite differs from "low (temperature)," hydrous cordierite in having lower refractive indices for the same compositional range. For low cordierite: $\alpha = 1.527$–1.560, $\beta = 1.532$–1.574, $\gamma = 1.538$–1.578, $\gamma - \alpha = 0.007$–0.015. Indices and birefringence increase with Fe^{2+} over Mg (Fig. 9.57) and also with the degree of distortion of the structure; those cordierites with a low distortion index (closer to indialite) having higher indices than those with a high distortion index (closer to "true" cordierite). Most hydrous (low) cordierites are in the range $\alpha = 1.53$–1.54, $\gamma = 1.54$–1.55, whereas for anhydrous (high) cordierite $\alpha = 1.52$–1.56, $\gamma = 1.53$–1.57. Heating low cordierites (1000°C for 10 min) reduces the refractive indices appreciably by drawing off H_2O, and these indices can be used to estimate the composition (Figs. 9.58 and 9.59).

Color. Usually colorless in thin section and crushed pieces. Detrital grains usually weakly pleochroic from $\alpha = $ pale yellow to $\beta = $ blue or violet

FIGURE 9.55 Multiple lamellar twinning in cordierite, cordierite-anthophyllite gneiss, Ruby Dam, Greenhorn Mountains, Montana. Polars crossed, ×74.

to $\gamma =$ blue. Fe^{2+}-rich types show pale pleochroism even in thin section, colorless to pale violet. Upon alteration a pale yellowish hue usually appears. Small opaque inclusions are common; pleochroic haloes envelope included zircons.

Form. Orthorhombic, pseudohexagonal. Cleavage (010) poor, usually without influence on orientation or shape of crushed fragments; nor does it show in thin section. Distinct (001) parting may develop upon alteration, and irregular fractures become prominent.

Orientation. Biaxial ($-$), much less commonly ($+$). $\alpha = c$, $\beta = a$, $\gamma = b$. Optic plane is (100) (Fig. 9.54). 2V: for anhydrous (high) cordierite ($+$) 80–89° and ($-$) 89–70°; for hydrous (low) cordierite ($+$) 75–89° and ($-$) 89–40°. Apparently composition has little effect on 2V. Generally,

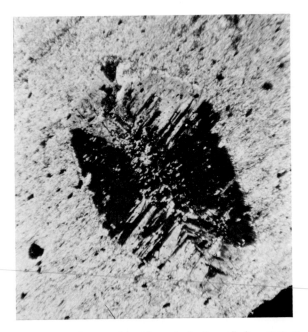

FIGURE 9.56 Sector and lamellar twinning in cordierite metacryst in spotted contact schist, Germany. Polars crossed, ×55.

$2V_\alpha$ increases as the distortion index increases and with heating. Dispersion usually $r < v$ weak but may change to $r > v$. Sign and 2V may vary from one grain to another in a single thin section.

Twinning in cordierite is common and resulting crystals may present a very complex aspect. Twinning is confined to the (110) and (130) planes. A complete classification of twinning in the two species is given in Table 9.4. Single and multilamellar twins (types 1–5) characterize many cordierites in coarse- to medium-grained schistose, gneissic, or granitoid rocks, in which sector twinning is rare (Fig. 9.55). Sector twins of uncomplicated appearance (types 6 and 7) occur in fine-grained hornfelses and contact slates (Fig. 9.56), although some examples have been recorded from pegmatites and gneisses. Cyclic twinning producing complex concentric and radial patterns apparently is characteristic of subdistortional cordierite, both synthetic and natural material from fused shales.

Occurrence. Chiefly in metamorphic rocks: gneisses, schists, and granulites, with such associates as quartz, plagioclase, biotite, muscovite, garnet, kyanite, and orthoclase. Also in amphibolites and some skarns with anthophyllite, in some instances associated with high-temperature Cu or Zn

TABLE 9.4   TWINNING IN CORDIERITE (MODIFIED FROM VENKATESH, 1954)

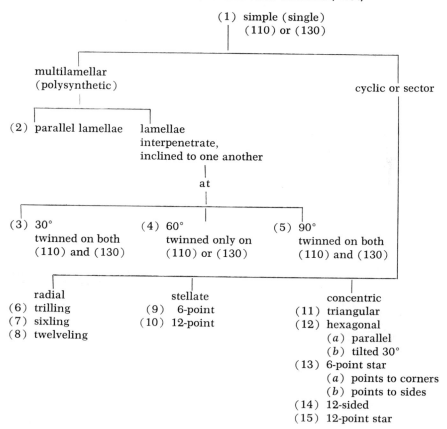

(1) simple (single)
    (110) or (130)

multilamellar
(polysynthetic)

cyclic or sector

(2) parallel lamellae

lamellae interpenetrate, inclined to one another

at

(3) 30°
twinned on both
(110) and (130)

(4) 60°
twinned only on
(110) or (130)

(5) 90°
twinned on both
(110) and (130)

radial
(6) trilling
(7) sixling
(8) twelveling

stellate
(9) 6-point
(10) 12-point

concentric
(11) triangular
(12) hexagonal
    (a) parallel
    (b) tilted 30°
(13) 6-point star
    (a) points to corners
    (b) points to sides
(14) 12-sided
(15) 12-point star

deposits. In hornfels and contact-metamorphic slates with biotite, muscovite, garnet, spinel, plagioclase, orthoclase, quartz, andalusite, sillimanite, hypersthene, and corundum. Uncommon in igneous rocks, including granite pegmatites, granite, quartz monzonite, and granodiorite, also some quartz veins. All of these occurrences are those of hydrous (low) cordierite (and the subdistortional type) which is apparently the more common type of cordierite and the most common form of $(Mg,Fe^{2+})_2Al_4Si_5O_{18}$.

High cordierite occurs as porphyritic crystals in volcanic rocks, andesite, and possible rhyolite (or in xenoliths in these rocks). Some is perdistortional, some subdistortional. It also occurs with indialite in fused sediments.

Detrital cordierite is retained in the light fraction but is rare. Cordierite alters readily to pinite, a fine-grained, usually green, variable, micaceous

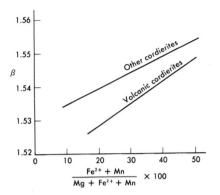

FIGURE 9.57 Variation of β index in cordierites with composition and manner of formation (Miyashiro et al., 1955).

mixture of muscovite and chlorite with minor biotite, zoisite, clinozoisite, garnet, and tourmaline. The partly altered mineral becomes pale yellowish in color, develops prominent (001) parting and irregular fractures that may be occupied by minute fillings of isotropic pinite.

Synthetic cordierites have been reported formed by reaction between Mg minerals and ceramic fire clays or in blast and glass furnace slags. Some of these may be indialite (if uniaxial); others are complicatedly twinned subdistortional cordierite.

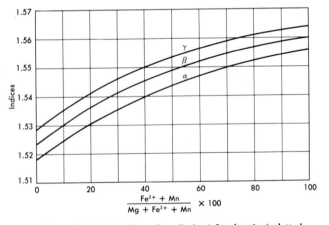

FIGURE 9.58 Refractive indices of cordierite (after heating) plotted against variation in composition (Iiyama, 1956).

BIAXIAL MINERALS

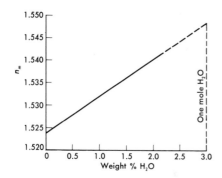

FIGURE 9.59 Variation in mean index of refraction (n_m) with water content in synthetic hydrous Mg-cordierite (Schreyer and Yoder, 1960).

Diagnostic features. If, upon heating, the indices decrease, the mineral was hydrous (low) cordierite; the indices of high cordierite do not change upon heating, and anhydrous (high) cordierite also differs in its occurrence. Cordierite differs from indialite in its biaxial, and in some examples, (+) character. (High) indialite occurs with subdistortional cordierite (in some cases, in single crystals with cores of indialite and margins of cordierite) in fused sediments (para-lavas) resulting from burning coal at Bokaro, India. These pyrometamorphic vitrophyres also contain glass, enstatite, and labradorite. Low indialite has been found neither in nature nor in artificial products, but its existence is probable on theoretical grounds. It could be recognized if, upon heating, its indices were to decrease to those of high indialite. To distinguish cordierite from indialite, optical tests alone may not be sufficient; this distinction may best be made by means of x-rays.

Osumilite, which was found in a hypersthene rhyodacite, is associated with tridymite and quartz. Its optical properties are uniaxial (+), $\omega = 1.545–1.547$, $\epsilon = 1.549–1.551$; $\epsilon =$ colorless, $\omega =$ light blue. Although osumilite and cordierite have similar refractive indices, the birefringence of the former is lower; and it is further distinguished by its uniaxial character, its distinctive pleochroism in thin section, and its occurrence. Osumilite probably is a relatively common mineral, usually mistaken for cordierite. It apparently is restricted in its occurrence to the matrixes of acid volcanic rocks.

Ordinary cordierite (hydrous, low, subdistortional cordierite), particularly that in coarse-grained metamorphic rocks and if it is unaltered and untwinned, greatly resembles quartz, from which it is distinguished by its biaxial character.

GENERAL   The epidote group consists of the following species:

Orthorhombic:
Zoisite, $Ca_2Al_3(SiO_4)_3OH$
Thulite, $Ca_2(Al,Fe^{3+},Mn)_3(SiO_4)_3OH$
Monoclinic:
Clinozoisite, $Ca_2Al_3(SiO_4)_3OH$
Epidote, $Ca_2(Al,Fe^{3+})_3(SiO_4)_3OH$
Piedmontite, $Ca_2(Al,Fe^{3+},Mn)_3(SiO_4)_3OH$
Allanite, $(Ca,Ce)_2(Al,Fe^{3+},Mn,Be,Mg)_3(SiO_4)_3OH$

Epidote is separated optically from clinozoisite by the change in optic sign, ($-$) in epidote to ($+$) in clinozoisite. However, Rapp (1960) found low-iron clinozoisite with both ($+$) and ($-$) forms in one specimen, and one ($+$) sample contained more Fe^{3+} than another that was ($-$).

ZOISITE

Composition. $Ca_2Al_3(SiO_4)_3OH$. Small amounts of Fe^{3+} may replace Al, but in general it seems that the addition of Fe^{3+} brings about the formation of the monoclinic form, clinozoisite. Mn^{3+} in minor amounts also substitutes for Al in thulite. Sr (as much as 2.50% SrO) and Ba (minor) substitute for Ca.

Indices.

	ZOISITE	ZOISITE no Fe^{3+} or Mn^{3+}	ZOISITE 2.50% SrO	THULITE	THULITE AVERAGE (16)
α	1.696–1.700	1.700	1.692	1.685–1.707	1.696
β	1.696–1.702	1.702		1.688–1.711	1.699
γ	1.702–1.718	1.706	1.698	1.698–1.725	1.707
$\gamma - \alpha$	0.005–0.018	0.006	0.006	0.007–0.022	0.010

The addition of Fe^{3+} increases the indices and the birefringence. In thulite the addition of Mn^{3+} alone lowers the indices, but many thulites also contain Fe^{3+}. The effect of even small amounts of Fe^{3+} upon the optical properties of zoisite is rather marked. Both low-iron zoisite and some thulite have abnormal interference colors in blue and brown, whereas zoisite with a higher iron content has normal interference tints of middle and upper first order in thin sections.

Color. Mn-free types are colorless in thin section or gray and gray green in thick grains. Even very small amounts of Mn^{3+} impart a pronounced pink color with pleochroism:

α = pale pink
β = colorless
γ = yellow

Crystals commonly show color zoning with the deep pink color more pronounced in the center of a radiating cluster and fading to colorless or pale pink away from the hub. Zoning is also very common in ordinary (Mn^{3+}-free) zoisite, owing to successive changes in Fe^{3+} content, and manifests itself by concentric variation in the birefringence, in some cases from abnormal to normal colors, and by differences in 2V.

FIGURE 9.60 Radial cluster of twinned and zoned zoisite crystals in marble, Axes Canyon, Dillon, Montana. Polars crossed, ×33.

Form. Orthorhombic anhedral to euhedral; usually prismatic, fibrous, bladed, or in columnar aggregates that may be parallel, subparallel, or radial (Fig. 9.60). The perfect (010) cleavage largely governs the shape of crushed material and also of many detrital grains. Detrital crystals tabular with (100) yield centered Bxa figures. Inclusions of minute amphibole needles are not uncommon, usually oriented parallel with (010). Both irregular and oriented intergrowths with epidote occur.

Orientation. Biaxial (+). Two types of zoisite may be contrasted (Fig. 9.61):

Nonferrian zoisite	Ferrian zoisite (ca. 5% Fe_2O_3 or more)
1. Lower indices	1. Higher indices
2. Lower birefringence and abnormal interference colors	2. Higher birefringence and normal interference colors
3. $\alpha = c, \beta = b, \gamma = a$ optic plane = (010)	3. $\alpha = b, \beta = c, \gamma = a$ optic plane = (001)
4. 2V = 30–0°	4. 2V = 0–60°
5. $r < v$ strong	5. $r > v$ strong
6. Parallel extinction	6. Parallel extinction
7. Length-fast in longitudinal sections	7. Length-fast or length-slow in longitudinal sections
8. Cleavage pieces yield a flash figure (α and γ)	8. Cleavage pieces yield a Bxo figure (β and γ)

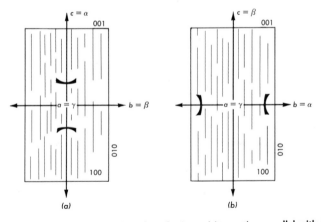

FIGURE 9.61 (*a*) Orientation of nonferrian zoisite, section parallel with (100). (*b*) Orientation of ferrian zoisite, section parallel with (100).

Since zoning may be revealed by variation in 2V or even by a shift of the optic plane, it is possible for contiguous zones to show opposing character of elongation. In some crystals outer zones show small extinction angles and thus grade into clinozoisite. With increasing Fe^{3+} content 2V decreases from 30 to 0° and then increases to 60°; some zoisite, therefore, is essentially uniaxial in character. Multilamellar twinning occurs in some specimens.

Occurrence. Occurs chiefly as an essential and accessory constituent of various metamorphic rocks and as a hydrothermal alteration product of mafic igneous rocks. Among metamorphic rocks that contain it are (1) nonmagnesian argillaceous limestones altered by thermal metamorphism and containing calcite, vesuvianite, anorthite, and grossularite; (2) nonmagnesian aluminous marbles of regional metamorphic origin with calcite, albite, quartz, biotite, and muscovite; (3) zoisite–biotite schists with some quartz and sphene; (4) zoisite granulites or zoisite–hornblende granulites containing quartz, plagioclase, biotite, and less abundant garnet, sphene and orthoclase; and (5) zoisite amphibolites with or without garnet. Accessory zoisite occurs in some eclogites and glaucophane schists.

In intermediate and mafic igneous rocks plagioclase may be altered to saussurite, a fine-grained, intimate intergrowth of zoisite (or clinozoisite-epidote or both) with albite and various of the following: orthoclase, garnet, chlorite, sericite, actinolite, prehnite, scapolite, and zeolites. This alteration may be developed to varying degrees, beginning with labradorite only slightly changed and proceeding to rocks completely transformed to coarse-grained zoisite–prehnite–actinolite aggregates. Zoisite appears less commonly in albite–corundum veins in peridotites; the associated minerals are chlorite, serpentine, margarite, scapolite, and sepiolite.

Thulite occurs in metamorphosed iron-poor carbonate rocks associated with epidote, calcite, quartz, corundum, actinolite, diopside, grossularite, chlorite, phlogopite, and calcic plagioclase; in granitic pegmatites and quartz veins; and in hydrothermally altered granites and granitic gneisses.

Zoisite is not as common as clinozoisite or epidote. Members of this group occur together as discrete grains, in intergrowths of epidote and zoisite, or in zoned zoisite–clinozoisite crystals.

Diagnostic features. Clinozoisite resembles zoisite but shows small to large extinction angles. Epidote also shows inclined extinction and has considerably higher birefringence. Melilite may resemble nonferrian zoisite in its abnormal birefringence but is always uniaxial and (−). Vesuvianite, which also may show abnormal interference colors, closely resembles zoisite in relief, form, optical sign, and some occurrences and associations; it is uniaxial and shows (110) cleavage. Careful index determination may be necessary to distinguish the two. Thulite grains may be mistaken for pleochroic andalusite, which has, however, lower indices and larger 2V and is $(-)$. Piedmontite is distinguished from thulite by its inclined extinction.

CLINOZOISITE

Composition. $Ca_2Al_3(SiO_4)_3OH$; with increasing substitution of Fe^{3+} for Al the mineral grades into epidote. The division point between the two is considered to coincide with the change in sign, $(+)$ in clinozoisite, $(-)$ in epidote, which takes place at between about 7–8% Fe_2O_3 or about 14–15% of the iron–epidote end-member molecule (see p. 190).

Indices. $\alpha = 1.706$–1.724, $\beta = 1.708$–1.729, $\gamma = 1.712$–1.735, $\gamma - \alpha = 0.004$–0.023. Indices and birefringence increase with Fe^{3+} (Fig. 9.62). The maximum interference colors vary from first order to lower second order in thin section. Interference tints in middle first order may be abnormal: blue gray, Berlin blue, green yellow. Crystals show zoning by birefringence variations.

Color. Colorless in thin section, but broken pieces or grains, especially of those types higher in Fe^{3+}, may show pleochroism in shades of yellow, pink, or green; for example:

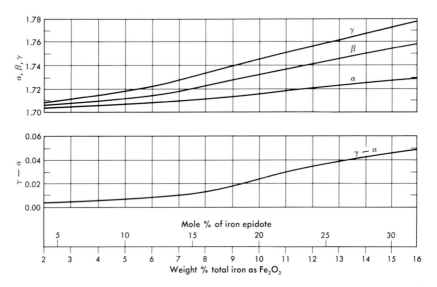

FIGURE 9.62 Variation of refractive indices and birefringence with composition in the epidote series (Johnston, 1949).

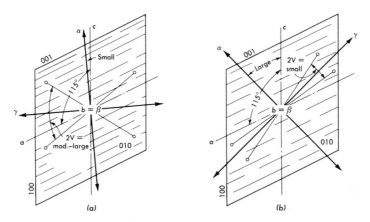

FIGURE 9.63 (a) Orientation of clinozoisite, section parallel with (010). (b) Orientation of iron-poor clinozoisite, section parallel with (010).

α = yellow green
β = green or pink
γ = green or red

Form. Monoclinic. Euhedral to subhedral. Bladed crystals or columnar aggregates of subparallel or radial arrangement are common. Also anhedral-granular. Cleavage (001) is perfect and governs the shape and orientation of crushed fragments and grains of detrital origin.

Orientation. Biaxial (+). $\alpha \wedge c = -5\text{--}85°$, $\beta = b$, $\gamma \wedge a = 30\text{--}59°$. Optic plane = (010) (Fig. 9.63). $2V = 14\text{--}90°$, $r > v$ or $r < v$. Iron-poor clinozoisites have very small 2V's and very large extinction angles. Below 5% Fe_2O_3 the change in orientation may not be due entirely to variations in the iron content, unless the sensitivity of the mineral to small changes in composition becomes much greater than in the upper part of the series. It has been suggested (Johnston, 1949) that the environment during crystallization may influence orientation in the low-iron mineral. Relationships between 2V, $\alpha \wedge c$, and dispersion are shown in Fig. 9.64. The change from $r < v$ to $r > v$ occurs when $\alpha \wedge c =$ about 45°; the dispersion is weak when the extinction angle is small and reaches a maximum when 2V is about 0°. Clinozoisite is elongate parallel with b, and elongate sections show parallel extinction and may be either length-fast or -slow. Multilamellar twinning on (100) is common; the composition surfaces can be irregular. Zoning also is common and, when present together with twinning, gives rise to individuals of great complexity. In some crystals outer zones have higher indices, a larger 2V, and smaller extinction angles,

CLINOZOISITE   **189**

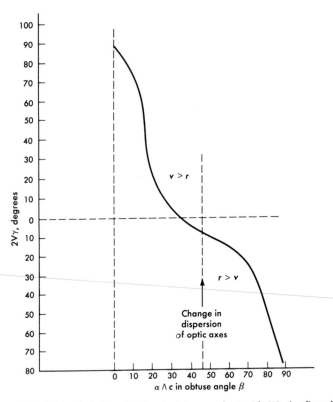

FIGURE 9.64 Variation of $\alpha \wedge c$ in obtuse angle β with $2V\gamma$ in clinozoisite (Johnston, 1949).

indicating a higher Fe^{3+} content than in the center. In other cases cores of epidote pass outward into clinozoisite. Overgrowths of clinozoisite on allanite also are known. In zoned iron-poor types the sign of elongation may vary from the center outward. This is due to the marked change in orientation, which brings either α or γ nearly across the elongation. Figures obtained from cleavage fragments vary considerably. With increasing $\alpha \wedge c$, the pieces show successively a centered or nearly centered optic-axis figure with a nearly straight isogyre, then an eccentric Bxo figure of increasing angle, and finally a slightly eccentric Bxa figure of moderate 2V. Thus the birefringence of these cleavage pieces also shows marked variation.

Occurrence. Epidote and clinozoisite both are formed by dynamothermal metamorphism, contact metamorphism, and metasomatism. Clinozoisite is found in aluminous calcareous metasediments with grossularite, vesuvianite, anorthite, microcline, quartz, and calcite; in amphibolites with hornblende and albite; in actinolite and hornblende schists; in glaucophane schists; in quartzites; and in slates. In some schists clinozoisite contains minute inclusions of amphibole. Also forms as a hydrothermal mineral in altered igneous rocks, usually mafic in character, associated with prehnite, sphene, actinolite, and albite. It may be a component of saussurite (see zoisite).

Diagnostic features. Differs from zoisite in showing inclined extinction; from epidote in sign, lower birefringence, and lower indices; from vesuvianite in biaxial ($+$) character.

EPIDOTE

Composition. $Ca_2(Al,Fe^{3+})_3(SiO_4)_3OH$. Separated from clinozoisite by the change in sign from ($+$) to ($-$). With substitution of Mn^{3+} as well as Fe^{3+} for Al, epidote grades into piedmontite. Very rarely some Cr replaces Al. The rare species manganepidote has Mn^{3+} replacing Al and minor Mn^{2+} replacing Ca.

Indices. $\alpha = 1.723$–1.751, $\beta = 1.730$–1.784, $\gamma = 1.736$–1.797, $\gamma - \alpha = 0.013$–0.046. Indices and birefringence increase with increasing Fe^{3+} (Fig. 9.62). The maximum interference colors in thin section vary from upper first order to middle second order, with the average nearer the latter. The birefringence commonly shows a peculiar mottled effect, varying in irregular patches or areas in a single grain to produce a vivid color contrast. Abnormal interference colors appear in middle first order.

Color. Colorless or pale yellow in thin section. Coarser grains or fragments may be greenish yellow, green, or brown. The color may be irregularly distributed. Pleochroism weak with:

$\alpha =$ colorless, pale green, pale yellow
$\beta =$ greenish yellow, greenish brown
$\gamma =$ pale green, yellow green

$\alpha < \gamma < \beta$. Zoning with clinozoisite either in the center or on the margin. Overgrowths, with allanite cores and epidote margins, also are known.

Form. Monoclinic. Commonly in euhedral fibrous or columnar crystals, elongate parallel with b, with six-sided cross sections and rectangular longitudinal sections. Also massive and finely to coarsely anhedral-granular. Perfect (001) cleavage governs shape of crushed material. Detrital grains may also be somewhat rounded (001) plates or irregular in shape. Inclusions of minute opaque dust may be present. Other included minerals are magnetite, quartz, chlorite, albite, and rutile.

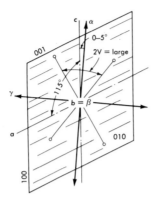

FIGURE 9.65 Orientation of epidote, section parallel with (010).

Orientation. Biaxial $(-)$. $\alpha \wedge c = -5\text{--}0°$, $\beta = b$, $\gamma \wedge a = 25\text{--}30°$. Optic plane is (010) (Fig. 9.65). $2V = 64\text{--}89°$, $r > v$ strong. With increasing Fe^{3+}, 2V decreases and the extinction angle increases. Elongate sections, which may be either length-slow or length-fast, show parallel extinction. Cleavage fragments (001) show a nearly centered optic-axis figure. Lamellar twinning on (100).

Occurrence. A common and widespread metamorphic and hydrothermal mineral. It may be of contact-metamorphic origin, as in altered aluminous limestones together with grossularite, chlorite, vesuvianite, anorthite, quartz, and calcite or in hornfels with diopside, actinolite, grossularite, and albite. Also in quartzites; quartz schists; slates; phyllites; epidote–muscovite schists with biotite, garnet, or plagioclase; actinolitic greenschists with chlorite, sphene, albite, quartz, calcite, magnetite, and rarely stilpnomelane; and epidote amphibolites with hornblende and some albite and chloritoid. In basalts epidote is found in cavities associated with augite, actinolite, grossularite, and sphene. Albitized basalts (spilites) contain it, along with albite, chlorite, actinolite, and calcite. In intrusive igneous rocks epidote forms veins of massive granular material. Vein material or replacement rocks containing epidote, quartz, and minor actinolite and chlorite are called epidosites. Also in igneous rocks epidote forms as a hydrothermal alteration product replacing plagioclase, as in saussurite (see zoisite), and also replacing hornblende, biotite, and augite. Unakite is an epidotized granite containing mainly epidote, orthoclase, and some quartz. In a few granites and syenites epidote forms as overgrowths on allanite and extends into surrounding biotite in symplectitic intergrowths. Not uncommon as an accessory heavy mineral in sandstones and sands.

Diagnostic features. Higher indices and birefringence, small extinction angle, $(-)$ sign, and, in some types, the pale color distinguish it from clinozoisite. The combination of mottled birefringence and weak pleochroism is distinctive. Cleavage and optic orientation differentiate it from olivine.

PIEDMONTITE

Composition. $Ca_2(Al,Fe^{3+},Mn)_3(SiO_4)_3OH$. Some Sr and a small amount of Mn^{2+} may replace Ca.

Indices. $\alpha = 1.725$–1.756, $\beta = 1.730$–1.789, $\gamma = 1.750$–1.832, $\gamma - \alpha = 0.028$–0.082. The considerable range in the indices and birefringence is due to the fact that piedmontite varies from manganiferous clinozoisite, very low in Fe^{3+}, to manganiferous epidote, high in Fe^{3+}. If the amount of Al is kept approximately constant, the indices increase with Fe^{3+} and decrease with Mn^{3+}. If Fe^{3+} remains constant, increasing substitution of Mn^{3+} for Al also raises the indices. Maximum interference tints in thin section vary from lower second order to upper third order but may be poorly defined owing to strong pleochroism.

Color. Pink, red, red violet. Strong pleochroism and variable absorption formulae, with $\gamma > \alpha > \beta$ or $\gamma > \beta > \alpha$ as the most common. Also reported are: $\alpha > \gamma > \beta$, $\gamma < \beta > \alpha$, $\alpha < \gamma < \beta$, $\gamma > \beta = \alpha$, and $\alpha > \beta > \gamma$:

α = pale yellow, yellow, lemon yellow, orange, gold, pale red, amethystine red

β = pale amethyst, amethyst, violet, rose, deep red

γ = deep purple red, brownish red, carmine, bright rose, pink

Marked color zoning may occur, especially in paler varieties. A core of pink piedmontite may be enclosed in a shell of colorless clinozoisite. Elongate crystals may consist in large part of colorless clinozoisite with one or both terminations of pink piedmontite. Only small amounts of Mn^{3+} are necessary to produce the characteristic color, and strong pleochroism appears with as little as 0.96% Mn_2O_3. In radial clusters the ends toward the center of the group may be deep pink which grades outward along each crystal into a brownish-pink hue, as in thulite. Individual blades may show a series of growth zones along their lengths parallel with terminations. Not uncommonly in some occurrences zoisite, thulite, clinozoisite, epidote, and piedmontite in various combinations appear together, either in zoned crystals or as individuals.

Form. Monoclinic. Commonly in euhedral crystals, elongate parallel with b, possibly with terminal faces on one or both ends. Cross sections are six-sided; longitudinal sections approach rectangular outlines. The crystals may be coarsely columnar in random or subparallel orientation or occur as minute fibers in radial aggregates or spherulites. Stellate groups of fibers occur in veins attached to the walls. In schists euhedra occur singly or concentrated in layers. Detrital grains usually are of irregular shape. Crushed material tends to lie on the perfect (001) cleavage; (100) cleavage has been reported; and subparallel fractures normal to b may be present. Inclusions of quartz occur in some crystals; in others liquid inclusions parallel with b appear.

Orientation. Biaxial (+). $\alpha \wedge c = -3 - -7°$, $\alpha = b$, $\gamma \wedge a = 28-32°$. Optic plane is (010). $2V = 50-86°$, $r > v$ strong. Parallel extinction in elongate sections. These are either length-fast or -slow, but the character of the elongation may be difficult to determine in deeply colored types. Cleavage fragments yield eccentric optic-axis figure. Twinning on (100) has been observed.

Occurrence. Chiefly (1) as a metamorphic mineral in schists, (2) as a hydrothermal mineral in altered volcanics, and (3) in manganese ore deposits of hydrothermal or metamorphic origin. Metamorphic rocks containing it include sericite phyllites; chlorite–sericite schists; piedmontite–sericite schists; glaucophane schists; and piedmontite–quartz or piedmontite–quartz–muscovite schists with colorless tourmaline, rutile, apatite, and spessartite. Piedmontite appears as a secondary constituent of felsites, rhyolites, rhyolite–porphyries, and andesites, usually as clusters of needles or spherulites. In other cases it may be found in shear zones and quartz veins. In manganiferous ore deposits a great variety of associates may be present, especially quartz, calcite, braunite, manganophyllite, and spessartite.

Diagnostic features. Piedmontite is distinguished from the rare manganepidote by the $(-)$ sign of the latter. Thulite has parallel extinction and lower indices of refraction. For the distinction from dumortierite, see under that mineral.

ALLANITE

Composition. $(Ca,Ce)_2(Al,Fe^{3+},Mn,Mg)_3(SiO_4)_3OH$. May show considerable variation in composition. La, other rare-earth elements including those of the Y subgroup, and Na also replace Ca; some Ti, Be, and Fe^{2+} may be present; O and F substitute for OH; and PO_4 (as much as 6.48% P_2O_5) proxies for SiO_4. Small amounts of Th and even U are present, and most allanites are weakly radioactive. Much allanite also is hydrated; analyses show more H_2O than required by the above formula.

Indices. $\alpha = 1.715–1.791$, $\beta = 1.718–1.815$, $\gamma = 1.733–1.822$, $\gamma - \alpha = 0.015–0.031$. This range is for the unaltered mineral. In general the indices and birefringence increase with Fe^{3+}, Ti, and rare-earth elements. Allanites that contain much Ca and Mg have indices near the lower end of the range. Because of the strong color of the mineral, interference tints are visible only rarely. Commonly altered to an isotropic (metamict) substance that contains variable amounts of loosely bound H_2O. For such material $n = 1.53–1.72$, and the index decreases in the general way with an increase in H_2O. However, as the mineral is altered, Fe^{2+} and Mn^{2+} become oxidized to Fe^{3+} and Mn^{3+}, which tends to increase the index and counterbalances somewhat the lowering effect of additional H_2O.

Color. Brown. If anisotropic, pleochroic with: $\beta > \gamma > \alpha$, $\beta = \gamma > \alpha$, $\gamma > \beta > \alpha$, or $\gamma = \beta > \alpha$.

$\alpha =$ colorless, pale yellow, yellow brown, pink, greenish yellow, greenish brown, pale green, greenish gray, brownish gray, brown, dark brown

$\beta =$ yellow brown, reddish brown, dark red brown, greenish brown, deep brown

$\gamma =$ pale yellow, brown yellow, brown green, green, brown, dark brown, red brown, gray brown, deep brown, black

Zoning is common, and outer zones normally have lower indices and lighter colors. Parallel overgrowths of zoisite, clinozoisite, epidote, and even piedmontite are known (Fig. 9.66). Irregular, nonzonal color variations also appear. These are in part due to incipient alteration, and isotropic and uniaxial portions may be scattered heterogeneously through anisotropic crystals. Different parts of single grains will be altered to varying degrees and thus show varied colors and different indices. With complete alteration the mineral becomes metamict. Ignition will usually bring about a partial recrystallization.

Form. Monoclinic. Occurs as euhedral, acicular crystals and needles (elongate parallel with b), as (100) tablets, or as irregular, anhedral grains. Where enclosed in biotite, chlorite, or hornblende, it is usually surrounded by a pleochroic halo. Cleavages (100) and (001) are irregularly developed; (100) twinning not uncommon. Detrital grains are irregular in shape.

Orientation. Biaxial $(-)$. $\alpha \wedge c = -22°\!-\!-\!47°$, $\beta = b$, $\gamma \wedge a = 47–72°$. Optic plane is (010). $2V = 40–80°$, $r > v$ distinct to strong, or $r < v$ distinct. Allanite with the optic plane normal to (010) has been reported, as has a $(+)$ variety. Because of the alteration and strong absorption the orientation can be determined only with difficulty. Crystals elongate with

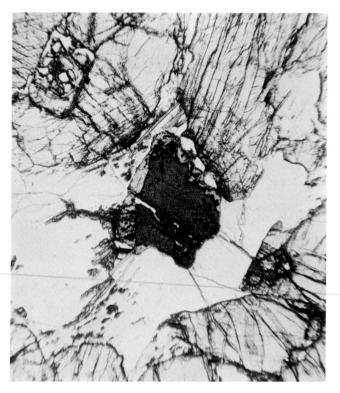

FIGURE 9.66   Allanite with rim of epidote in gneissic granite, Spring Creek, Ruby Mountains, Montana. Polars not crossed, ×55.

b show parallel extinction and are length-slow or -fast. Plates tabular with (100) are length-fast and also have parallel extinction.

Occurrence. An accessory mineral of granites, aplites, syenites, nepheline syenites, granodiorites, diorites, their pegmatites, and also in some of their extrusive equivalents. Occurs in some contact marbles and skarns or calcareous blocks ejected from volcanoes and in biotite schist, granitic gneiss, migmatite, hornblende gneiss, eclogite, glaucophane gneiss, and amphibolite. Rare as a detrital species.

Diagnostic features. Resembles brown hornblende but lacks the amphibole cleavage. The crystal habit and orientation also are different. The relief, dark colors, and single-crystal variation are distinctive. Commonly radioactive. When metamict, distinguished from gadolinite, euxinite, samarskite, and fergusonite by *x*-ray powder patterns obtained on recrystallized material.

Composition. $Ca_2(Al,Mg,Fe)_3(SiO_4)_3(OH) \cdot H_2O$. Al predominates greatly over Mg, Fe^{2+}, and Fe^{3+}. Composition is close to that of clinozoisite but differs in having an extra molecule of water.

Indices. $\alpha = 1.674–1.748$, $\beta = 1.675–1.754$, $\gamma = 1.688–1.764$, $\gamma - \alpha = 0.002–0.020$. Both the birefringence and refractive indices increase with increasing Fe^{2+} or $Fe^{2+} + Fe^{3+}$ content. (Fig. 9.67.) Zoned crystals usually have an outer zone with higher birefringence than the core.

Color. Colorless to dark green, rarely brownish in detrital grains. Usually somewhat pleochroic:

α = colorless, yellow, pale brownish yellow, pale greenish yellow
β = light green, green, deep bluish green, blue green, brownish yellow
γ = colorless, yellow, pale yellow, brownish yellow

$\beta > \gamma > \alpha$. Higher indices, generally stronger birefringence, and stronger dispersion are associated with darker colors and more intense pleochroism.

Form. Monoclinic. Elongated parallel with b. Usually bladed, lathlike, acicular, or fibrous; commonly in radial aggregates, rosettes, or spherulites and also in parallel or subparallel groups. The better cleavage is (001); another is (100). Detrital grains are subangular to subrounded aggregates of fibers.

Orientation. Biaxial $(+)$; rarely $(-)$. Orientation varies (Fig. 9.69): (1) $\alpha \wedge a = -4--25°$, $\beta = b$, $\gamma \wedge c = 24-45°$. Optic plane is (010). 2V =

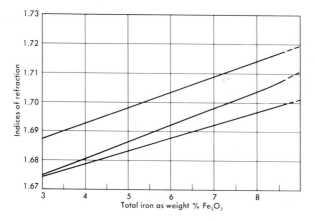

FIGURE 9.67 Variation in indices of refraction with composition in pumpellyite (Coombs, 1953).

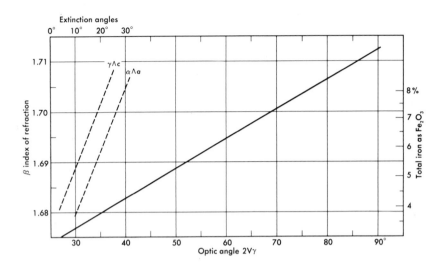

FIGURE 9.68   Variation in 2V and extinction angles with β index of refraction and composition in pumpellyite (Coombs, 1953).

80–0°, decreasing with increasing extinction angle $\alpha \wedge a$ (Fig. 9.68); $r < v$ strong. (2) $\alpha \wedge a = -28$––31°, $\beta \wedge c = 48$–51°, $\gamma = b$. Optic plane normal to (010). 2V = 0–38°, $r > v$ strong (Fig. 9.69). Twinning on (001) may be present. Complex crystals in which both zoning and twinning occur are not uncommon.

Occurrence.   A rather widespread low-grade metamorphic and hydrothermal mineral. Occurs in the amygdaloidal cavities of basalts, as a con-

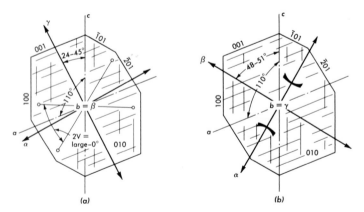

FIGURE 9.69   (a), (b) Orientations of pumpellyite, sections parallel with (010).

stituent of lawsonite veins cutting glaucophane schists and in the schists themselves, in altered diabases, greenstones, albitophyres, metadolerites, diabasic tuffs, spilites, lime–silicate rocks, skarns, and hornfelses and in quartz–albite veins cutting altered mafic rocks. It occurs as detrital species (Netherlands Quaternary sands), in which form it may have been overlooked elsewhere.

Diagnostic features. May be confused with members of the epidote group. However, zoisite is orthorhombic. Epidote has generally higher indices, a smaller extinction angle $\alpha \wedge c$, and usually a larger 2V and is ($-$). May be difficult to tell from clinozoisite. Some detrital grains resemble those of glauconite.

CHAPTER TEN

BIAXIAL MINERALS: CHAIN SILICATES

PYROXENE GROUP

Orthopyroxenes

The orthopyroxenes form a series whose main chemical variation can be expressed in terms of two end-member molecules, $MgSiO_3$ (En) and $Fe^{2+}SiO_3$ (Of). The following subdivisions of the series may be useful in some instances:

$En_{100}–En_{88.5}$	enstatite
$En_{88.5}–En_{80}$	bronzite
$En_{80}–En_{50}$	hypersthene
$En_{50}–En_{12}$	ferrohypersthene
$En_{12}–En_{0}$	orthoferrosilite

In the following descriptions bronzite, hypersthene, and ferrohypersthene are included under hypersthene. Orthoferrosilite is found in nature only very rarely. Volcanic orthopyroxenes contain some Ca, which is largely exsolved as clinopyroxene in intrusive orthopyroxenes (Bushveld and Stillwater types).

ENSTATITE

Composition. $(Mg,Fe^{2+})SiO_3$ or, in terms of percentage of the enstatite molecule, $En_{100}–En_{88.5}$. The division between enstatité and hypersthene is so chosen that all enstatite is optically ($+$) and nearly all hypersthene is optically ($-$). This change takes place when the FeO content becomes about 7%. Enstatite may also contain minor Fe^{3+}, Al, and Ca.

Indices. $\alpha = 1.650\text{–}1.668$, $\beta = 1.652\text{–}1.673$, $\gamma = 1.659\text{–}1.679$, $\gamma - \alpha = 0.009\text{–}0.011$. The indices increase rapidly and the birefringence increases slightly with additional Fe^{2+} (Figs. 10.2 and 10.3). In thin section the maximum interference colors are of middle first order. Al increases indices slightly—about 0.001 per weight percent Al_2O_3.

Color. Colorless in thin section. Coarser crushed or larger detrital pieces are colorless to pale green.

Form. Orthorhombic. Anhedral to euhedral. Crystals are prismatic with four- or eight-sided cross sections, on which the characteristic (110) cleavages appear as two sets of lines at angles of 93 and 87°. Longitudinal sections show one direction of cleavage lines. In pyroxenites large irregular crystals may be poikilitic with olivine inclusions. Radial fibrous enstatite forms kelyphitic borders on garnet. Fibrous enstatite also occurs in reaction shells between olivine and plagioclase, although hypersthene is usually more common in these rims. Enstatite may have well-developed partings, (010) and (100). Detrital pieces or crushed grains tend to be oriented by the (110) cleavage, although usually a small to moderate percentage of broken pieces lies on a parting face. Radiating or lattice aggregates (chondrules) occur in meteorites. Inclusions are of gas, magnetite, and apatite.

Orientation. Biaxial ($+$). $\alpha = a$, $\beta = b$, $\gamma = c$. Optic plane is (010)

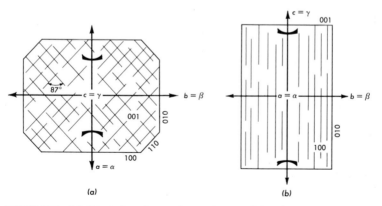

(a) (b)

FIGURE 10.1 (a) Orientation of enstatite, section parallel with (001). (b) Orientation of hypersthene, section parallel with (100).

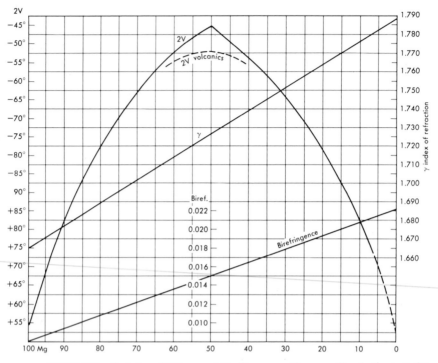

FIGURE 10.2 Variation of 2V, γ, and birefringence with composition in the enstatite–hypersthene series (Hess, 1952).

(Fig. 10.1a).[1] 2V = 54–90°, increasing with Fe^{2+}; $r < v$ weak. Ca increases 2V slightly. Longitudinal sections show extinction parallel with cleavage traces, which are length-slow. Cross sections have symmetrical extinction.

ORIENTATION	FIGURE	INDICES
(110) cleavage	eccentric flash figure	γ,α'
(010) parting	flash figure	γ,α
(100) parting	Bxo figure	γ,β

Twins are rare, appearing as lamellae; (014) is the twin plane.

Occurrence. As a primary mineral mainly in peridotites, pyroxenites, and serpentinites and in both stony and metallic meteorites. Less common in

[1] This orientation follows the American usage in the orthorhombic system of choosing the b axis as intermediate in length between a and c. Thus $(110) \wedge (1\bar{1}0) = 87°$.

gabbros. Rare in some orthogneisses, granulites, and eclogites. Alters readily to antigorite (variety called bastite), with parallelism of the serpentine and pyroxene cleavages. Uralitization of enstatite is not common. Talc may replace enstatite in rare cases. Not uncommon as a detrital species. Clinoenstatite forms in magnesite bricks attacked by SiO_2 and in talc-derived porcelains.

Diagnostic features. From hypersthene it is separated by (+) sign and by lower indices: from diopside and augite by parallel extinction. The alteration is also characteristic. Crushed pieces lying on (110) are not easy to distinguish from cleavage fragments of the orthorhombic amphibole, anthophyllite, with which enstatite may be closely associated, but anthophyllite has higher birefringence and may show pleochroism in pale brown colors.

HYPERSTHENE

Composition. $(Fe^{2+},Mg)SiO_3$. In terms of the enstatite end member: $< En_{87.5}$. Included here are the series members: bronzite, hypersthene in the narrow sense, ferrohypersthene, and orthoferrosilite. Mn may be present in considerable amounts, especially in some iron-rich hypersthenes. Fe^{3+}, Al, and Ca usually are present. Hypersthenes from charnockites have as much as 6–9% of Al_2O_3, and those of extrusive rocks may have higher than normal amounts of Fe_2O_3. Ti^{4+} is commonly found as a minor constituent. Some of the Ti^{4+} and Fe^{3+} recorded in various analyses is from inclusions. Between En_{85}–En_{65} as much as 9% $CaSiO_3$ may be retained in solid solution. The maximum recorded content of the orthoferrosilite molecule in a natural hypersthene is Of_{88} (FeO $= 41.65\%$ and MnO $= 5.02\%$).

Indices. $\alpha = 1.669$–1.755, $\beta = 1.674$–1.763, $\gamma = 1.680$–1.773, $\gamma - \alpha = 0.011$–0.018. Both the indices and the birefringence increase with Fe^{2+} (Figs. 10.2 and 10.3). The maximum interference colors range from middle to upper first order in thin section. Some crystals may show zoning, with thin margins having higher birefringence and higher indices.

Color. Colorless to pink or pale green in thin section. The pleochroism is:

α = pink, brownish pink, pale yellow, pale red, brownish red
β = pinkish yellow, greenish yellow, yellow, light green, greenish gray
γ = light green, bluish green, grayish green, blue

Not all hypersthenes are pleochroic; thus the absence of this property does not place the mineral in the enstatite range. Nor does the intensity of the pleochroism increase with the Fe^{2+} content; for many high-iron hyper-

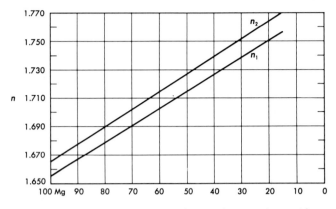

FIGURE 10.3 Variation in n_1 and n_2 of (110) cleavage pieces with composition in the enstatite–hypersthene series (Burri, 1941).

sthenes, including natural orthoferrosilite (Of_{88}), show only very weak pleochroism. Furthermore, there is no regular variation of the pleochroism with Mn. Possibly the content of Fe^{3+} or Ti^{4+}, where they are not due to included ilmenite or magnetite, may account for the variation in pleochroism.[1] The pleochroism varies in different hypersthenes closely associated in the same rock mass. Detrital grains and coarse crushed pieces usually show darker pleochroic colors than those listed above, such as brown, olive green, and deep red.

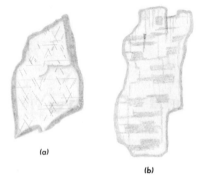

(a)

(b)

FIGURE 10.4 Detrital hypersthene. (a) North Fairhaven, New York. (b) Lake Erie shore sand, Cedar Point, Ohio (Krumbein and Pettijohn, 1938).

[1] Howie (1963) has also suggested that this pleochroism is a physical effect resulting from contraction of cell dimensions with increasing Al content.

Form. Orthorhombic. Anhedral to euhedral. Phenocrysts of extrusive rocks are usually euhedral to subhedral with a prismatic habit and nearly square cross sections on which the (110) cleavage forms two sets of diagonal lines intersecting at 87 and 93°. Parting (010) and rarely (100) also occur. In intrusive rocks subhedral to anhedral, typically in stumpy prisms. These may show resorption or replacement by augite or pigeonite with by-product dusty magnetite. In diabases usually interstitial (ophitic to subophitic texture), with triangular and wedge-shaped sections. In reaction rims around olivine, as fibers normal to the olivine contact and perhaps in turn surrounded by a shell of amphibole or clinopyroxene. In diabases hypersthene may be overgrown on pigeonite with the latter forming either a euhedral core or a central group of irregular, corroded blebs.

Orthorhombic pyroxenes of the composition En_{85}–En_{65} that formed in intrusive rocks and cooled slowly show optical anomalies, which may be due to the presence of exsolved Ca in the form of diopside or augite. In diabases this exsolution is usually incomplete or on a very fine scale, giving rise to a peculiar patchy or uneven extinction. This grades over into a graphic intergrowth of hypersthene and diopside, which may extend over the entire crystal or occur either in the core or in a marginal zone. In plutonic bodies, i.e., in norites, peridotites, and in some eclogites, the exsolution intergrowth commonly is coarser and more regular, and thin plates form a lamellar intergrowth with host hypersthene. One, two, or even three sets of clinopyroxene lamellae may be present. This results in a finely striated or "polysynthetically-twinned" appearance in the ortho-pyroxene and gives rise to an anomalous extinction angle from the aggregate. Where one set of lamellae occurs, the lamellae are commonly oriented parallel with (010) of the hypersthene host (Bushveld type), but within them their optic plane (010) is parallel with (100) of the hypersthene (Fig. 10.18). Plates parallel with (110) and (100) of hypersthene also appear.[1] Not all of such fine lamellae are formed by exsolution. In some cases they have the same composition as the host but are slightly disoriented with respect to its optical directions. This type may represent twin lamellae that formed owing to deformation during crystallization. Parallel growths of hypersthene with cummingtonite are recorded from some metamorphic rocks.

In coarser-grained rocks hypersthene commonly contains brown to

[1] In some cases these hypersthenes are inverted pigeonites and the exsolved augite plates are parallel with a plane that was (001) in the parent clinopyroxene.

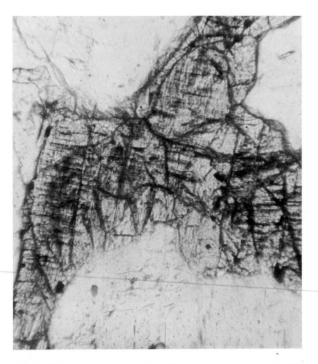

FIGURE 10.5 Hypersthene with schiller inclusions, in norite, Hitterö, Norway. Polars not crossed, ×94.

opaque inclusions of ilmenite, brookite, or titaniferous magnetite, which appear as minute, thin, often six-sided plates or rods or blades (Fig. 10.5). These lie parallel to (010) or (001) or are inclined 30° to c. They cause the schiller of bronzite and hypersthene. They may be rather uniformly distributed throughout the crystal or concentrated in the central part. They represent material exsolved with slow cooling.

Detrital hypersthene appears as prisms with ragged ends, although euhedral crystals occur where source lavas are nearby (Fig. 10.4).

Orientation. Biaxial $(-)$. Orthoferrosilite is $(+)$. $\alpha = a$, $\beta = b$, $\gamma = c$. Optic plane is (010) (Fig. 10.1b). 2V variable (Fig. 10.2):

$En_{87.5}$ $2V = 90° \ (-)$
En_{58} $2V = 46° \ (-)$
En_{15} $2V = 90° \ (-)$
En_{12} $2V = 83° \ (+)$

In zoned crystals 2V and the sign may be variable. In the enstatite–hypersthene series the dispersion (over α) is as follows (weak to strong):

En_{100}–En_{82} $r < v$
En_{82}–En_{58} $r > v$
En_{58}–En_{12} $r < v$

Twins on (101) are rare. Cross sections display symmetrical extinction.
Occurrence. In both extrusive and intrusive, basic to intermediate igneous rocks. Common in norite, hyperite, gabbro, diabase, basalt, andesite, and dacite. Less common in peridotites and meteorites, in which enstatite predominates. Also in diorites and monzonites and to an even lesser extent in some granites, syenites, and trachytes, in which iron-rich varieties may appear. In metamorphic rocks hypersthene occurs in charnockites with diopside, garnet, and antiperthite; in hypersthene granulites with quartz, orthoclase, plagioclase, and diopside or garnet; in eulysite with fayalite or hedenbergite, garnet, grunerite, magnetite, and apatite, in which the hypersthene may be rich in both Fe^{2+} and Mn; in hypersthene eclogites; in hypersthene amphibolites; and in hypersthene hornfels with biotite, quartz, cordierite, magnetite, and locally orthoclase or cummingtonite.

Relatively rare as a detrital mineral. It may be replaced by amphibole (uralite), antigorite (bastite), chlorite, and interlayered smectite–chlorite (Wilshire, 1958).

Diagnostic features. From enstatite it differs in its usual $(-)$ sign and higher indices; from the clinopyroxenes by parallel extinction. Detrital pleochroic grains may be confused with those of andalusite, which have, however, α = pale green, β = pale green, and γ = rose red.

Clinopyroxenes

Within the clinopyroxene subgroup the two main divisions are:

1. The augite subgroup
   a. The diopside–hedenbergite series
   $CaMgSi_2O_6$–$CaFe^{2+}Si_2O_6$
   b. The augite–ferroan augite series
   $Ca(Mg,Fe^{2+},Al)(Si,Al)_2O_6$–$Ca(Fe^{2+},Mg,Al)(Si,Al)_2O_6$
   c. The clinoenstatite–pigeonite–clinohypersthene series
   $Mg_2Si_2O_6$–$(Ca,Mg)(Mg,Fe^{2+})Si_2O_6$–$(Fe^{2+},Mg)_2Si_2O_6$
2. The alkali pyroxenes
   a. The aegirine (acmite)–jadeite series
   $NaFe^{3+}Si_2O_6$–$NaAlSi_2O_6$
   b. The aegirine–augite–aegirine series
   $(Ca,Na)(Fe^{2+},Mg,Fe^{3+},Al)(Si,Al)_2O_6$–$NaFe^{3+}Si_2O_6$

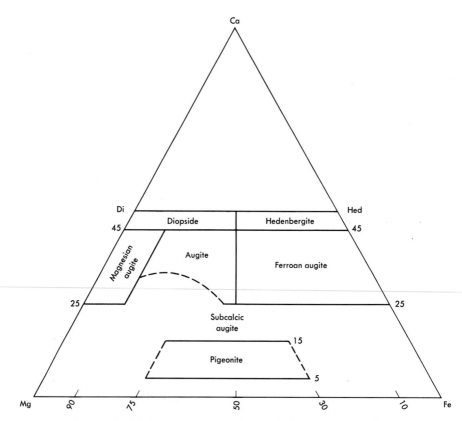

FIGURE 10.6 Subdivisions and nomenclature of the Ca–Mg–Fe–($\pm$Al) clinopyroxenes. (Modified from Hess, 1941.)

Other pyroxenes of minor significance as rock-forming minerals are schefferite, $Ca(Mg,Fe^{2+},Mn,Zn)Si_2O_6$ and jeffersonite, $Ca(Mg,Mn,Fe^{2+}, Zn)Si_2O_6$.

The composition of the augite subgroup can be expresed adequately, at least for purposes of optical determination, in terms of the three elements: Mg, Fe^{2+}, and Ca (Fig. 10.6). This representation does not affect the common definition of augite as an aluminous clinopyroxene, for nearly all the clinopyroxenes that appear in the Mg–Fe^{2+}–Ca field defined for augite also contain Al. Pigeonites also contain about the same proportion of Al as augites. The diopside–hedenbergite series is defined as having a $Ca:Mg+Fe^{2+}$ ratio of about 1:1.

Pigeonite apparently does not form a complete miscible series with augite. Relatively few pyroxenes that fall in the range 15–25% Ca (subcalcic augite) have been discovered. Pyroxenes that lie in the field $> 50\%$ Ca are rare and also contain so much Ti^{4+} and Al that they cannot be discussed adequately in terms of Mg–Fe^{2+}–Ca alone.

DIOPSIDE

Composition. $CaMgSi_2O_6$–$Ca(Mg,Fe^{2+})Si_2O_6$. In terms of the three main variables in the augite group: 25–55% Mg, 0–27.5% Fe^{2+}, 45–50% Ca. The variation in Ca is small. The lower boundary for the diopside–hedenbergite series is placed at 45% Ca because this provides a natural separation between most clinopyroxenes of contact-metamorphic origin and igneous clinopyroxenes. The specialized nomenclature that has been suggested for divisions of the diopside–hedenbergite series based on the Mg:Fe^{2+} ratio is:

100–90% Mg	diopside
90–50% Mg	salite
50–10% Mg	ferrosalite
10–0% Mg	hedenbergite

Little advantage is gained in this further subdivision and the names as here employed. In terms of the Mg:Fe^{2+} ratio, 100–50% Mg = diopside and 50–0% Mg = hedenbergite. If desired, adjectival modifiers may be employed: ferroan diopside or magnesian hedenbergite. Some Mn enters into the composition, usually in those members in which $Fe^{2+} \geqq Mg$. The end member of the diopside range tends to occur as the relatively pure Ca–Mg silicate, whereas the intermediate members usually contain Al, sometimes in considerable amounts. Fe^{3+} also may be present in variable quantities, usually small. Minor elements are Ti, Cr, and Na.

Indices. $\alpha = 1.663$–1.699, $\beta = 1.671$–1.705, $\gamma = 1.693$–1.728, $\gamma - \alpha = 0.028$–0.031. The indices increase with increasing Fe^{2+} (Fig. 10.8a). If Fe^{3+} is present in moderate amounts, the indices are somewhat above those of the curves. Thus a ferroan diopside, near the limit of the species range, with 3.9% Fe_2O_3 has $\alpha = 1.707$, $\beta = 1.714$, $\gamma = 1.733$ (measured) instead of $\alpha = 1.698$, $\beta = 1.705$, $\gamma = 1.727$ (from curves). The birefringence increases slightly toward the magnesian end. Maximum interference colors in thin section are of middle second order.

The substitution of Al for Si in the central part of the diopside–hedenbergite series has only slight effect on the indices but does cause a noticeable decrease in birefringence. The ferroan diopside listed above, which contains 3.95% Al_2O_3, has $\gamma - \alpha = 0.0255$ (measured) and $\gamma - \alpha = 0.028$ (from curves).

Color. In thin section colorless for the magnesian end to pale green and green at the ferroan end. With increasing Fe^{2+}, pleochroism also appears:

α = pale bluish green, dark blue green
β = brownish green, dark blue green
γ = yellow green, yellowish blue green

Form. Monoclinic. Euhedral prismatic crystals of stubby habit are common. These have eight- or four-sided cross sections showing the pyroxene cleavage (110) in two directions at angles of 87 and 93°. Granular anhedral masses in a mosaic texture also occur. Detrital pieces are tabular to irregular. In addition to the cleavage, diopside may show well-developed parting on (100) and (001). The majority of crushed pieces tend to lie on the (110) cleavage, but a few lie on the partings, usually (100).

Orientation. Biaxial (+). $\alpha \wedge a = -23$--$29°$, $\beta = b$, $\gamma \wedge c = 38$--$45°$. Optic plane is (010) (Fig. 10.7). $2V = 54$--$58°$, $r > v$ weak to moderate. The extinction angle increases with the Fe^{2+} content as does 2V, slightly (Fig. 10.8a,b). Slightly higher values for the extinction angles may be obtained when the diopside contains unusually large amounts of Fe^{3+} or Al.

SECTION	EXTINCTION	FIGURE	INDICES
(100)	parallel extinction	eccentric optic-axis figure	β
(110)	inclined extinction	eccentric flash figure	n_1, n_2
(010)	inclined extinction	flash figure	γ, α
(001)	symmetrical	slightly eccentric optic-axis figure	β

Multiple twins occur with (100) as the twin plane. Symmetrical extinction is characteristic of cross sections.

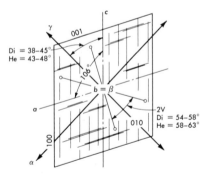

FIGURE 10.7 Orientation of diopside–hedenbergite, section parallel with (010).

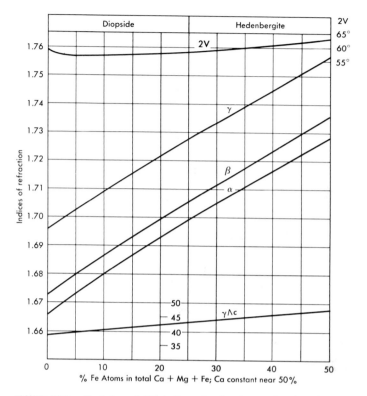

FIGURE 10.8a Variation of 2V, indices of refraction, and γ ∧ c with composition in the diopside–hedenbergite series (Hess, 1949).

Occurrence. The chief occurrence is as a metamorphic mineral in (1) impure magnesian marbles, calc schists, skarns, and lime–silicate rocks of both contact and regional metamorphic origin, together with various of the following: calcite, quartz, forsterite, grossularite, vesuvianite, tremolite or actinolite, epidote, wollastonite, phlogopite, scapolite, chondrodite, spinel, graphite, and apatite; (2) pyroxene hornfels associated with plagioclase, grossularite, hypersthene, biotite, quartz, and orthoclase; (3) amphibolite and hornblende gneiss with epidote, hypersthene, almandite, and plagioclase; and (4) pyroxene granulites along with plagioclase, hypersthene, quartz, and orthoclase.

In igneous rocks it is found to a minor extent in some intrusive types such as bronzitite (chrome diopside) with hypersthene and also in some anorthosites. Clinopyroxenes of many lamprophyric dikes tend to be diopsidic in composition. It is a rare constituent of meteorites and some

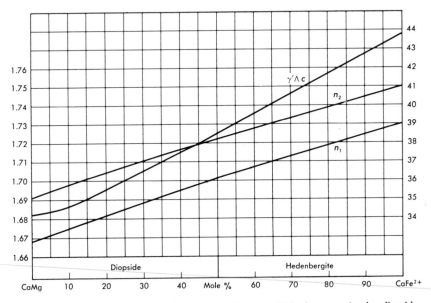

FIGURE 10.8b Variation in $\gamma' \wedge c$, n_2, and n_1 on (110) cleavages in the diopside–hedenbergite series (Parker, 1961; Hess, 1949).

slags. Many of the clinopyroxene phenocrysts of extrusive rocks, particularly basalts, alkali olivine basalts, andesites, latites, and quartz latites, have a $Ca:Mg+Fe^{2+}$ ratio that approaches 50:50 rather than 40:60, and they also are relatively rich in Al_2O_3 and Fe_2O_3. Thus by the classification employed here, these pyroxenes are diopside rather than augite. Normal augite may form phenocrysts in these rocks, as does hypersthene commonly, but augite is more common in the intrusive counterparts, in which diopside is rare. Diopside may be altered to talc, serpentine, tremolite, or chlorite. It may also show uralitization.

Diagnostic features. Distinguished from augite with difficulty:

	DIOPSIDE	AUGITE
2V	55–58°	40–60°
$\gamma - \alpha$	0.028–0.031	0.020–0.029
$\gamma \wedge c$	39–45°	41–57°
occurrence	metamorphic, phenocrysts of mafic extrusives	mainly in mafic and ultramafic intrusives

Pigeonite has a smaller 2V, lower birefringence and, in part, a different optical orientation. The last property is also characteristic of clinoenstatite.

Hedenbergite is separated from diopside by higher refractive indices. Tremolite commonly forms acicular prisms and in thin section shows the amphibole cleavage. Cleavage fragments of tremolite also have lower extinction angles than those of diopside. Olivine can be distinguished from diopside by the absence of cleavage, shape of crystals, and characteristic alteration.

HEDENBERGITE

Composition. $Ca(Fe^{2+},Mg)Si_2O_6-CaFe^{2+}Si_2O_6$. As expressed in the three variables of the augite subgroup: 25–55% Fe^{2+}, 0–27.5% Mg, 45–50% Ca. As in diopside, the end member, $CaFe^{2+}Si_2O_6$, tends to occur in rather pure form, but magnesian hedenbergite usually contains Al and Fe^{3+}. Mn may also be present in considerable amounts; and with increasing amounts of

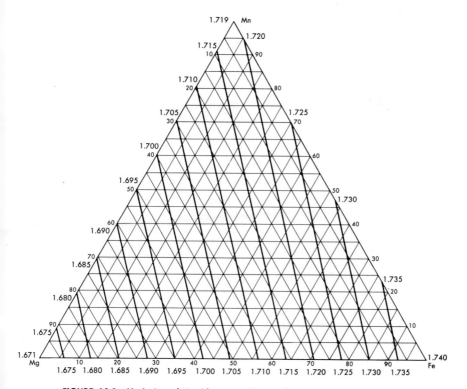

FIGURE 10.9 Variation of β with composition in the diopside–hedenbergite–johannsenite group (Schaller-Kennedy, 1947).

Mn, hedenbergite grades into johannsenite, $CaMnSi_2O_6$. Na and Ti are usually minor.

Indices. $\alpha = 1.699$–1.739, $\beta = 1.673$–1.745, $\gamma = 1.728$–1.757, increasing with Fe^{2+} (Fig. 10.8a, b). $\gamma - \alpha = 0.028$–0.029. The presence of considerable Fe^{3+} will cause an abnormal rise in the indices. A decrease in the birefringence occurs with high Al content and toward the Fe^{2+} end of the range. Johannsenite with from 72 to 97% $CaMnSi_2O_6$ has $\alpha = 1.708$–1.713, $\beta = 1.717$–1.721, $\gamma = 1.735$–1.740, $\gamma - \alpha = 0.026$–0.029. The maximum interference colors in thin section range from upper first to middle second order.

Color. Colorless to pale green or pale brown in thin section. May be pleochroic with:

α = purplish green, pale blue green, dark blue green, deep green
β = pinkish purple, pale green, blue green
γ = pale yellowish green, yellow, brownish yellow

Form. Monoclinic. Columnar or prismatic subparallel aggregates, masses of randomly oriented blades, or radial to spherulitic fibrous groups or sheaves are typical for iron-rich hedenbergite and johannsenite. Magnesian hedenbergite tends to be more like diopside in form. Crushed pieces lie on the (110) cleavage, which in cross section forms two sets of lines intersecting at 93 and 87°; (010) and (100) parting planes also may be present. In some specimens exsolution lamellae can be observed parallel with (001) and/or (100) (Fig. 10.18). Inclusions of calcite and quartz are common. Minute grains of sphene also occur as inclusions, probably the result of exsolution.

Orientation. Biaxial (+). $\alpha \wedge a = -28$–$-33°$, $\beta = b$, $\gamma \wedge c = 43$–$48°$. The extinction angle increases with Fe^{2+} (Fig. 10.8a). Slightly higher values (as much as $\gamma \wedge c = 51°$) occur if the Fe^{3+} and/or Al contents are abnormally high. The optic plane is (010) (Fig. 10.7). The general range in $2V = 58$–$63°$, but values as low as 53° and as high as 67° result from large amounts of elements other than Fe^{2+} and Mg. $r > v$ weak to strong. Cross sections show symmetrical extinction. Paired and multiple twinning on (100) are common in some specimens. Johannsenite is (+) and has $2V = 70°$, $r > v$ and $\gamma \wedge c = 48°$; the optic plane is (010); multiple twin lamellae are common.

Occurrence. Typically of pyrometasomatic origin in lime–silicate rocks, tactites, or skarns; associated with ore deposits of magnetite or sulfides of Fe, Zn, Pb, or Cu. Minerals that may be found with it are andradite, vesuvianite, wollastonite, epidote, actinolite, anthophyllite, cummingtonite, ilvaite, plagioclase, magnetite, quartz, and calcite. Johannsenite is also commonly of contact origin and occurs with ore deposits of various metals.

It alters readily to rhodonite, and the calcium released in this alteration may form xonotlite.

Hedenbergite appears in certain iron-rich metamorphic rocks—various greenstones and eulysites. Associated constituents are grunerite, actinolite, anthophyllite, hypersthene, fayalite, almandite, and magnetite. Hedenbergite also is a rare constituent of intrusive igneous rocks ranging from gabbro (iron-rich with fayalite) to syenite. It has been found altered to nontronite. Has been found in high-silica slags.

Diagnostic features. From diopside and most augite it is distinguished by higher indices. It also has a slightly higher 2V than augite, and the occurrences and associated minerals are distinctive.

AUGITE

Composition. $Ca(Mg,Fe^{2+},Al)(Si,Al)_2O_6$. Ferroan augite contains more Fe^{2+} than Mg. Fe^{3+}, Ti, and Cr may be present in significant amounts; Mn, Ni, and Na are minor constituents. Most augites fall into the range 27.5–65% Mg, 10–37.5% Fe^{2+}, 25–45% Ca. Augites with less than 10% Fe^{2+} are relatively rare (magnesian augites). Most natural augites tend to be concentrated along the 40% Ca line (Fig. 10.6), whereas ferroan augites lie near the line, 30% Ca. Al is variable from low to moderate.

Indices. $\alpha = 1.680$–1.703, $\beta = 1.684$–1.711, $\gamma = 1.706$–1.729 for ordinary augites. For ferroan augite $\alpha = 1.699$–1.712, $\beta = 1.706$–1.718, $\gamma = 1.728$–1.742. Cr decreases the indices and Fe^{3+} increases them above the values of the graphs (Figs. 10.14–10.17). $\gamma - \alpha = 0.024$–0.030, increasing for

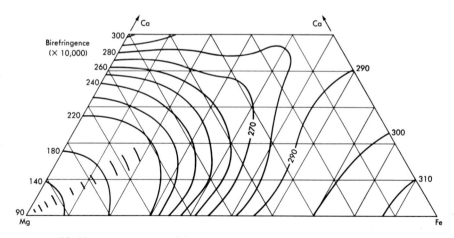

FIGURE 10.10 Variation of birefringence in Ca–Mg–Fe clinopyroxenes with composition (Hess, 1949).

the ferroan augites (Fig. 10.10) but modified by Al, which causes a marked decrease in the birefringence. The maximum interference tints in thin section fall in the lower and middle second order.

Color. Colorless, gray, pale green, pale brown, and pale purple in thin section. Pleochroism:

$\alpha =$ pale green, pale brownish green, pale blue green, blue green, pale purple grown

$\beta =$ pale brownish green, pale greenish brown, pink, yellow green

$\gamma =$ pale green, pale brownish green, pale blue green, yellow green, pale purple brown

The absorption surface does not coincide with the three directions of the optical indicatrix in many pyroxenes. However, the absorption formulae are stated as $\alpha > \gamma > \beta, \alpha = \gamma < \beta, \gamma > \alpha > \beta$.

Detrital pieces are yellow green, green, dark green, and brown. Green color is related to Fe content. The purple and purple-brown shades, which are caused by Ti, may be zonally arranged in wedge-shaped growth sectors ("hourglass" structure) owing to selective absorption of Ti ions during growth onto certain faces, generally the pinacoids, (111) and (101). In other crystals the sectors are of less regular shape. The differences in color are accompanied by variation in the extinction angles and 2V. Hourglass type of zoning may be combined with a concentric zoning, which also occurs separately.

Form. Monoclinic. Phenocrysts and microphenocrysts in lavas are euhedral to subhedral (Fig. 10.24) and may show embayment or corrosion. The crystals are stubby or elongated parallel with c, with eight-sided cross sections on which the two diagonal sets of cleavage traces (110) intersect at 87 and 93°. Matrix augite is anhedral and interstitial. In diabases augite is subophitic to ophitic. Poikilitic augite includes olivine or plagioclase. Various parallel growths are formed between augite and other pyroxenes (Fig. 10.20):

CORE	MARGIN	ROCK
augite	pigeonite	diabase
augite	hypersthene	hornblende andesite
pigeonite	augite	diabase
hypersthene in ragged relicts	augite	olivine andesite
hypersthene with oriented plates of augites	augite	diabase
clinoenstatite	augite	slag

Irregular, ragged intergrowths with hypersthene also occur. In glassy rocks augite may form microlites of varied form: fernlike branches, tufts with fanlike terminations, curved and plumose groups, rectangular branching aggregates of needles and prisms knobbed at both ends. Augite shows the partings (100) (diallage) and (001). Crushed material tends to lie mainly on (110) cleavage faces and to a lesser extent on (100) partings. Detrital pieces are subrounded and prismatic grains or are cleavage fragments with irregular outline (Fig. 10.23). Blebs of volcanic glass and specks of magnetite occur in augites of some extrusives. In augites of intrusive rocks thin plates or rods of ilmenite may be arranged along one or more of several crystal planes, commonly (100) as in hypersthene. The inclusions of augite or diopside as oriented lamellae in hypersthene are described under that mineral. Similarly, augite may contain exsolution lamellae of enstatite or hypersthene, commonly along (100) and less commonly along (001) (Fig. 10.18); also (001) pigeonite lamellae and a combination of (001) pigeonite and (100) hypersthene lamellae (Brown and Gay, 1959).

Orientation. Biaxial $(+)$. $\alpha \wedge a = 4\text{--}9°$, $\beta = b$, $\gamma \wedge c = 39\text{--}52°$ (Fig. 10.19). Many augites have extinction angles in the range $\gamma \wedge c = 41\text{--}48°$. In a few types angles as high as $57°$ are recorded. The size of the angle is probably affected somewhat by Al, Fe^{3+}, and Ti as well as Mg, Fe^{2+}, and Ca. Optic plane is (010). $2V = 39\text{--}63°$. For augites of extrusive and hypabyssal rocks the normal range is $2V = 40\text{--}52°$, averaging about $46°$. For augites with exsolved hypersthene from plutonic rocks, $2V = 45\text{--}60°$. The size of 2V is also somewhat increased by Ti and Fe^{3+} as well as

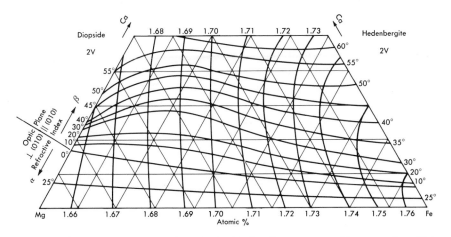

FIGURE 10.11 Variation of 2V in Ca–Mg–Fe clinopyroxenes with composition (Muir, 1951).

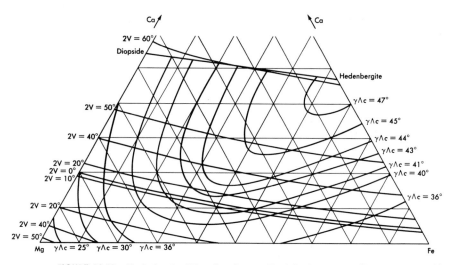

FIGURE 10.12 Variation in 2V and $\gamma \wedge c$ on (010) in Ca–Mg–Fe clinopyroxenes with composition (Tomita, 1934; modified by Deer and Wager, 1938).

variable with the Mg–Fe^{2+}–Ca ratio (Fig. 10.11), and a 2V as high as 70° has been recorded for ferroan augite. Dispersion: $r > v$ weak to moderate, strong in varieties containing much Ti. Cross sections display symmetrical extinction.

Zoning is not uncommon, especially in phenocrysts. In normal zoning

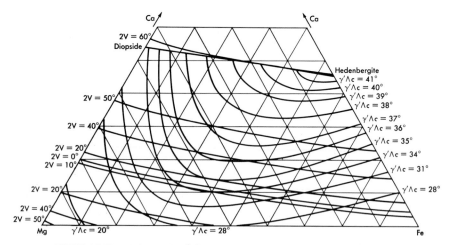

FIGURE 10.13 Variation in $\gamma' \wedge c$ on (110) and 2V in Ca–Mg–Fe clinopyroxenes with composition (Tomita, 1934).

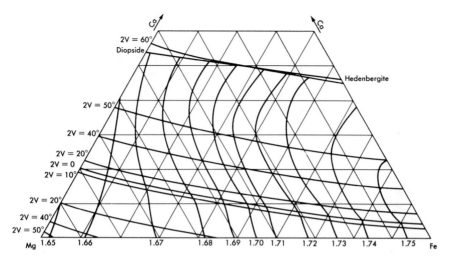

FIGURE 10.14 Variation in n_1 on (110) and 2V in Ca–Mg–Fe clinopyroxenes with composition (Tomita, 1934).

the outer layers have higher refractive indices and a smaller 2V, indicating increasing Fe^{2+} contents in outer layers. Reverse zoning may also be present. The extinction angle, $\gamma \wedge c$, also varies in zoned crystals, usually decreasing with a decrease in the indices but in some cases increasing with

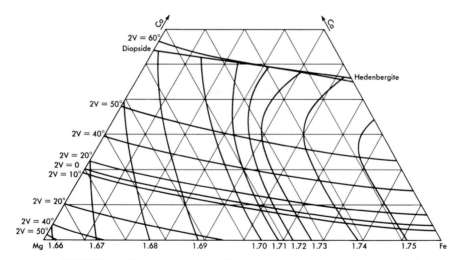

FIGURE 10.15 Variation in n_2 on (110) and 2V in Ca–Mg–Fe clinopyroxenes with composition (Tomita, 1934).

AUGITE

them. In a few varieties both the extinction angle and 2V decrease outward, whereas the indices increase. Zonal variation in birefringence also occurs. Contacts between zones are straight to extremely irregular. In some cases a zone of irregular patches of varying composition borders the crystal, or the core may show a ragged, embayed outline and display mottled extinction. In crystals that show both hourglass and concentric types of zoning the variation in properties is progressive in each sector.

Paired or multiple twins on (100) are common. Fine multiple twinning on (001) may also be present. Crystals in which paired (100) twins are combined with either multiple (001) twinning or (001) parting display herringbone structure when viewed normal to (010). Crystals of complicated aspect are produced by combinations of twinning and both types of zoning.

Occurrence. A common igneous mineral, especially of the ultramafic to intermediate rock types. In mafic lavas augite forms phenocrysts, although diopside is more common. Hypersthene may appear together with augite. In the groundmass of basalts pigeonite is common with or without augite, or ferroaugite alone may be present. Extrusive rocks in which augite occurs are basalts, andesites, and quartz latites.

Diabases and dolerites contain ophitic augite, in combination with pigeonite and hypersthene. In peridotites augite is common, usually occur-

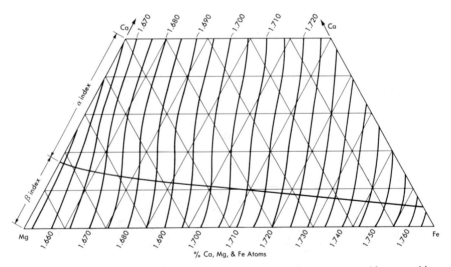

FIGURE 10.16 Variation in α and β in Ca–Mg–Fe clinopyroxenes with composition (Hess, 1949).

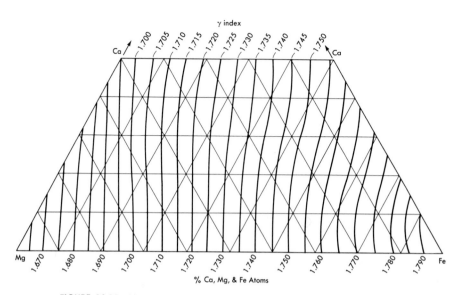

FIGURE 10.17 Variation in γ in Ca–Mg–Fe clinopyroxenes with composition (Hess, 1949).

ring with hypersthene that contains oriented plates of exsolved augite. Augite is also an important constituent in olivine gabbros, gabbros, hyperites, diorites, and tonalites; less common in granodiorites, granites, and syenites.

In extrusive rocks phenocrysts of hornblende or biotite commonly show resorption effects in which these minerals are peripherally or completely replaced by a fine-grained mixture of pyroxene (diopside or augite), magnetite, hematite, and plagioclase or epidote. Augite phenocrysts in extrusives may display a reaction rim that consists of granules of pigeonite with fine-grained magnetite and tridymite. In intrusive rocks augite may be replaced by amphibole through uralitization. In this reaction, hornblende replaces augite, usually along cleavages; in other cases hornblende forms parallel overgrowths. Augite may also be replaced by serpentine, chlorite, epidote, and calcite.

Among metamorphic rocks, augite appears in some dark-colored gneisses and in pyroxene granulites, in which it is associated with hypersthene, garnet, calcic plagioclase, and magnetite. Relict augite rimmed by actinolite or chlorite is found in greenschists formed by the metamorphism of mafic igneous rocks. Augite is moderately common as a detrital mineral.

Diagnostic features. For the distinction between augite and diopside see under the latter. Pigeonite has $2V \leqq 32°$. Enstatite and hypersthene show

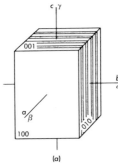
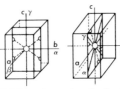
(a)

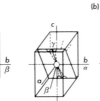

(b)

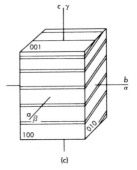
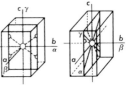

(c)

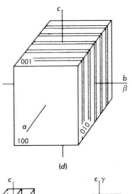

Host hypersthene Lamellae diopside   Host pigeonite   Lamellae augite   Lamellae hypersthene   Lamellae augite

(d)

(e)

Host augite   Lamellae hypersthene   Host augite   Lamellae pigeonite

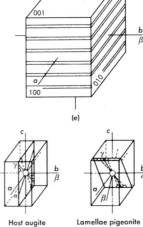

FIGURE 10.18 Common orientations of exsolution lamellae in pyroxenes (Poldervaart and Hess, 1951). (*a*) Fine lamellae of diopsidic pyroxene parallel with (100) in orthopyroxene. (*b*) Broad lamellae of augite parallel with (001) in pigeonite. (*c*) Broad lamellae of augite parallel with the relict monoclinic (001) plane in orthopyroxene inverted from pigeonite. (*d*) Broad lamellae of orthopyroxene parallel with (100) in magnesian augite. (*e*) Broad lamellae of pigeonite parallel with (001) in more ferroan augite.

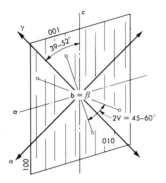

FIGURE 10.19 Orientation of augite, section parallel with (010).

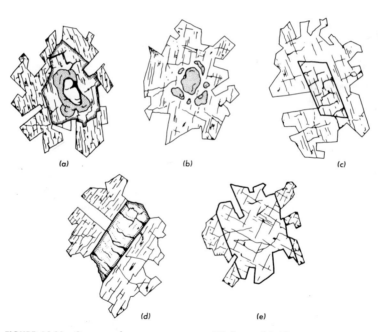

FIGURE 10.20 Overgrowths among pyroxenes (Walker and Poldervaart, 1949). (*a*) Core of olivine, surrounded by successive mantles of pigeonite, hypersthene, and augite. (*b*) Partly resorbed core of pigeonite surrounded by hypersthene. (*c*) Columnar core of pigeonite surrounded by hypersthene. (*d*) Columnar core of pigeonite surrounded by augite. (*e*) Core of augite surrounded by ferroan pigeonite.

AUGITE

223

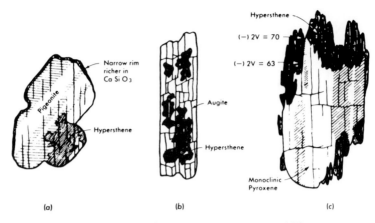

FIGURE 10.21   (a)-(c) Intergrowths among pyroxenes (Kuno, 1936).

parallel extinction in all longitudinal sections, and the latter is $(-)$ as well. Furthermore, front pinacoid (100) sections of hypersthene yield a Bxa figure, whereas those of augite give an optic-axis figure. Olivine shows higher double refraction in some sections than does augite. Crushed material that lies on (110) cleavage faces may be difficult to tell from cleavage fragments of colorless or light-colored monoclinic amphiboles, but the latter have lower extinction angles.

PIGEONITE

Composition. $Ca(Mg,Fe^{2+},Al)(Si,Al)_2O_6$. The pigeonite field includes: 30–70% Mg, 25–65% Fe^{2+}, 5–15% Ca. There apparently is not a complete gradation between natural pigeonite and augite; few natural clinopyroxenes (subcalcic augite) fall into the range 15–25% Ca. Fe^{3+} and Ti are generally

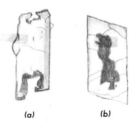

FIGURE 10.22   (a) Zoned, corroded augite, Vesuvius, Italy. (b) Augite with corroded core, from nephelinite, Kaiserstuhl, Bavaria (Wülfing and Mügge, 1925).

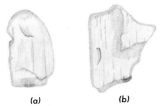

(a)   (b)

FIGURE 10.23   Detrital augite. (a) River sand, Austria. (b) Cape Verde Islands, Portugal.

present as minor elements. The members of the series, clinoenstatite [1], $Mg_2Si_2O_6$, clinohypersthene, $(Mg,Fe^{2+})_2Si_2O_6$, and clinoferrosilite, Fe^{2+}_2 Si_2O_6, contain very little or no Ca and Al and have few natural representatives.

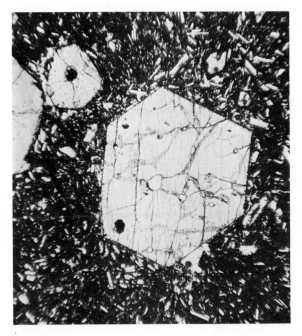

FIGURE 10.24   Phenocryst of augite in basalt, Brohl, Rhineland, Germany. Polars not crossed, $\times 28$.

[1] Series between synthetic clinoenstatite and diopside and between clinoferrosilite and hedenbergite have been demonstrated but are not known in nature.

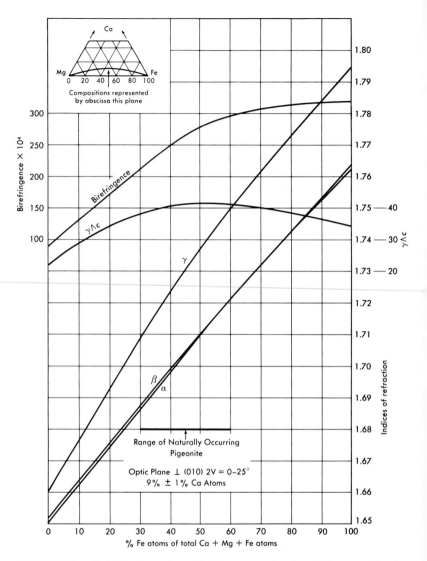

FIGURE 10.25a Variation in α, β, γ, birefringence, and $\gamma \wedge c$ with composition in pigeonites (Hess, 1949).

Indices. $\alpha = 1.683–1.722$, $\beta = 1.684–1.722$, $\gamma = 1.704–1.752$, increasing with Fe^{2+} (Figs. 10.14–10.17, 10.25). $\gamma - \alpha = 0.021–0.028$, also increasing with Fe^2 (Fig. 10.10). Maximum interference colors in thin section range from lower to middle second order. $\beta - \alpha$ is very low ($0.000–0.002$). For

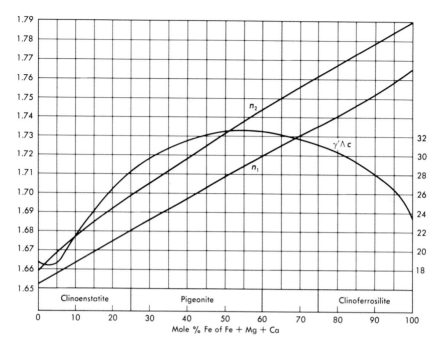

FIGURE 10.25b Variation in $\gamma' \wedge c$, n_2, and n_1 on (110) cleavages in pigeonite (Parker, 1961; Hess, 1949).

clinoenstatite $\alpha = 1.651$, $\beta = 1.652$, $\gamma = 1.661$, and for clinoferrosilite $\alpha = 1.763$, $\beta = 1.764$, $\gamma = 1.794$.

Color. Colorless to faintly pleochroic in thin section. Some titaniferous types show pale purple colors; others have:

α = faint pink, pink, pale greenish brown
β = faint pink, brownish pink, pale greenish brown
γ = faint or pale green, pale reddish brown

Absorption is $\gamma = \alpha < \beta$ or $\alpha = \gamma < \beta$.

Form. Monoclinic. Euhedral to anhedral. Very rarely as well-formed phenocrysts. Euhedral crystals are prismatic, may be considerably elongate with c, and their eight-sided cross sections show the pyroxene cleavage (110) in two directions at 87 and 93°. Parting on (100) also may be present. Much pigeonite is anhedral-interstitial or anhedral-ophitic. Overgrowths betwen augite and pigeonite are described under augite. Pigeonite may contain oriented (001) lamellae of hypersthene or augite. Hypersthene or augite shells on pigeonite are formed either as overgrowths or by

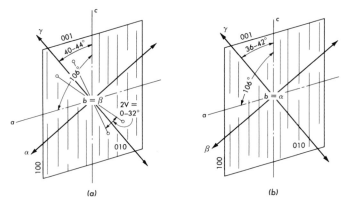

FIGURE 10.26 Orientations of pigeonite. (a) Section parallel with (010), optic plane parallel with (010). (b) Section parallel with (010), optic plane normal to (010).

inversion (Fig. 10.20); pigeonite bordering hypersthene is also known. Inclusions of magnetite and ilmenite may be abundant.

Orientation. Biaxial ($+$). Two orientations occur (Fig. 10.26):

1. $\alpha \wedge a = -23$--$27°$, $\beta = b$, $\gamma \wedge c = 40$–$44°$. Optic plane is (010), $2V = 32$–$0°$, $r < v$ distinct to weak
2. $\alpha = b$, $\beta \wedge a = -19$--$25°$, $\gamma \wedge c = 36$–$42°$. Optic plane is normal to (010), $2V = 0$–$30°$

The latter orientation is apparently more common. The change from the optic plane parallel with (010) to normal to (010) takes place when the Ca falls below 12–8% (Fig. 10.11). Clinoenstatite, clinohypersthene, and clinoferrosilite all have their optic planes normal to (010).

The characteristic twinning is either multiple or paired, with (100) as the twin plane. Zoning is not uncommon; usually, outer zones contain more Fe^{2+}. The Ca content may increase or decrease toward the margin, whereas Mg tends to stay constant. The zoning is expressed optically by variations in size of 2V, shift of the optic plane, and changes in extinction angle, indices, and birefringence. Variations in 2V can occur in different grains in the same thin section.

Occurrence. Principally as a matrix mineral in rapidly cooled basalts and andesites; very rarely as phenocrysts in these rocks. Commonly augite occurs with pigeonite in the matrix of extrusive rocks. Also common in diabases with augite or hypersthene. Rare in meteorites. Rarely it forms a reaction rim around olivine in some extrusives. Hypersthene–pigeonite and augite–pigeonite intergrowths and overgrowths are known. Rarely uralitized, although rims of secondary amphibole are known.

The most widespread and abundant mineral that forms microlites in glassy volcanic rocks is clinopyroxene, which exhibits a variety of forms: prisms, radial groups of curving filaments (trichites), coiled, looped, spiraled, and segmented types (Fig. 10.27). The identification of these microlites poses special problems in microscopy, inasmuch as some prisms are but 0.005 mm in length and 0.0005 mm in diameter (Ross, 1962). The

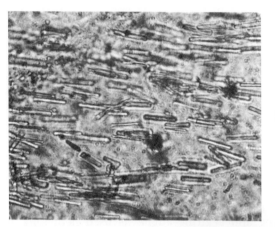

(a)

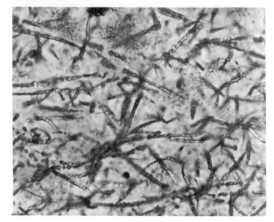

(b)

FIGURE 10.27 (a) Obsidian from Arroyo Hondo, Valles Mountains, New Mexico. Prismatic type of pyroxene microlites showing alignment imposed during emplacement. (b) Obsidian from the north side of Metztitlán Barranca on the road to Zacualtipán, Hidalgo, Mexico. Abundant pyroxene microlites made up of beadlike segments with disordered arrangement. Each figure is about 0.10 mm wide (Ross, 1962).

large and diagnostic extinction angle may be determined by using an oil-immersion lens system and the full intensity of a 500-watt bulb.

Diagnostic features. Augite has $2V > 39°$; pigeonite has $2V < 32°$. The occurrence is so restricted that if its presence is suspected, pigeonite may readily be determined. Clinoenstatite has lower indices, $\beta = 1.652$–1.680, lower birefringence, $\gamma - \alpha = 0.010$–0.020, and lower extinction angles, $\gamma \wedge a = 22$–$38°$. Clinoferrosilite has higher indices, $\beta = 1.730$–1.764, higher birefringence, $\gamma - \alpha = 0.029$–0.031, and extinction angles $\gamma \wedge c = 34$–$40°$, and it has been discovered in lithophysae in an obsidian. Clinohypersthene is much like pigeonite in many of its optical properties but has a somewhat smaller extinction angle. With the exception of clinoferrosilite, the occurrence of clinoenstatite or clinohypersthene in terrestrial rocks has not been demonstrated conclusively, although the two occur in meteorites and slags and have been prepared in melts. Some iron-rich pyroxenes with the optic plane parallel with (010) and containing small amounts of Ca are known from slags. These can be classified as ferroan pigeonites. A few rare Al- and Ti-rich pyroxenes with $Ca:Mg+Fe^{2+}<1:1$ are known with $2V = $ small–$0°$.

AEGIRINE

Composition. $NaFe^{3+}(SiO_3)_2$ to $(Na,Ca)(Fe^{3+},Fe^{2+},Mg,Al)(Si,Al)_2O_6$. The division between aegirine and aegirine–augite as used here falls at $Na_{0.45}Fe^{3+}_{0.45}$, or at about 38% of the aegirine molecule, at which point

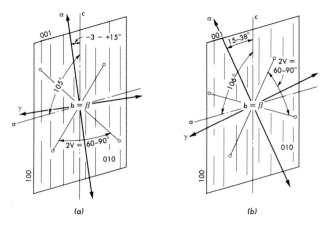

FIGURE 10.28 (a) Orientation of aegirine, section parallel with (010). (b) Orientation of aegirine–augite, section parallel with (010).

$2V = 90°$ and the sign changes from $(-)$ in aegirine to $(+)$ in aegirine–augite. This usage (Sabine, 1950) is analogous to the subdivision in the orthorhombic pyroxenes. K is commonly present in minor amounts, replacing Na; V may be an important constituent (nearly 4% V_2O_3); and Ti^{4+} (rarely Ti^{3+}), Zr, Mn, Be, Cr, and rare earths can be present in small amounts.

Indices. $\alpha = 1.720–1.778$, $\beta = 1.740–1.819$, $\gamma = 1.757–1.839$, $\gamma - \alpha = 0.032–0.060$. Both the refractive indices and the birefringence increase with increasing percentage of the aegirine molecule end member (Fig. 10.29). The maximum thin-section interference colors, which range from upper second order to fourth order, may be somewhat masked by the color of the mineral itself.

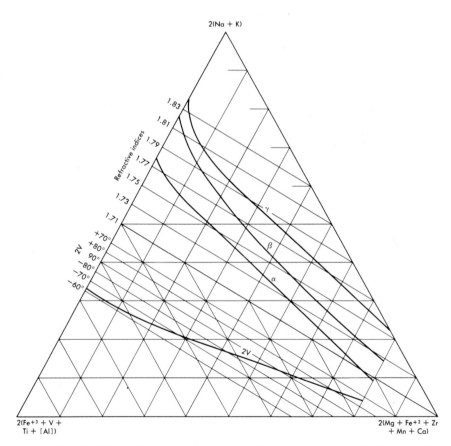

FIGURE 10.29   Variation in α, β, γ, and 2V with composition in aegirine (Sabine, 1950).

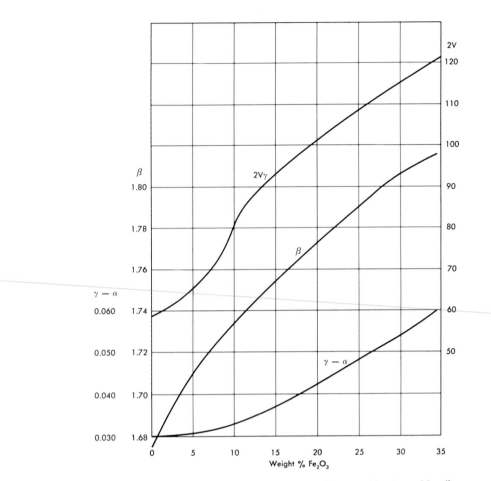

FIGURE 10.30a Variation in β, $\gamma - \alpha$, and $2V_\gamma$ with composition in aegirine (**Larsen, 1942**).

Color. Green or brown in thin section. Pleochroism weak to strong with:

α = olive green, pale green, pale bluish **green**, dark grass green, **deep** green
β = light blue green, yellowish green, **grass** green, emerald **green**
γ = yellowish green, honey yellow, light yellow

The absorption is $\alpha > \beta \geqq \gamma$.

Acmite is very similar in its optical properties and in general composition except that it is brown with pleochroic colors:

α = pink, grass green, olive green, dark brown
β = greenish blue, brownish green, brown
γ = blue, brown yellow brown, brownish green

The difference in color between acmite and aegirine has been ascribed to a higher Mn content in acmite (Grout, 1946). Zoned crystals are common in the augite–aegirine series; examples are (Fig. 10.33):

CORE		MARGIN
augite		aegirine-augite
augite	aegirine-augite	aegirine
aegirine-augite		aegirine
augite		aegirine
diopside		aegirine
aegirine		acmite
augite, aegirine-augite, aegirine	acmite	

Examples of reverse zoning, in which aegirine or aegirine–augite is surrounded by augite, are rare.

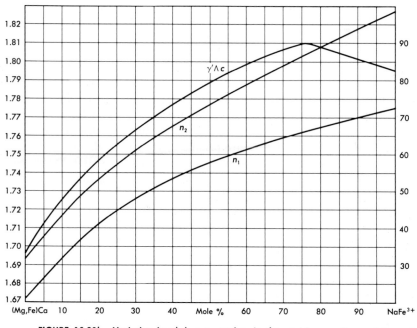

FIGURE 10.30b   Variation in $\gamma' \wedge c$, n_2, and n_1 in the aegirine–augite–aegirine series (Parker, 1961).

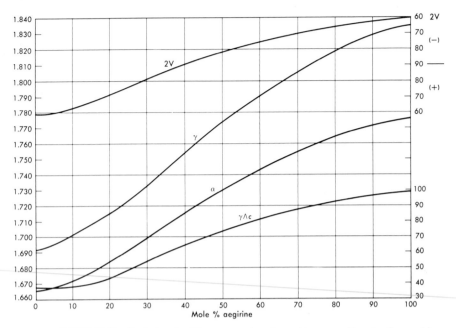

FIGURE 10.31 Variation in 2V, α, γ, and γ ∧ c with composition in the aegirine–diopside series. (Based on data by Ostrovsky, 1946, for synthetic pyroxenes.)

Form. Monoclinic. Typically in euhedral to subhedral, long prismatic, bladed, or needle-like crystals (Fig. 10.32). Cross sections are eight- or four-sided, commonly with the trace of (100) longer than that of (010). The (110) cleavage makes two sets of lines at 87 and 93°; (001) and (010) partings may be developed. Radial and subparallel felted aggregates also are found. Detrital pieces are prismatic with ragged ends and vertical striations. Crushed material tends to lie on (110) cleavage faces. Small hematite flakes appear as inclusions in acmite. Aegirine may enclose nepheline or albite in poikilitic fashion.

Orientation. Biaxial (−). $\alpha \wedge c = -3–15°$, $\beta = b$, $\gamma \wedge a = 20–2°$. Optic plane is (010) (Fig. 10.28). $2V = 60–90°$, $r > v$. 2V and $\alpha \wedge c$ increase with increasing percentage of the aegirine molecule (Fig. 10.30). Cross sections show symmetrical extinction. Elongated sections are length-fast. Twinning on (100) is common.

Occurrence. In alkalic granites, syenites, monzonites, shonkinites, feldspathoidal syenites, and feldspar-free feldspathoidal rocks and in their extrusive equivalents: soda rhyolites, trachytes, phonolites, and alkalic basalts. In alkali granites it is associated with an alkali amphibole, riebeck-

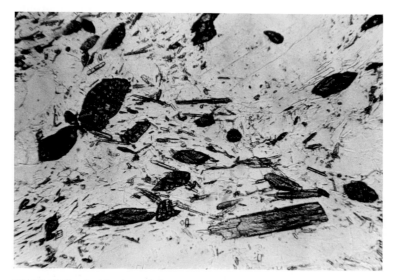

FIGURE 10.32 Aegirine in nepheline syenite, Sarna, Sweden. Polars not crossed, ×94.

ite or arfvedsonite, by which it may be replaced in uralitic fashion. Hastingsite also forms replacement rims on aegirine. Graphic intergrowths with microperthitic potash feldspar are known. Aegirine is commonly a late magmatic mineral in soda-rich intrusives, forming as a rim on augite or aegirine–augite. In extrusives aegirine can occur both as phenocrysts or as minute matrix crystals, even microlites. Alkali amphibole phenocrysts can be partly or entirely replaced in extrusives by an aggregate of small aegirine crystals, a process similar to the replacement of hornblende or biotite phenocrysts by augite in calc-alkalic extrusives.

In some extrusives the phenocrysts are augite rimmed by aegirine–augite, whereas aegirine alone appears as the smaller matrix crystals. Leucite-bearing lavas may have small aegirine needles tangential to leucite phenocrysts (ocellar texture).

In metamorphic rocks aegirine is formed by pyrometasomatic action where alkalic intrusives have encountered carbonate rocks, just as diopside may be produced by calc-alkalic rocks under similar conditions. Aegirine also appears in a variety of relatively rare soda-rich metamorphic rocks: metasomatized diabases with lawsonite and albite, alkalic gneisses and granulites with biotite and arfvedsonite or garnet and riebeckite, and some alkali pyroxene schists. Acmite, in addition to forming as a narrow rim around aegirine in various igneous rocks, appears as a metamorphic mineral in taconite, banded iron schists, and manganese ores; associated minerals are magnetite, hematite, braunite, quartz, siderite, adularia, riebeckite, grunerite, cummingtonite, nontronite, and apatite. Aegirine has been synthesized in glass furnaces and in boiler scale. Aegirine alters to limonite.

Diagnostic features. Differs from aegirine–augite and augite in deeper color, $(-)$ sign, higher refractive indices, higher birefringence, and smaller extinction angle. Green amphiboles generally show larger extinction angles, stronger pleochroism, and length-slow elongation.

AEGIRINE–AUGITE

Composition. $(Na,Ca)(Fe^{3+},Fe^{2+},Mg,Al)(Si,Al)_2O_6$, with less than 38% of the aegirine molecule, to $(Ca,Na)(Mg,Fe^{2+},Fe^{3+},Al)(Si,Al)_2O_6$, in which Na and Fe^{3+} becomes minor. Varieties toward the aegirine end of the series may also contain small quantities of K, Ti, V, and Mn. Varieties toward the augite end are characterized by either high Mg or Fe^{2+} contents; thus the minerals grouped here under aegirine–augite might be considered as sodian augite, sodian ferroan augite, or varieties richer in Ca such as sodian diopside or sodian hedenbergite.

Indices. $\alpha = 1.673–1.720$, $\beta = 1.679–1.744$, $\gamma = 1.691–1.759$, $\gamma - \alpha = 0.018–0.039$. The range in the indices and in the birefringence is considerable, owing to the bivariant nature of the series (Fig. 10.31). At the augite end the variation is due mainly to the $Mg–Fe^{2+}$ substitution, whereas the series as a whole varies in terms of $(Mg–Fe^{2+})–Fe^{3+}$ and Ca–Na. Maximum interference colors in thin section range from upper first order to upper second order.

Color. Colorless to green. Pleochroism slight:

$\alpha =$ pale yellow, green, olive green
$\beta =$ pale yellow, green
$\gamma =$ pale green, pale brownish green, yellow

Zoning is common (see under aegirine). Normally, outer zones are more strongly colored and have higher contents of aegirine.

Form. Monoclinic. Commonly euhedral stubby prisms. Cross sections are eight-sided and show the two directions of (110) cleavage at 87 and 93°. The (010) and (001) partings also may be present. Crushed grains tend to orient themselves on (110) cleavages.

Orientation. Biaxial (+). $\alpha \wedge c = 15\text{--}38°$, $\beta = b$, $\gamma \wedge a = 2\text{--}{-}21°$. Optic plane is (010) (Fig. 10.28b). $2V = 60\text{--}90°$, $r > v$. Cross sections are characterized by symmetrical extinction. Twins with (100) as the twin plane are not uncommon.

Occurrence. In soda-rich igneous rocks: alkalic granites, syenites, feldspathoidal syenites, soda rhyolites, trachytes, and phonolites. It can form as phenocrysts or in the matrix. Commonly shows zonal relations with augite or aegirine. Also found to a much lesser extent in some soda-rich pyroxene schists with riebeckite or glaucophane, in iron–quartz schists

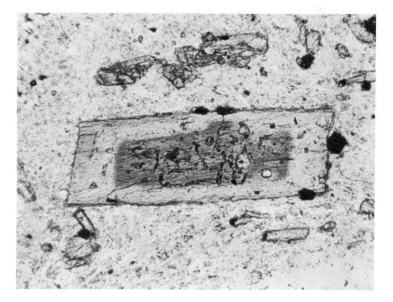

FIGURE 10.33 Zoned aegirine–augite phenocryst in tinguaite porphyry. Lookout Peak, Judith Mountains, Montana. Polars not crossed, ×50 (S. B. Wallace).

with glaucophane, magnetite, and hematite, and in pyrometasomatic rocks adjoining alkalic intrusives. In carbonatites.

Diagnostic features. Aegirine is $(-)$ and has smaller extinction angles and higher indices. Augite has a somewhat lower 2V, different extinction angles, generally lower indices, and lacks the marked green color. Green amphiboles show different cleavage and stronger pleochroism and have the slow direction closer to the elongation. Members of the diopside–jadeite series likewise have $\gamma \wedge c$ as the smaller extinction angle.

JADEITE

Composition. $NaAlSi_2O_6$. The formula for most natural jadeite can be written $(Na,Ca)(Al,Mg)(Si,Al)_2O_6$, and the mineral grades toward soda-diopside $(Ca,Na)(Mg,Fe^{2+},Al)(Si,Al)_2O_6$. Fe^{3+}, Ti, and Cr also may be present. Omphacite lies in this series near the diopside end, although some varieties of omphacite also contain considerable Fe^{3+}. Some types of jadeite that contain much Fe^{3+} can best be considered as part of a poorly represented aegirine–jadeite series.

Indices. $\alpha = 1.654$–1.673, $\beta = 1.659$–1.679, $\gamma = 1.667$–1.693, $\gamma - \alpha = 0.012$–0.027. Indices and birefringence decrease from diopside to jadeite. Omphacite has indices near the upper end of the range and higher (β as high as 1.685). The maximum interference colors vary from middle first to middle second order in thin section.

Color. Colorless to light green. The depth of color and pleochroism, in shades of green and yellow, increase with Fe^{3+} and Cr in jadeite. Omphacite usually is pale green and nonpleochroic.

Form. Monoclinic. Anhedral to subhedral. Fine- to coarse-grained masses consist of stubby prisms, irregular granules, or fibrous or needle-like aggregates. The cleavage is (110) in two directions at 93 and 87°; (100) parting has been observed. Omphacite, which is anhedral and equigranular, may contain rutile or kyanite as inclusions and may show parallel overgrowths of green amphibole.

Orientation. Biaxial $(+)$. $\alpha \wedge a = -13$–$-24°$, $\beta = b$, $\gamma \wedge c = 30$–$41°$. $\gamma \wedge c$ increases with decreasing content of the jadeite molecule. Jadeite usually has $\gamma \wedge c = 30$–$36°$; omphacite commonly has $\gamma \wedge c = 39$–$41°$. The optic plane is (010) (Fig. 10.34). $2V = 70$–$80°$ (jadeite), 60–$67°$ (omphacite). $r < v$ (jadeite); $r > v$ (omphacite). Twinning in jadeite is finely lamellar: (100) and less commonly (001).

Occurrence. Jadeite is the main mineral in jade, a nearly monomineralic semiprecious rock formed in association with some altered ultramafic rocks. Other minerals rarely encountered in jade in minor amounts are

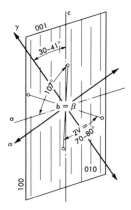

FIGURE 10.34 Orientation of jadeite, section parallel with (010).

albite, actinolite, wollatonite, nepheline, zoisite, muscovite, picotite, and quartz. Also occurs in quartz–jadeite rocks (metagraywackes) with glaucophane and lawsonite. Omphacite occurs in eclogites, together with garnet, plagioclase, quartz, green hornblende and accessory kyanite, enstatite, glaucophane, zoisite, muscovite, rutile, apatite, ilmenite, and magnetite. It may show varying degrees of replacement by hornblende.

Diagnostic features. Distinguished from the fibrous amphibole nephrite (a variety of tremolite or actinolite) by higher indices and larger extinction angles. Diopside is much like omphacite. The occurrences of both jadeite and omphacite are distinctive.

SPODUMENE

Composition. $LiAl(SiO_3)_2$. Some minor Fe^{3+}, Mn, Na, and K.

Indices. $\alpha = 1.648$–1.661, $\beta = 1.655$–1.670, $\gamma = 1.662$–1.679, $\gamma - \alpha = 0.014$–0.027. Small amounts of Fe^{3+} increase the indices and the birefringence. Upper first to middle second order maximum interference colors in thin section.

Color. Colorless in thin section and in crushed pieces. Pink (kunzite) and green (hiddenite) varieties are rare.

Form. Monoclinic. In tabular (100) crystals elongated with the c axis. Euhedral to rounded corroded. Usually in large to very large crystals. Distinct (110) cleavages, (010) and (100) partings. Much of the broken material will lie on (110) cleavage faces. Inclusions of quartz, albite, and muscovite are not uncommon.

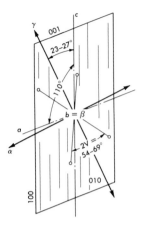

FIGURE 10.35 Orientation of spodumene, section parallel with (010).

Orientation. Biaxial (+), $\alpha \wedge a = -3$--$7°$, $\beta = b$, $\gamma \wedge c = 23$-$27°$. Optic plane is (010) (Fig. 10.35). $2V = 54$-$69°$, $r < v$ slight. Parallel extinction in sections normal to c. Twinned with (100) as the twin plane. Cleavage pieces yield an eccentric flash figure.

Occurrence. Almost entirely in granitic pegmatites, typically in crystals too large to be encompassed by thin sections. Alters to or is replaced by fine-grained muscovite, quartz, albite, cookeite, eucryptite, kaolinite, and other clay minerals.

Diagnostic features. Occurrence and association are distinctive. Has a smaller extinction angle than diopside.

RHODONITE

Composition. $(Mn,Ca,Fe^{2+})SiO_3$. Minor Mg may also be present and rarely a little Fe^{3+}. Zn replaces Mn in fowlerite.

Indices. $\alpha = 1.711$-1.738, $\beta = 1.714$-1.741, $\gamma = 1.724$-1.751, $\gamma - \alpha = 0.011$-0.014. The indices of refraction increase with Fe^{2+} and decrease with rise in Ca (Fig. 10.36). The maximum interference colors are of middle first order in thin section. A synthetic rhodonite from slag with 64% $FeSiO_3$ has $\alpha = 1.750$, $\beta = 1.754$, $\gamma = 1.767$.

Color. Colorless.

Form. Triclinic. Anhedral to euhedral crystals commonly tabular parallel with (001). Cleavages are (110) and (1$\bar{1}$0) perfect and (001) poor. Inclusions of other minerals, such as calcite, quartz, diopside, grunerite and other amphiboles, and bustamite, may be present. Both parallel and

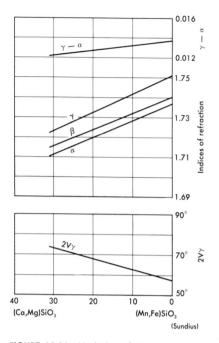

FIGURE 10.36 Variation of 2V$_\gamma$, α, β, γ, $\gamma - \alpha$ with composition in rhodonite (Sundius, 1931).

subgraphic intergrowths with bustamite are known. A type of "perthitic" structure consisting of small diopside lamellae along (001) probably results from exsolution.

Orientation. Biaxial (+) or (−). Positions of optical directions vary with composition. $\gamma' \wedge c$, measured on (010) = 2–11°; $\gamma' \wedge c$ on (110) = 17–30°; $\gamma' \wedge c$ on (1$\bar{1}$0) = 14–20°. In (+) rhodonite α approaches the normal to (100) and the trace of the optic plane in (100) makes an angle of 39° with the (100):(1$\bar{1}$0) edge, inclined toward (110). 2V = 61–76°; $r < v$ weak, with marked crossed dispersion. For some synthetic, high-Fe^{2+} rhodonites 2V = 37°. Multiple (010) twinning is common. Inclined extinction in all sections of the vertical zone. Some crystals are zoned with lighter-colored and untwinned borders around a darker core that shows fine lamellar twinning.

Occurrence. In ore deposits of iron, copper, zinc, and manganese of the vein or replacement type, also in pyrometasomatic deposits. Usually with rhodochrosite or other manganese minerals. May alter to pyrolusite or rhodochrosite. A rare constituent of some unusual manganiferous metamorphic rocks with grunerite, spessartite, pyroxmangite, and magnetite.

Diagnostic features. Differs from pyroxmangite, $(Mn,Fe^{2+},Mg)SiO_3$, in its larger 2V.

WOLLASTONITE

Composition. $(Ca,Fe^{2+})SiO_3$. Both natural and synthetic iron wollastonites are known, although natural iron wollastonites are much less common than iron-free types. A small amount of Mn may also be present. Wollastonites containing as much as 3.2% MgO have been synthesized, and others have been prepared with excess Ca or Si. The compound $(Ca,Fe^{2+},Mn,Mg)SiO_3$, vogtite, found in slags, also is isomorphous with wollastonite and emphasizes the relationship of wollastonite to bustamite.

Indices. The indices increase with increasing Fe^{2+} (Fig. 10.37):

	NATURAL	SYNTHETIC 70% $FeSiO_3$	SYNTHETIC 17% $CaMg(SiO_3)_2$
α	1.615–1.646	1.716	1.619
β	1.627–1.659	1.729	1.631
γ	1.629–1.662	1.734	1.634
$\gamma - \alpha$	0.013–0.017	0.018	0.015

In thin section maximum interference colors are of upper first order.

Color. Colorless or turbid owing to many small inclusions (gas?). Coarse pieces are colorless, gray, or pale yellow.

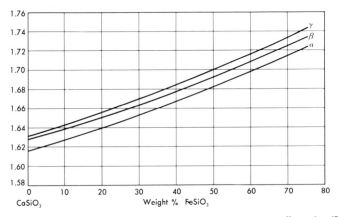

FIGURE 10.37 Variation of α, β, γ with composition in wollastonite (Bowen et al., 1933).

Form. Triclinic, pseudomonoclinic. Rarely monoclinic (parawollastonite). Typical crystals are subhedral to euhedral, columnar to fibrous, usually elongated parallel with b, and tabular with (100). May be poikiloblastic. Cleavages: (100) perfect, ($\bar{1}02$) and (001) good, and (101) and ($10\bar{1}$) poor. The shape of crushed and detrital material is governed principally by (100) and (001) cleavages.

Orientation. Biaxial $(-)$. Optic plane almost parallel with (010), and β nearly coincides with b. $\alpha \wedge c$ (almost in the acute axial angle β) $= -32$–$-44°$, increasing with Fe^{2+} content; $\beta \wedge b$ (in the plane normal to c) $= 1\frac{1}{2}$–$4°$; $\gamma \wedge a$ (nearly in the obtuse axial angle β) $= 37$–$49°$. $2V = 35$–$63°$ in natural wollastonites, increasing with Fe^{2+} content, but may become as large as $85°$ in synthetic iron-rich wollastonite. $r > v$ distinct. Parawollastonite has the orientation $\alpha \wedge c = -34°$, $\beta = b$, $\gamma \wedge a = 39°$. Optic plane is (010). $2V = 44°$, $r > v$, perceptible. Cleavages and twinning as in wollastonite.

Extinction may be somewhat irregular or mottled. Zoning may be present in cores having lower indices, a smaller 2V, and a lesser $\alpha \wedge c$ extinction angle. Contacts between zones may be smooth or ragged. Cleavage pieces (100), (001), and (101) yield eccentric optic-axis figures; ($\bar{1}02$), a nearly centered optic-axis figure; and (101), a nearly centered Bxo figure. Elongate sections show small extinction angles and are either length-slow or -fast. Multiple twinning on (100) is common.

Occurrence. Mainly a contact-metamorphic mineral in both endogenic and exogenic zones. Common associates are calcite, diopside, grossularite, tremolite, epidote, vesuvianite, monticellite, and calcic plagioclase; less common are melilite, leucite, graphite, fluorite, spinel, scheelite, magnetite, periclase, perovskite, xonotlite, pectolite, bustamite, and the rare Ca– and Ca–Mg–silicates, larnite, merwinite, rankinite, scawtite, spurrite, tilleyite, and cuspidine (Table 9.2). Rarely of igneous origin, as in some ijolites. Wollastonite is common also in slags and boiler scale. Rare as a detrital mineral. Parallel overgrowths of pectolite occur, and wollastonite blades are locally enveloped in sheaths of calcite or xonotlite. Both pectolite and calcite vein and replace wollastonite. Pseudowollastonite is a devitrification product in window glass melts and occurs in some silica bricks and in some blast furnace slags. Graphic intergrowths of wollastonite with tilleyite, gehlenite, vesuvianite, and merwinite are known.

Diagnostic features. Tremolite has the diagnostic amphibole cleavage, lower extinction angles, and larger 2V. Pectolite has smaller extinction angles, higher birefringence, and the optic plane normal to (010) and is always length-slow. Pseudowollastonite, the high-temperature modification (above $1180°C$), which occurs as a synthetic mineral and very rarely in nature, has $\alpha = 1.610$–1.614, $\beta = 1.611$–1.615, $\gamma = 1.648$–1.651, $\gamma - \alpha = 0.034$–0.041; colorless; monoclinic, pseudohexagonal; irregular to equant grains or prismatic; (001) cleavage; biaxial ($+$); $\alpha \wedge a = $ small, $\beta = b$, $\gamma \wedge c = $ small; $2V = 0$–$6°$. Lamellar twinning appears on basal sections.

The rare species bustamite, $CaMn(SiO_3)_2$, is closely related in its properties to wollastonite, but compounds intermediate between the two are not known in nature. It has $\alpha = 1.662$–1.708, $\beta = 1.674$–1.716, $\gamma = 1.676$–1.724, $\gamma - \alpha = 0.013$–0.022; triclinic; perfect (010) and ($1\bar{1}0$) cleavages; biaxial ($-$) or less commonly ($+$); optic plane and α nearly normal to (010); $\gamma \wedge c$ (nearly in (010)) $= 40°\pm$; $2V = 44$–$85°$, $r < v$ weak, strong crossed dispersion.

PECTOLITE

Composition. $NaCa_2Si_3O_8(OH)$. Ca may be replaced by Mg, Fe^{2+}, and Mn^{2+}. Magnesian pectolite has been called walkerite; manganoan pectolite, schizolite.

Indices. $\alpha = 1.594$–1.610, $\beta = 1.603$–1.614, $\gamma = 1.631$–1.642, $\gamma - \alpha = 0.032$–0.038. In thin section the maximum interference tints are of middle and upper second order. Indices increase and birefringence decreases with Mn.

Color. Colorless.

Form. Triclinic. Perfect (100) and (001) cleavages at 85 and 95°, resembling pyroxene cleavage. Thin crystals commonly elongate parallel with b. Stellate or radial masses of acicular or needlelike crystals are typical, less commonly fine-grained and massive.

Orientation. Biaxial ($+$). Optic plane and γ are nearly normal to (010), but the b axis does not lie exactly in the optic plane. Thus on (100), $\alpha \wedge (001)$ cleavage trace $= 2°$. $\alpha \wedge c = 10°$ nearly in the acute axial angle β. Similar to wollastonite in the pseudomonoclinic symmetry of its optical elements. $2V = 50$–$63°$, decreasing with Mn, $r > v$ slight. Cleavage pieces yield either a somewhat eccentric flash figure (001) or a nearly centered Bxo figure (100). Twinning on (100) may be present.

Occurrence. In veins and cavities in mafic igneous rocks such as diabases and basalts. Common associates are prehnite, datolite, calcite, and zeolites. Less commonly in phonolites and nepheline syenites and rarely in contact-metamorphic marbles, in which it may replace calcite. Recognized as a constituent of boiler scale.

Diagnostic features. Wollastonite has lower birefringence and is length-fast.

GENERAL   The amphiboles form an exceedingly complex group in which many members and varieties have been reported, some based on minor chemical differences and others on optical variations alone. Several classifications of the amphiboles have been made, including some involving the calculation of various end-member molecules. Such a classification becomes unsatisfactory, however, when the number of end-member molecules is large, as in the hornblende series, in which as many as eight have been proposed. Furthermore, many of the end members or even approximations of them do not exist in nature. In the following classification, although the formulae for the various species are given in somewhat generalized form, the listed compositions do not represent theoretical end members, but indicate the range in composition of the natural species.

A. Orthorhombic amphiboles
   1. Anthophyllite
      $(Mg,Fe^{2+},Al)_7(Si,Al)_8O_{22}(OH)_2$

B. Monoclinic amphiboles
   1. Cummingtonite–grunerite series
      $(Mg,Fe^{2+})_7Si_8O_{22}(OH)-(Fe^{2+},Mg,Mn)_7Si_8O_{22}(OH)$
   2. Tremolite–actinolite series
      $Ca_2Mg_5Si_8O_{22}(OH)_2-Ca_2(Mg,Fe^{2+})_5Si_8O_{22}(OH)_2$
   3. Hornblende series
      a. Edenite
         $Ca_2Na(Mg_{4-5},Al_{1-0})(Si_{6-8}Al_{2-0})O_{22}(OH,F)_2$
      b. Hornblende
         $Ca_2Na(Mg,Fe^{2+},Fe^{3+},Al)_5(Si_{6-7},Al_{2-1})O_{22}(OH,F)_2$
   4. Oxyhornblende (basaltic hornblende)
      $Ca_2Na(Mg,Fe^{3+},Fe,Al,Ti)_5(Si_{6-7},Al_{2-1})O_{22}(O,OH)_2$
   5. Hastingsite
      $(Ca_{2-1.5},Na+K_{1-1.5})(Fe^{2+},Mg)_{4-3}Fe^{3+}_{1-2}$
      $(Si_{5.5-6.5},Al_{2.5-1.5})O_{22}(OH)_2$
   6. Alkali amphiboles [1]
      a. Arfvedsonite
         $(Na_{2-2.5},Ca_{1-0.5})(Fe^{2+}_{4-2.5},Mg_{0.5-1.5})Fe^{3+}_{1.5-0.5}Si_8O_{22}(OH)_2$
      b. Riebeckite
         $Na_2Fe^{2+}_3Fe^{3+}_2Si_8O_{22}(OH)_2$

[1] For more detailed classifications and descriptions of the alkali amphiboles including other, rarer species, see Miyashiro (1957*b*) and Deer et al. (1963*a*, p. 209).

c. Glaucophane
$$Na_2(Mg,Fe^{2+})_3Al_2Si_8O_{22}(OH)_2$$

ANTHOPHYLLITE

Composition. $(Mg,Fe^{2+})_7Si_8O_{22}(OH)_2$. Al may be present up to a maximum of $(Mg,Fe^{2+})_5Al_2(Si_6,Al_2)O_{22}(OH)_2$. Varieties rich in Al have been called gedrite. Usually Fe^{2+} replaces Mg from 5 to 50% on an atomic basis. Fe^{3+}, Mn, Na, and Ti are present in very minor amounts. Small quantities of F replace OH. A rare ferroan aluminian variety also is known (Seki and Yamasaki, 1957). Anthophyllite and cummingtonite are not isomorphous.

Indices. $\alpha = 1.598\text{--}1.674$, $\beta = 1.605\text{--}1.685$, $\gamma = 1.615\text{--}1.697$, $\gamma - \alpha = 0.013\text{--}0.025$. The indices increase in general with an increase in Fe^{2+} (or $Fe^{2+}+Fe^{3+}+Ti+Mn$) (Fig. 10.39). For a Mg-free type: $\alpha = 1.694$, $\beta = 1.710$, $\gamma = 1.722$. The birefringence, on the other hand, increases generally with a rise in Mg. Al also influences the indices significantly; points lying above the index curve represent mainly anthophyllites high in Al. The birefringence apparently is not changed significantly by Al variations. In thin section the maximum interference colors range between middle first and lower second orders.

Aluminian and ferroan anthophyllites show weak to moderate pleochroism in shades of tan, increasing in depth with Fe^{2+}. Magnesian anthophyllite is colorless in thin section. The absorption is $\alpha = \beta < \gamma$, and some of the pleochroic colors are:

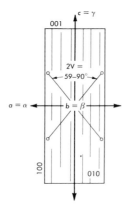

FIGURE 10.38 Orientation of (+) anthophyllite, section parallel with (010).

	Mg-free
α = pale tan, clove brown, greenish yellow	pale green
β = pale tan, clove brown, greenish yellow	brownish green
γ = tan, smoky gray, dark brown, grayish green	green blue

Form. Orthorhombic. Typically in elongated prismatic, euhedral to sub-hedral crystals (Fig. 10.40). May be columnar, bladed, fibrous, or asbesti-form. Commonly subparallel, but radial aggregates also are common. Cross sections are four-sided (diamond-shaped) or six-sided, with (010) subordinate, and show the typical amphibole cleavage (110) in two directions intersecting at 54 and 126°. Crushed coarse grains tend to lie on cleavage faces.

Orientation. Biaxial ($+$) or ($-$). $\alpha = a$, $\beta = b$, $\gamma = c$. Magnesian antho-phyllites are generally ($-$), and most aluminian types are ($+$). Optic plane is (010) (Fig. 10.38). $2V = -57°-90°-59°$ and is usually 80° or larger; ($+$) anthophyllite apparently is more common than ($-$) antho-phyllite. Dispersion $r > v$ or $r < v$ distinct to weak. Longitudinal sections are length-slow and exhibit parallel extinction. Cross sections show sym-metrical extinction. Asbestiform amphiboles, both orthorhombic and monoclinic, have their c axes parallel with the fiber length, but their a and b axes are randomly arranged around the long direction. Since the smallest fiber that is usable microscopically still consists of a bundle of submicro-

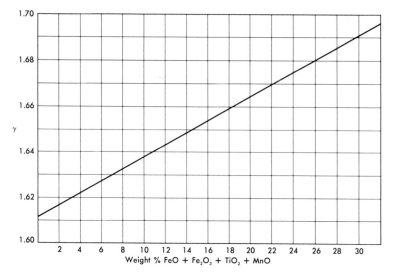

FIGURE 10.39 Variation of γ with composition in anthophyllite (Rabbitt, 1950).

scopic fibers in which the a and b axes are randomly placed, such aggregates appear to show parallel extinction. From microscopic evidence alone such material has therefore been classed as orthorhombic. Asbestiform amphiboles must be studied by x-ray methods in order to determine their orthorhombic or monoclinic symmetry, and some asbestiform amphiboles, previously determined as anthophyllite, are in reality tremolite, actinolite, soda–tremolite, and cummingtonite. Anthophyllite shows twinning very rarely, on (100).

Occurrence. Forms as a secondary, hydrothermal mineral in serpentinites and altered peridotites, commonly with talc, actinolite, magnesite, vermiculite, and chlorite. Also common in metamorphic rocks such as anthophyllite schists, anthophyllite–talc schists, anthophyllite amphibolite, anthophyllite–biotite gneiss ± sillimanite, anthophyllite–cordierite gneiss ± cummingtonite and rarely staurolite, anthophyllite–andalusite–plagioclase gneiss, anthophyllite–cordierite hornfels with biotite or cummingtonite, and granulites with biotite, hypersthene, and cordierite. Found with ore deposits in metamorphic rocks and associated with pyrite, chalcopyrite, pyrrhotite, quartz, and cordierite. Parallel intergrowths occur with cummingtonite, usually as alternating laths, as if "twinned" on (100). In stream sands

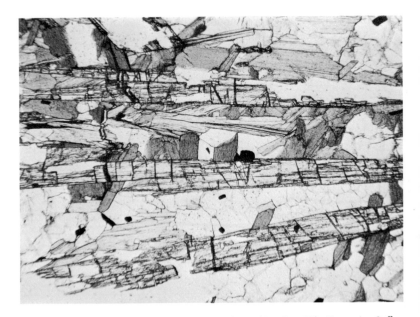

FIGURE 10.40 Anthophyllite in anthophyllite schist, Thirty-One Mile Mountain, Guffey, Colorado. Polars not crossed, ×32 (J. E. Bever).

BIAXIAL MINERALS: CHAIN SILICATES

from areas underlain by metamorphic rocks anthophyllite may be an important detrital mineral. Anthophyllite may be replaced by biotite, and it alters to talc.

Diagnostic features. The asbestiform types are not distinguishable from monoclinic amphiboles by optical methods. Otherwise, the parallel extinction of (010) sections or (110) cleavage pieces distinguishes anthophyllite from cummingtonite, actinolite, and hornblende.

CUMMINGTONITE

Composition. $(Mg,Fe^{2+})_7Si_8O_{22}(OH)_2$. Most cummingtonites have $Fe > Mg$, although in a few $Mg > Fe$. Cummingtonite grades into grunerite with an increase in Fe^{2+} to nearly the pure iron silicate. However, the pure magnesium silicate is not known, and the maximum reported content of $MgO = 22.11\%$. Cummingtonite is separated from grunerite by the change in optical sign from $(+)$ to $(-)$, analogously to the division between enstatite and hypersthene. This change in sign takes place when $FeO + MnO = $ ca. 40% or the content of $(Fe^{2+},Mn)_7Si_8O_{22}(OH)_2 = $ ca. 78%. Minor amounts of Fe^{3+}, Mn, Ti, Ca, Na, and K may be present, usually as less than 1% of the oxides. Al is low, and a maximum of 5.02% Al_2O_3 is recorded. A rare variety contains Zn.

Indices. $\alpha = 1.639\text{--}1.671$, $\beta = 1.647\text{--}1.689$, $\gamma = 1.664\text{--}1.708$, $\gamma - \alpha = 0.025\text{--}0.038$. The indices and birefringence increase with Fe^{2+} (Fig.

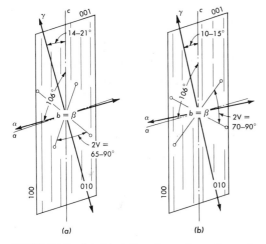

FIGURE 10.41   (a) Orientation of cummingtonite, section parallel with (010). (b) Orientation of grunerite, section parallel with (010).

10.42a). Cummingtonite with several percent Al_2O_3 usually will have indices that lie slightly above those of the nonaluminous varieties   Al causes a decrease in the birefringence; with 5% Al_2O_3, $\gamma - \alpha = 0.020$. In general in thin section the maximum interference colors range from lower to uppermost second order. The substitution of Mn for Fe^{2+} causes a decrease in birefringence and in γ. The substitution of F for OH also lowers the refractive indices.

Color. Normally colorless or faint grayish brown in thin section; may show weak pleochroism:

α = colorless, pale yellow

β = pale yellow, pale brown

γ = pale greenish yellow, pale violet

Form. Monoclinic. Prismatic bladed or fibrous in subparallel, parallel, or radial aggregates. Curved crystals and asbestiform varieties are found.

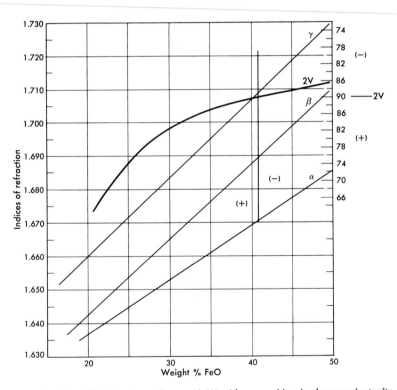

FIGURE 10.42a Variation in α, β, γ, and 2V with composition in the cummingtonite–grunerite series.

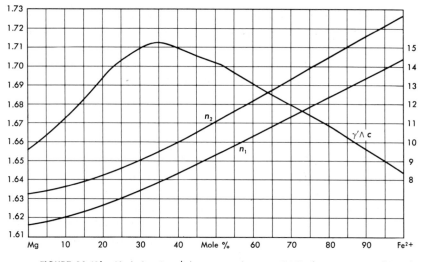

FIGURE 10.42b Variation in $\gamma' \wedge c$, n_2, and n_1 on (110) cleavages, cummingtonite–grunerite series (Parker, 1961).

Cross sections of larger crystals can be six-sided with (010) subordinately developed, showing the two directions of (110) cleavage at angles of 56 and 124°. Intergrowths with anthophyllite are not uncommon. Parallel intergrowths with hornblende also occur, and overgrowths of green hornblende have been noted. Because of the usual fibrous habit, the cleavage does not noticeably influence the orientation of crushed material.

Orientation. Biaxial (+). $\alpha \wedge a = -5$–$2°$, $\beta = b$, $\gamma \wedge c = 21$–$14°$. Optic plane is (010) (Fig. 10.41a). $2V = 65$–$90°$. $\gamma \wedge c$ decreases with increasing Fe^{2+}; 2V increases with Fe^{2+}. F for OH increases $\gamma \wedge c$ slightly. Dispersion $r < v$ for the magnesian varieties, $r > v$ for the ferroan. Cross sections show symmetrical extinction; elongate sections are length-slow. Twinning is very common, sometimes paired but usually multiple. The twin plane is (100), and the typically very narrow lamellae should not be confused with parallel intergrowths with anthophyllite.

Occurrence. Almost exclusively in metamorphic rocks: cummingtonite–anthophyllite gneiss with cordierite; cummingtonite hornfels, some varieties with cordierite and anthophyllite or plagioclase and biotite; cummingtonite schists; cummingtonite amphibolites; and cummingtonite granulites. Some skarn rocks associated with iron or copper ores also contain cummingtonite with anthophyllite and cordierite. Mafic xenoliths in intrusive and extrusive igneous rocks may contain it. Manganiferous cummingtonite has been identified in eulysites.

Diagnostic features. Grunerite is $(-)$; cummingtonite is $(+)$. Grunerite also has higher indices and a slightly lower extinction angle. For the distinction between cummingtonite and anthophyllite see under anthophyllite. In general, cummingtonite has higher refractive indices than tremolite, and tremolite is $(-)$. Hornblende has pleochroic colors in green and brown and also is $(-)$.

GRUNERITE

Composition. $(Fe^{2+},Mg,Mn)_7Si_8O_{22}(OH)$ with $FeO + MnO > 40\%$ or $(Fe^{2+},Mn)_7Si_8O_{22}(OH)_2 > 78\%$. MnO may attain 7%. Fe^{3+} is present usually in amounts less than 2% Fe_2O_3. Al is minor, in contrast to some cummingtonites; Ti, Ca, K, and F present in small amounts.

Indices. $\alpha = 1.663-1.686$, $\beta = 1.680-1.709$, $\gamma = 1.696-1.729$. The slight overlap in the index ranges of cummingtonite and grunerite is due to the modifying influences of Mn and Al. $\gamma - \alpha = 0.038-0.045$, with maximum interference colors in thin section ranging between uppermost second and lower third orders. The indices and the birefringence increase with Fe^{2+} (Fig. 10.42a). Mn for Fe^{2+} and F for OH decrease the indices and birefringence slightly.

Color. Usually colorless or pale brown in thin section. May be slightly pleochroic with:

α = colorless, pale yellow
β = pale yellow, yellowish gray
γ = pale green, pale yellow, pale greenish yellow

Absorption $\beta \geqq \gamma > \alpha$.

Form. Monoclinic. Fibrous, asbestiform, or columnar in random, parallel, or radiating aggregates. Cross sections tend to be rhombic and show the amphibole cleavage traces (110) intersecting at 56 and 124°. Longitudinal sections may show cross striations parallel with (001). Parallel intergrowths with actinolite and blue-green soda amphiboles occur.

Orientation. Biaxial $(-)$. $\alpha \wedge a = 1-6°$, $\beta = b$, $\gamma \wedge c = 15-10°$. Optic plane is (010) (Fig. 10.41b). $2V = 90-70°$, $r > v$. 2V and $\gamma \wedge c$ decrease with Fe^{2+}. The extinction angle is increased slightly by F for OH. Interference figures may be difficult to secure because of the thin crystals. Fine multiple twinning with (100) as the twin plane is very common.

Occurrence. A metamorphic mineral common in iron-rich rocks such as magnetite–hematite schist; magnetite–grunerite schist in some cases with a blue-green sodic amphibole, actinolite, garnet, or fayalite; grunerite schist; grunerite quartzite; and eulysite with fayalite, hedenbergite, antho-phyllite, and garnet. Alters to limonite.

Diagnostic features. Cummingtonite has a $(-)$ sign, lower indices, and higher extinction angles. Tremolite and actinolite have lower indices, and hornblende has marked pleochroism. Anthophyllite is orthorhombic and normally lacks twinning.

TREMOLITE

Composition. $Ca_2Mg_5Si_8O_{22}(OH)_2$. Unlike the cummingtonite–grunerite series, there is no natural optical division between tremolite and actinolite. Both optically and chemically, tremolite grades imperceptibly into actino-lite. The previous arbitrary division point, which has been taken by Winchell (1924) at 4% FeO or about 10% of the actinolite end-member molecule, is here applied. In tremolite Al is usually present but in amounts generally below 2% Al_2O_3. Na + K may substitute for Ca, the former in large amounts to form sodatremolite, which is isomorphous with glauco-phane and is considered under that mineral. Small quantities of F may replace OH. In some rare varieties of tremolite Mn proxies for Mg; another rare type contains 2% Cr_2O_3.

Indices. $\alpha = 1.599$–1.612, $\beta = 1.613$–1.626, $\gamma = 1.625$–1.637, $\gamma - \alpha = 0.022$–0.027. The indices increase with Fe^{2+} (Fig. 10.44). The maximum interference tints in thin section are in the lower second order. F for OH causes a decrease in the indices.

Color. Colorless in thin section, but coarse detrital pieces may show very pale green or pale pink colors (Mn).

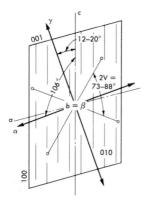

FIGURE 10.43 Orientation of tremolite–actinolite, section parallel with (010).

Form. Monoclinic. Crystals are elongate parallel with c, columnar to fibrous; some types are asbestiform. May be in rather stubby crystals or in granular aggregates; otherwise, randomly oriented, subparallel, or radially arranged. Cross sections usually are four-sided and show the two-directional amphibole cleavage (110) at 56 and 124°. Partings parallel with (100) and more rarely parallel with (010) may be present. Detrital pieces are prismatic with ragged terminations. Biotite or carbonaceous spots may be included.

Orientation. Biaxial $(-)$. $\alpha \wedge a = -6$–$-1°$, $\beta = b$, $\gamma \wedge c = 20$–$15°$. Optic plane is (010) (Fig. 10.43). $2V = 88$–$80°$, $r < v$. $\gamma \wedge c$ decreases with increasing Fe^{2+}, as does $2V$ (Fig. 10.43). F for OH increases the extinction angle. Elongate sections are length-slow. Cross sections display symmetrical extinction. Twins with (100) as the twin planes are common. Very fine multiple twinning parallel with (001) is present in some types.

Occurrence. A product of contact and regional metamorphism of impure magnesian limestones, associated with calcite, phlogopite, garnet, apatite, graphite, or even some wollastonite, diopside, forsterite, scapolite, chondrodite, and dravite. Also occurs in tremolite schists, some of which contain talc. Alters to talc.

Diagnostic features. Tremolite differs from actinolite in its lack of pleochroism, lower indices, larger 2V, and larger extinction angle. Hornblende has marked pleochroism. Cummingtonite is $(+)$; grunerite has higher refractive indices. Edenite is $(+)$ and has lower birefringence. Sillimanite is orthorhombic, as is anthophyllite. If difficulty is encountered in determining the orthorhombic nature of an amphibole in thin section

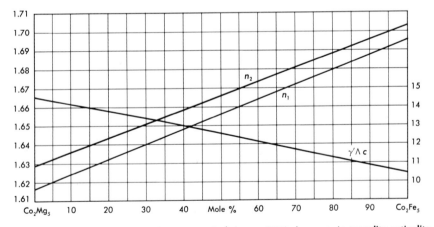

FIGURE 10.44 Variation in n_2, n_1, and $\gamma' \wedge c$ on (110) cleavages in tremolite–actinolite (Parker, 1961).

owing to lack of longitudinal sections, the crystal system can be determined readily by an observation on crushed fragments, in which (110) sections predominate. Wollastonite lacks the amphibole cleavage and is also distinguished by its low 2V and different orientation.

ACTINOLITE

Composition. $Ca_2(Mg,Fe^{2+})_5Si_8O_{22}(OH)_2$ to $Ca_2(Fe^{2+},Mg)_5Si_8O_{22}(OH)_2$. The pure-iron member does not occur in nature, and the maximum reported iron content is about 20% $FeO + Fe_2O_3$. In high-iron varieties some Ti may be present. With an increase in Na for Ca and an accompanying increase in Al, actinolite grades into glaucophane. Normally actinolite does not contain much Al, usually less than 2% Al_2O_3; similar amounts of Fe_2O_3 may be present. Some varieties associated with manganiferous ores contain important amounts of Mn (5–8% MnO), and in other types F substitutes for OH to a limited extent.

Indices. $\alpha = 1.613$–1.628, $\beta = 1.627$–1.644, $\gamma = 1.638$–1.655, $\gamma - \alpha = 0.024$–0.028. The indices increase with Fe^{2+}; the birefringence increases slightly with Fe^{2+} (Fig. 10.44). This variation is also somewhat influenced by the substitution of Na for Ca and F for OH, which tend to lower the indices, and by the replacements Fe^{3+}, Al, and Mn, which increase the indices. The maximum interference colors in thin section are in the lower and middle second order. Some rare iron-manganese actinolites have indices above the ranges listed.

Color. Colorless to pale green in thin section. May be pleochroic, increasing with Fe^{2+}:

α = very pale yellow, colorless, pale brown, light green
β = greenish yellow, pale yellow green, pale brown, bluish green
γ = pale green, green, dark green, pale bluish green

$\gamma > \beta \geqq \alpha$.

Form. Monoclinic. Crystals are either short- or long-bladed, parallel with c (Fig. 10.45). Also columnar to fibrous and asbestiform, in parallel, subparallel or radial aggregates. Rarely massive and granular. Nephrite is a variety of tremolite or actinolite that consists of very fine interlocking or felted fibers. Cross sections of actinolite are diamond-shaped and show the amphibole (110) cleavage at 56 and 124°. A (100) parting may also be present. Inclusions of biotite, magnetite, or dark carbonaceous material occur. Crushed pieces tend to lie on cleavage surfaces. Detrital grains are slender prisms with splintered ends. Uncommon zoned crystals contain cores of actinolite and outer zones of soda–actinolite. May be pseudomorphous after or replace augite.

Orientation. Biaxial $(-)$. $\alpha \wedge a = -3\text{-}2°$, $\beta = b$, $\gamma \wedge c = 17\text{-}12°$. Optic plane is (010) (Fig. 10.43). $2V = 84\text{-}73°$, $r < v$. $\gamma \wedge c$ and 2V decrease with increasing Fe^{2+} (Fig. 10.44). Longitudinal sections are length-slow; cross sections show symmetrical extinction. Twins with (100) as the twin plane are common.

Occurrence. Chiefly in metamorphic rocks: calc-silicate hornfels with actinolite, epidote or zoisite, and albite; calc schists with calcite, epidote, actinolite, and locally diopside; actinolite–anthophyllite schists with chlorite; actinolitic greenschists in which actinolite, epidote, albite, chlorite, and talc occur. In some lime-silicate rocks of contact-metamorphic origin actinolite forms instead of tremolite. Also forms by the hydrothermal alteration of mafic and ultramafic rocks, along with anthophyllite, chlorite, serpentine, talc, and carbonate. Forms pseudomorphs after igneous pyroxenes (usually augite) by deuteric replacement, or after metamorphic pyroxene by hydrothermal alteration. Alters to chlorite.

Diagnostic features. Tremolite has lower indices, a larger extinction angle, and a larger 2V and is colorless. Cummingtonite is $(+)$, and grunerite has higher indices; anthophyllite is orthorhombic. May be difficult to distinguish from pale green hornblende, but the latter usually has larger extinction angles. Edenite is $(+)$ and has lower birefringence.

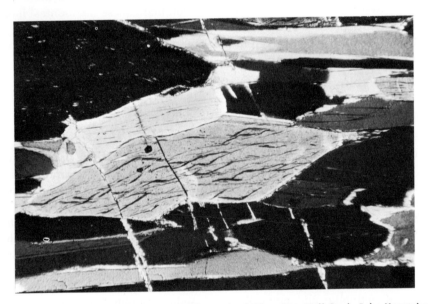

FIGURE 10.45 Actinolite in actinolite–anthophyllite schist, Wolf Creek, Ruby Mountains, Montana. Polars crossed, $\times 55$.

Composition. $Ca_2Na(Mg_{4-5},Al_{1-0})(Si_{6-8},Al_{2-0})O_{22}(OH,F)_2$. Fe^{2+} and Fe^{3+} usually are low in edenite but increase in the variety pargasite, in which as many as $2Fe^{2+}$ may replace 2Mg; most pargasites have about $1Fe^{2+}$. The number of Al atoms varies from about ½ to nearly 3 per formula weight and is generally near 2. Na (and K) may be low to absent. Pargasite may contain relatively large amounts of F replacing OH; edenite may also have small amounts of F. Edenite and pargasite grade into hornblende.

Indices. $\alpha = 1.606-1.649$, usually < 1.640; $\beta = 1.617-1.660$, usually < 1.645; $\gamma = 1.631-1.672$, usually < 1.655; $\gamma - \alpha = 0.016-0.023$. The indices increase in a general way with Fe^{2+}, but at the lower end of the range the variation in Al also influences the indices markedly. F for OH tends to lower the indices considerably. The maximum interference colors in thin section vary from upper first to lower second order.

Color. Colorless to pale green. May be pleochroic, especially the iron-bearing type, pargasite:

α = pale gray, colorless, greenish yellow, pale green, pale yellow
β = pale brown, light green, emerald green, green, light violet blue, yellowish brown
γ = pale bluish gray, light brownish green, greenish blue, greenish pink, light violet blue, light greenish yellow

$\alpha < \beta \leqq \gamma$.

Form. Monoclinic. In prisms with four-sided or six-sided cross sections; (010) absent to subordinate. Cleavage (110) in two directions at 56 and 124° appears in cross sections. Longitudinal sections show one direction of cleavage lines and commonly a cross fracture normal to them. Crushed material tends to rest on (110) cleavages.

Orientation. Biaxial $(+)$, $\alpha \wedge a = -2--12°$, $\beta = b$, $\gamma \wedge c = 17-27°$, usually 20° or more. Optic plane is (010) (Fig. 10.46). $2V = 52-83°$, usually $< 75°$; $r > v$ weak to distinct. Cross sections display symmetrical extinction. Twins with (100) as the twin plane may occur.

Occurrence. Chiefly in magnesian marbles of pyrometasomatic origin, associated with some of the following: phlogopite, scapolite, dravite, diopside, wollastonite, serpentine, dolomite, quartz, and graphite. Rarely in amphibolite, pyroxenite or augite gneiss, and amphibole gneiss. Pargasite is a rare mineral in cavities of some dacites and andesites.

Diagnostic features. Hornblende is $(-)$ and usually shows stronger pleochroism. Members of the tremolite–actinolite series are $(-)$ and have a

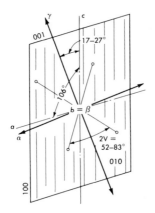

FIGURE 10.46 Orientation of edenite, section parallel with (010).

larger 2V and generally a smaller extinction angle. Cummingtonite, which is also (+), normally has higher indices, stronger birefringence, a lower extinction angle, and multiple twinning. Anthophyllite is orthorhombic and may be (−). Some amphiboles of edenitic composition are (−) and thus not distinguishable by optics readily from hornblende.

HORNBLENDE

Composition. $Ca_2Na(Mg,Fe^{2+},Fe^{3+},Al)_5(Si_{6-7},Al_{2-1})O_{22}(OH,F)_2$. The composition varies more or less continuously with respect to the ratios $Mg:Fe^{2+}$, $Al:Fe^{3+}$, and $Al:Si$. The Ca:Na ratio also is variable but less so, and generally Ca = 2. The OH:F ratio also fluctuates to some extent. K substitutes for Na in minor amounts; Mn and Ti may be present also in small amounts, and Cl replaces F.

Indices. The indices of refraction of nearly 200 hornblendes, many of them analyzed chemically, fall into the following groups:

α	1.602–1.615 1.2%	1.616–1.645 18%	1.646–1.665 72.7%	1.666–1.680 7.5%	1.681–1.690 0.6%
β	1.613–1.625 0.7%	1.626–1.655 15.3%	1.656–1.675 71.5%	1.676–1.695 10.5%	1.696–1.702 2.0%
γ	1.624–1.635 0.6%	1.636–1.660 10.2%	1.661–1.685 84.1%	1.686–1.700 3.4%	1.701–1.705 1.7%

$\gamma - \alpha = 0.015$–0.034, but most hornblendes have $\gamma - \alpha = 0.018$–0.028, and the average is 0.022. Thus maximum interference colors range from upper first to lower second order in thin section. However, the color of the mineral masks or alters the interference tints.

An approximation of the composition from optical properties is difficult because of the number of variables involved. The most important factor affecting the refractive indices is the MgO:FeO ratio, and in a general way the indices rise with increasing Fe^{2+}. In ordinary hornblende changes in this ratio have a more marked effect on the indices than changes in the Fe_2O_3:FeO ratio have. The latter ratio appears to affect the optical properties significantly only when Fe^{3+}:Fe^{2+} is at least $2:1$ (Deer, 1937). SiO_2 and Al_2O_3 are the most difficult to estimate from optical constants. In hornblendes of diorites increasing refractive indices indicate a decrease in SiO_2 and an increase in Al_2O_3. However, in hornblendes of granites the increase in indices is due usually to large increases in FeO. The CaO:$Na_2O + K_2O$ ratio in hornblendes is less variable than the other ratios and thus has little influence on the indices. The substitution of F for OH tends to lower the indices, whereas the replacement of OH by Cl will raise them. A chlorhornblende with MgO:FeO $= 0.18$ and 7.24% Cl has $\alpha = 1.728$, $\gamma = 1.751$, $\gamma \wedge c = 11°$, $2V = 15°$.

Color. Various shades of green and brown with marked pleochroism:

Green varieties:

α	β	γ
light green	dark green	very dark green
light bluish green	deep green	deep bluish green
yellowish green	dark yellow green	dark bluish green
very light green	light gray green	gray green
straw yellow	yellow green	blue green
greenish yellow	green	blue green
pale yellow	olive green	deep green
yellow	greenish yellow	indigo blue

Brown varieties:

α	β	γ
greenish yellow	yellowish brown	black
pale yellow brown	yellow brown	brownish yellow
yellow	pale brown	grayish brown
pale yellow	sepia brown	sepia brown
greenish yellow	yellowish brown	deep reddish brown
greenish brown	reddish brown	red brown

$\gamma \geqq \beta > \alpha$ or rarely $\beta > \gamma > \alpha$.

HORNBLENDE

The two varieties intergrade. When ordinary green hornblende is heated, Fe^{2+} becomes Fe^{3+} and the color changes to brown. Thus in oxyhornblendes or in heated hornblendes the $Fe_2O_3 : FeO$ ratio is an important factor in determining the depth of the brown color and the pleochroism. In naturally occurring ordinary hornblendes, however, the change from a green to a brown color is independent of this ratio. In some cases a general correlation between brown color and an increased Ti content can be established, but the Ti variation is not the only deciding factor in this change. Probably the color of hornblende cannot be correlated with the percentage of any one constituent but may depend on whether Ti exists as Ti^{3+} or Ti^{4+} (Deer, 1937). In general, green hornblende is characterized by higher SiO_2, lower Al_2O_3, lower $Na_2O + K_2O$, and lower TiO_2 when compared with brown hornblende.

The hourglass type of zoning is found in brown varieties. Rare zoning with edenite is reported in crystals in which optically $(+)$ and $(-)$ zones occur together. In some igneous rocks brown hornblende may show a peripheral zone of green color; rarely the reverse relations are observed (Fig. 10.49). Zoning in shades of green also is recorded, with either rims or cores of darker tints. Another type of zoning is represented by a frayed core containing a lattice work of ilmenite inclusions surrounded by an inclusion-free margin. Other combinations include margins of green hornblende on nearly colorless hornblende cores and overgrowths of green hornblende on cores of fibrous hypersthene. Phenocrysts of hornblende in porphyries may lack a core, whose position is occupied by fine-grained matrix material. Pleochroic haloes in hornblende occur around zircon and allanite inclusions.

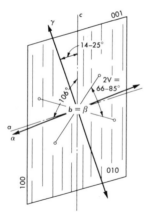

FIGURE 10.47 Orientation of hornblende, section parallel with (010).

Form. Monoclinic. Crystals are prismatic, usually with poorly developed terminations but with well-outlined, six-sided cross sections in which (010) is subordinately developed (Fig. 10.49). These sections show the cleavage (110) in two directions at 56 and 124°. Longitudinal sections show one set of cleavage lines and rarely (001) parting. Hornblende usually is anhedral to subhedral, and in some peridotites it is markedly poikilitic with included olivine and/or augite. Other minerals included are zircon, apatite, magnetite, ilmenite, sphene, rutile, quartz, epidote, and biotite. Hornblende may be fibrous, usually in reaction rims around pyroxene or garnet. Detrital grains are typically elongated (110) cleavage plates with frayed or ragged ends and may show prismatic striae; rounded to angular grains also occur (Fig. 10.48). Crushed pieces tend to lie on (110) cleavage faces.

Orientation. Biaxial ($-$). $\alpha \wedge a = 1$-$-10°$, $\beta = b$, $\gamma \wedge c = 14$-$25°$. Optic plane is (010) (Fig. 10.47). $2V = 66$-$85°$, dispersion is generally $r < v$ weak, but may be $r > v$. For nearly 200 hornblendes, many of which are analyzed, the distribution of 2V and $\gamma \wedge c$ values are:

2V		$\gamma \wedge c$	
36–55°	2.5%	8–13°	5.6%
56–65°	11.6%	14–25°	87.2%
66–85°	79.3%	26–28°	5.6%
86–89°	6.6%	29–37°	1.6%

In a general way 2V increases with Fe^{2+} content; Al for Si decreases 2V. Cross sections show symmetrical extinction. Paired or lamellar twins with (100) as the twin plane are common (Fig. 10.49).

Occurrence. One of the most common of rock-forming minerals in igneous and metamorphic rocks. Also widely distributed in detrital sediments. In igneous rocks more common in intrusive types. Found as a primary mineral in peridotites to granites, including cortlandite and hornblendite, hornblende gabbro and norite, diorite and tonalite, syenite, nepheline syenite (rarely), monzonite, granodiorite, quartz monzonite, and granite. In

(a)   (b)

FIGURE 10.48   (a), (b) Detrital hornblende, North Fairhaven, New York.

FIGURE 10.49   Euhedral twinned and zoned hornblende in hornblende gabbro, Cuya-maca, southern California. Polars crossed, ×35.

diorites and tonalites it is the most common mafic constituent. Horn-blende of secondary, usually of deuteric origin, is common in peridotites and gabbros. It forms reaction rims (coronas), usually as fine needles normal to the contact, around olivine, pyroxene, ilmenite, magnetite, and pyrope (kelyphytic borders). The hornblende may lie directly against the olivine; in other cases a layer of pyroxene or a double layer of ortho- then clinopyroxene intervenes between the amphibole and the olivine. Not un-commonly the hornblende of these reaction rims is intergrown with small vermicular granules of pleonaste. Some gabbros and anorthosites contain myrmekitic intergrowths of hornblende and plagioclase.

Secondary hornblende that replaces pyroxene (usually augite or diop-side) is called uralite, and this replacement, which is widespread in pyroxene-bearing intrusive igneous rocks and in some metamorphic rocks, is known as uralitization. The replacement begins at the borders of the pyroxene grain and proceeds inward, typically along cleavage planes, leav-ing relicts of pyroxene throughout the amphibole. This alteration also may be accompanied by formation of primary hornblende as an overgrowth on the pyroxene. The replacing amphibole can form a single grain or a group of grains in nearly parallel position. In each case, however, the c axis of

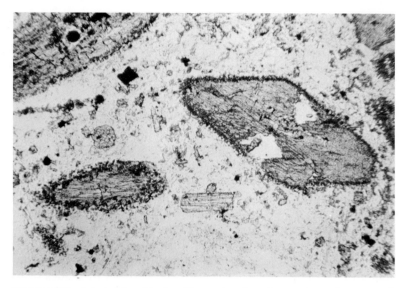

FIGURE 10.50 Euhedral hornblende with reaction rim of magnetite and pyroxene in monzonite porphyry, Cone Butte, Judith Mountains, Montana. Polars not crossed, ×50 (S. B. Wallace).

the amphibole is parallel or nearly parallel with that of the pyroxene. Much uralite is actinolite rather than hornblende.

Hornblende also occurs as phenocrysts in extrusive igneous rocks, although it is not as common as in intrusive types. Not uncommonly the amphibole of lavas is oxyhornblende, but green and brown ordinary hornblende also occur. Rock types that contain it are basalt (rarely), andesite, dacite, quartz latite, latite, and rhyolite. These hornblendes show resorption or reaction effects and are partly replaced by fine-grained augite and magnetite (Fig. 10.50).

In metamorphic rocks hornblende is found in: marbles; hornblende schists with biotite and garnet, or albite and epidote, or talc and chlorite; amphibolites with garnet, plagioclase, or epidote; hornblende gneisses with plagioclase, biotite, and garnet; hornblende eclogites with relict omphacite; granulites with zoisite, in which the hornblende forms large poikiloblastic crystals; and some quartzites. Parallel intergrowths and overgrowths with cummingtonite occur.

In intrusive rocks hornblende may be replaced by biotite. Hornblende also alters to chlorite plus magnetite or epidote plus quartz and shows bleaching and consequent production of secondary magnetite dust. The change of brown hornblende to a green type may bring about the production of granules of secondary sphene. Some hornblendes low in Al are replaced by serpentine.

Diagnostic features. Edenite and pargasite are distinguished optically by their ($+$) optical character. Cummingtonite also is ($+$). Grunerite has higher refractive indices and lower extinction angles. Anthophyllite shows parallel extinction in sections parallel with c. All of these amphiboles show little or no pleochroism. Tremolite also is colorless and has lower indices. Actinolite and hornblende are distinguished with difficulty, but hornblende normally has higher refractive indices, lower birefringence, deeper colors, stronger pleochroism, and usually a larger extinction angle. Their occurrences also differ considerably, for actinolite does not form as a primary igneous mineral and is a lower-grade metamorphic mineral. Oxyhornblende has stronger pleochroism, higher refractive indices, stronger birefringence, and a smaller extinction angle than hornblende. Fibrous brown hornblende has its maximum absorption nearly parallel with c, whereas fibrous tourmaline shows maximum absorption at right angles to c. Biotite also shows strong pleochroism in shades of brown but has only one direction of perfect cleavage, will show "birdseye maple" structure near extinction and has parallel or nearly parallel extinction.

OXYHORNBLENDE [1]

Composition. $Ca_2Na(Mg,Fe^{3+},Fe^{2+},Al,Ti)_5(Si_{6-7},Al_{2-1})O_{22}(O,OH)_2$. Fe_2O_3 usually predominates over FeO. TiO_2 attains a maximum of about 10% and is particularly abundant in the variety kaersutite (titanian oxyhornblende). OH may be very low to nearly absent, for the oxidation of Fe^{2+} to Fe^{3+} is consequent on the escape of H from the structure, leaving "extra" O ions. Ordinary iron-bearing hornblendes can be transformed upon heating into oxyhornblende, and the process is reversible by heating oxyhornblende in an atmosphere of H. The Ca:Na(K+) and Al:Si ratios are generally similar to those in ordinary hornblendes.

Indices. $\alpha = 1.650-1.702$, usually > 1.675; $\beta = 1.683-1.769$, usually > 1.695; $\gamma = 1.689-1.796$, usually > 1.700; $\gamma - \alpha = 0.020-0.094$, usual range $= 0.031-0.072$. Thus in thin section maximum interference colors lie between those of middle second order and the pale tints of the fourth order. However, because of the very strong color and pleochroism, the interference colors ordinarily cannot be observed. The indices increase with an increase in the $Fe^{3+}:Fe^{2+}$ ratio, the $Fe^{2+}:Mg$ ratio, and Ti.

[1] The name "oxyhornblende" (Winchell) is considered more appropriate than the double name of basaltic hornblende; for many of these amphiboles do not occur in basalts.

Color. Brown and strongly pleochroic. May be nearly opaque in position of maximum absorption, or contains opaque borders of numerous inclusions of magnetite and hematite.

α = pale yellow, greenish yellow, pale brown, yellowish green
β = dark brown, brownish green, dark brown, reddish brown
γ = dark olive green, dark blue, dark brown, dark reddish brown

$\alpha < \beta < \gamma$. The depth of color and strength of the pleochroism increase with increasing oxidation of Fe^{2+} to Fe^{3+}.

Form. Monoclinic. Euhedral crystals display six-sided cross sections and short to long prismatic habit. Not uncommonly with a corroded margin. Common as phenocrysts and may be in very large crystals. Inclusions are of magnetite, hematite, augite, zircon, apatite, and glass. Around included zircon crystals pleochroic haloes may be prominent. Cross sections show (110) cleavage at angles of 56 and 124°. Prismatic sections show a single cleavage direction. Partings (100) and (010) may be present. Detrital pieces are mainly prismatic (110) tablets with irregular ends. Crushed pieces tend to be oriented on cleavage planes.

Orientation. Biaxial $(-)$. $\alpha \wedge a = 4\text{--}16°$, $\beta = b$, $\gamma \wedge c = 12\text{--}0°$. Optic plane is (010). $2V = 56\text{--}88°$, usually $> 70°$ (Fig. 10.51). $r < v$ or $r > v$ weak to moderate. $\gamma \wedge c$ decreases with increasing oxidation of Fe^{2+} to Fe^{3+}. Zoning may be present, with outer zones usually having smaller extinction angles. Elongated sections are length-slow, but the dark color may make this determination difficult. Cross sections display symmetrical extinction. Lamellar twinning with (100) as the twin plane may be present.

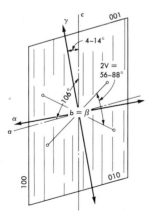

FIGURE 10.51 Orientation of oxyhornblende, section parallel with (010).

Occurrence. Mainly as phenocrysts in extrusive and hypabyssal igneous rocks. Normally not found as a groundmass mineral in these rocks. Rock types in which it occurs include monchiquite, camptonite, basalt, andesite, trachyte, quartz latite, and rhyolite. Shows resorption effects and is peripherally replaced by a fine-grained mixture of various minerals, including magnetite, hematite, feldspar, pigeonite, augite, aenigmatite, and biotite. The replacement may go to completion, in which case only the outline of the fine-grained replacement aggregate indicates the former presence of the amphibole. Also occurs as a detrital mineral.

Diagnostic features. Has higher indices, higher birefringence, lower extinction angles, and deeper color and pleochroism than hornblende. These properties, as well as its specialized occurrence, suffice to distinguish it from most other amphiboles. Biotite has only one direction of cleavage, has lower relief, lower birefringence, and different crystal outline. Tourmaline shows maximum absorption at right angles to c. Oxyhornblende resembles the soda amphibole barkevikite, which also has relatively high indices of refraction and shows deep brown pleochroic colors but has a smaller 2V and lower double refraction.

HASTINGSITE

Composition. $(Ca_{2-1.5},Na+K_{1-1.5})(Fe^{2+},Mg)_{4-3}Fe^{3+}_{1-2}(Si_{5.5-6.5},Al_{2.5-1.5})O_{22}(OH)_2$. Hastingsite is characterized by a low Si content for amphiboles (43–37% SiO_2), by an approach toward a 1 : 1 ratio of Ca:Na in some types, and generally by a high iron content. Most hastingsite is ferrohastingsite according to the classification (Billings, 1928) in which:

ferrohastingsite   FeO:MgO > 2
femaghastingsite   2 > FeO:MgO > ½
magnesiohastingsite   FeO:MgO < ½

Ti and Mn are present usually in small amounts, but in a few varieties Ti becomes a significant constituent. Types in which Ca:Na = 1:1 or Na > Ca are called alkali hastingsite. The variety of amphibole, barkevikite, also is low in SiO_2 and may be grouped with hastingsite, usually with ferrohastingsite, rarely with femaghastingsite.

Indices.

TYPE	α	β	γ	$\gamma - \alpha$
ferrohastingsite	1.679–1.705	1.694–1.731	1.703–1.732	0.014–0.029
femaghastingsite	1.669–1.680	1.682–1.695	1.684–1.705	0.015–0.033
magnesiohastingsite	1.653–1.670	1.661–1.690	1.669–1.700	0.016–0.030

The indices vary with the FeO:MgO ratio; as Fe^{2+} increases, the indices rise (Fig. 10.53). Na for Ca in large amounts decreases the indices if the FeO:MgO ratio remains the same. An alkali femaghastingsite with Na:Ca = ca. 2:1 has $\alpha = 1.639$, $\beta = 1.658$, $\gamma = 1.660$. Varieties called barkevikite have $\alpha = 1.687$–1.694, $\beta = 1.700$–1.707, $\gamma = 1.701$–1.712, $\gamma - \alpha = 0.014$–0.023.

Color. Shades of green, blue green, and brown; pleochroism marked:

$\alpha = $ dull yellow, deep brown, greenish yellow, pale greenish brown, pale green, yellow

$\beta = $ liver brown, straw yellow, yellowish brown, dark greenish brown, dusky green, deep greenish blue

$\gamma = $ greenish yellow, yellow, reddish brown, bluish green, dark blue, deep olive green

$\alpha < \beta \leqq \gamma$ or $\alpha < \gamma \leqq \beta$. The depth of color and intensity of pleochroism increase with Fe^{2+}. Barkevikite usually shows pleochroism in shades of brown.

Form. Monoclinic. Varies in shape from subhedral, rounded grains to euhedral, elongated prisms. Cross sections are six-sided with two directions of cleavage (110) at 56 and 124°. Spheroidal aggregates of radial needles also occur. In some rocks a poikilitic texture is characteristic, with quartz and orthoclase included in the more silicic rocks and hypersthene and diopside in mafic types. Also may form as a late mineral, either interstitially

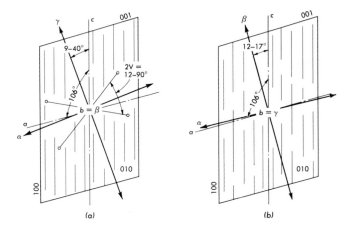

FIGURE 10.52 (a) Orientation of hastingsite, section parallel with (010), optic plane parallel with (010). (b) Orientation of hastingsite, section parallel with (010), optic plane normal to (010).

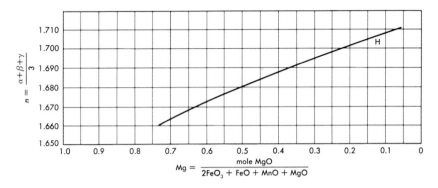

FIGURE 10.53 Variation of median index of refraction with composition in hastingsite (Foslie, 1945).

or as borders around augite or hedenbergite. Broken material tends to be oriented on cleavage faces (110).

Orientation. Biaxial $(-)$. Most hastingsite has an orientation in which the optic plane is parallel with (010) (Fig. 10.52a). In a few ferrohastingsites, very high in FeO, the optic plane is normal to (010) (Fig. 10.52b). Common orientation: $\alpha \wedge a = -25{-}6°$, $\beta = b$, $\gamma \wedge c = 40{-}9°$. $2V = 12{-}85°$, $r < v$ or $r > v$ weak to strong. Orientation for some ferrohastingsite: $\alpha \wedge a = 3{-}{-}2°$, $\beta \wedge c = 12{-}17°$, $\gamma = b$. $2V = 16{-}25°$, $r < v$ strong. The size of 2V and $\gamma \wedge c$ vary with the composition as follows:

	2V	DISPERSION	$\gamma \wedge c$
ferrohastingsite, optic plane $\perp$ (010)	16–25°	$r < v$ strong	$(\beta \wedge c)$ 12–17°
ferrohastingsite (incl. most barkevikite), optic plane = (010)	12–76°	$r < v$ strong to medium, and $r > v$ medium	9–20°
femaghastingsite	40–81°	$r > v$ distinct	9–22°
magnesiohastingsite	60–90°	$r > v$ slight	27–40°

Dispersion may differ for the two optic axes in a single specimen: it may be reversed and is usually of different intensity. In a general way with increasing Fe^{2+}, 2V decreases, $\gamma \wedge c$ decreases, and the dispersion changes from $r > v$ weak to $r < v$ strong. Fe^{3+}, Ti, and Na variations modify this general trend somewhat. Zoning occurs in some hastingsite, in which cores have a larger 2V than peripheries. Outer zones of hastingsite on hornblende have been reported. Twinning with (100) as the twin plane may be present.

Occurrence. Usually in alkalic igneous rocks or in marbles and limestones metasomatized by such rocks.

Variety	Rock types
1. Ferrohastingsite with optic plane ⊥ (010)	Nepheline syenite, albite–nepheline syenite, syenite, amphibole granite, amphibole granite gneiss
2. Ferrohastingsite with optic plane = (010), including most barkevikite	Nepheline syenite, foyaite, nordmark-ite, alkali syenite, alkali granite, rapakivi granite, contact marble, fer-rohastingsite schist (rare)
3. Femaghastingsite	Akerite, theralite, essexite, campton-ite, pyroxene granulite, quartz syenite gneiss, amphibole granite, amphibole granite gneiss, contact marble
4. Magnesiohastingsite	Gabbro, diorite, pyroxenite (rare), contact marble

Hastingsite may be a product of the uralitic alteration of augite.

Diagnostic features. The combination of strong pleochroism and high refractive indices serves to separate members of the hastingsite series from grunerite and tremolite–actinolite. Anthophyllite is orthorhombic. Cummingtonite and edenite are optically $(+)$. Ferrohastingsite is distinguished from hornblende by generally higher refractive indices, usually a smaller 2V, a smaller extinction angle, deeper and different pleochroic colors, and in some cases by a difference in optical orientation. It may be very difficult, however, to distinguish between some pale-colored femaghastingsite or magnesiohastingsite and some types of hornblende on optical properties alone. The chief cause of the optical variation in both hastingsite and hornblende is the MgO:FeO ratio, and in this magnesiohastingsite and hornblende may be similar. The variations in the Ca:Na and Si:Al ratios, which distinguish the two chemically, influence the optical constants to a lesser extent. Oxyhornblende is distinctive in its occurrence and in having a small $\gamma \wedge c$ angle combined with a large to moderate 2V, whereas ferro-hastingsite has a small $\gamma \wedge c$ angle usually with a small 2V.

ARFVEDSONITE

Composition. $(Na_{2-2.5},Ca_{1-0.5})(Fe^{2+}_{4-2.5},Mg_{0.5-1.5})Fe^{3+}_{1.5-0.5}Si_8O_{22}(OH)_2$. In several rare varieties $Mg > Fe^{2+}$. Some types contain considerable K in place of Na; others have some Ti and Al. The latter element normally replaces Fe^{3+}, but there may be a minor replacement of Si as well. OH may be replaced by F and by extra O. A large number of superfluous names have been applied to minor chemical or optical varieties.

Indices. $\alpha = 1.652$–1.699, $\beta = 1.660$–1.705, $\gamma = 1.666$–1.708, $\gamma - \alpha = 0.005$–0.014. In thin section maximum interference colors lie in the lower to upper first order range. Interference colors are strongly modified by the color of the mineral. The indices increase in general with increase in Fe^{2+}, and the birefringence generally decreases with increasing Fe^{2+}. Variations in Fe^{3+} and to a lesser extent in Ca also affect the indices.

Color. Brown and green. Strongly pleochroic, may be almost opaque in the γ direction:

$\alpha = $ yellow, dark green, dark bluish green, deep bluish green
$\beta = $ green, pale brown, dark bluish green, bluish green
$\gamma = $ blue gray, black, brownish green, dirty green

$\gamma > \alpha > \beta$, rarely $\alpha = \beta > \gamma$, or $\alpha > \beta > \gamma$. Varieties with high iron contents show stronger absorption. Zonal structures with arfvedsonite around barkevikite or around magnesian arfvedsonite are known. Where zircon is included a brown pleochroic halo is developed.

Form. Monoclinic. As elongated crystals or broad laths, with euhedral prisms and anhedral terminations. Poikilitic structure is not uncommon, with included albite, potash–soda feldspar, and apatite. Interstitial network aggregates of arfvedsonite also occur; may be fibrous. Cross sections of larger crystals are rhombic and show (110) cleavages at 56 and 124°. Rarely an (010) parting is present. Detrital pieces are long blades or fibers. Crushed material tends to orient itself on (110) cleavage faces.

Orientation. Biaxial $(-)$. $\alpha \wedge c = -7$–$-36°$, $\beta \wedge a = 23$–$52°$, $\gamma = b$. Optic plane is normal to (010) (Fig. 10.54). $2V = 30$–$70°$, $r > v$ strong. This is the normal orientation for arfvedsonite. Varieties rich in Mg have the orientation: Biaxial $(-)$; $\alpha \wedge a = -8$–$-54°$, $\beta = b$, $\gamma \wedge c = 24$–$70°$; optic plane is (010); $2V = $ small to medium; $r > v$. The strong dispersion leads to incomplete extinction in white light. Cross sections display symmetrical extinction. Longitudinal sections are length-fast. Twins with (100) as the twin plane are uncommon.

Occurrence. Chiefly in alkalic igneous rocks: alkalic granite, alkalic syenite, nepheline syenite and nepheline syenite pegmatite, pulaskite, shonkinite, alkalic rhyolite, trachyte, and phonolite. Rare as a detrital. May be in parallel intergrowths with aegirine. Has been found replaced by a mixture of acmite and iron-rich biotite. Alters to limonite and siderite.

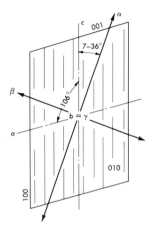

FIGURE 10.54 Orientation of arfvedsonite, section parallel with (010), optic plane normal to (010).

Diagnostic features. Arfvedsonite is length-fast; glaucophane is length-slow. Riebeckite also is length-fast but normally has (010) as the optic plane. Riebeckite also has a large 2V and a small extinction angle and may be (+). From ferrohastingsite (and barkevikite) arfvedsonite differs in having lower birefringence, length-fast character, and in general a different orientation.

RIEBECKITE

Composition. $Na_2Fe^{2+}{}_3Fe^{3+}{}_2Si_8O_{22}(OH)_2$. Among the amphiboles riebeckite is characterized chemically by high Si, Fe^{2+}, Fe^{3+}, and Na, and very low Al, Mg, and Ca. K, Mn, and Ti are minor elements. Several investigators have sought to establish the presence of a continuous series between glaucophane and riebeckite: $Na_2Mg_3Al_2Si_8O_{22}(OH)_2$—$Na_2Fe^{2+}{}_3$ $Fe^{3+}{}_2Si_8O_{22}(OH)_2$, but intermediate types appear to be very rare. There are, however, varieties of riebeckite (rhodusite) in which Mg replaces Fe^{2+} to a variable extent, but in which Fe^{3+} is not replaced appreciably by Al. This is also true of some of the asbestiform types, crocidolite.

Indices. $\alpha = 1.680\text{--}1.698$, $\beta = 1.683\text{--}1.700$, $\gamma = 1.685\text{--}1.706$, $\gamma - \alpha = 0.003\text{--}0.009$. The indices decrease and the birefringence increases with increasing substitution of Mg for Fe^{2+}. Magnesian riebeckite, in which $Fe^{2+}:Mg = 2:1$, has $\gamma - \alpha = 0.018\text{--}0.022$. The first-order interference colors normally are masked by the mineral color.

Color. Deep blue in thin section. Strongly pleochroic:

α = green blue, deep blue, smoky green, yellow, indigo

β = gray blue, light yellowish green, yellow, deep violet, blue

γ = yellow brown, deep blue, blue violet, dark smoky green, black, yellowish green

$\alpha > \beta > \gamma$, $\gamma \gtrless \alpha > \beta$, $\alpha > \gamma > \beta$. Crocidolite and other varieties containing Mg have somewhat lighter colors.

Form. Monoclinic. Anhedral to euhedral. In slender prismatic, elongated parallel with c, fibrous, felted, or asbestiform (crocidolite). The fibrous type may show radial arrangement. Also interstitial or in very irregular, spongelike crystals. Forms microlites in feldspar and replaces aegirine in

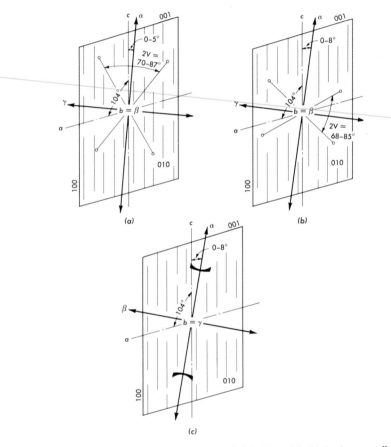

FIGURE 10.55 (a)–(c) Orientations of riebeckite. (a), (b) Sections parallel with (010), optic planes parallel with (010). (c) Section parallel with (010), optic plane normal to (010).

uralitic fashion. Ragged, irregular forms with striations parallel with c characterize rare detrital grains. Cross sections show the two directions of (110) cleavage at 56 and 124°. For some large pegmatitic crystals (010) and (001) partings are recorded.

Orientation. Biaxial $(-)$ or $(+)$. Three orientations are established; in probable order of abundance (Fig. 10.55):

a. $(-)$, optic plane parallel with (010); $\alpha \wedge c = 0\text{--}5°$, $\beta = b$, $\gamma \wedge a = 14\text{-}19°$; $2V = 70\text{-}87°$, $r > v$ strong to extreme

b. $(+)$, optic plane parallel with (010); $\alpha \wedge c = 0\text{--}8°$, $\beta = b$, $\gamma \wedge a = 14\text{-}22°$; $2V = 68\text{-}85°$, $r < v$ strong.

c. $(+)$, optic plane normal to (010); $\alpha \wedge c = -3\text{--}10°$, $\beta \wedge a = 17\text{-}24°$; $\gamma = b$; $2V = 50°\text{-large}$, $r > v$ strong

The relation between chemical variation and the shift in the position of the optic plane is not well known. Mg for Fe^{2+} increases the size of $\alpha \wedge c$. Some magnesian crocidolite has $\alpha \wedge c$ as large as 20°. Elongated crystals are length-fast.

Occurrence. Typically in soda-rich igneous rocks including alkalic granite, alkalic syenite, syenite, alkalic rhyolite, alkalic trachyte, and phonolite. In the metamorphic group it appears in alkalic granite gneiss, riebeckite granulite, nepheline syenite gneiss, aegirine–riebeckite quartzite, banded ferruginous quartzite (may be crocidolite), ferruginous hornfels (crocidolite), quartz–magnetite schist with cummingtonite, riebeckite–tremolite schist, and lawsonite schist. It is also a rare hydrothermal mineral in some quartz veins and an uncommon detrital species. It alters to limonite or siderite. Crocidolite may be replaced by fine-grained quartz.

Diagnostic features. From glaucophane it is distinguished by its length-fast character, by higher indices, lower birefringence, and in some cases by a $(+)$ sign. Also differs from hastingsite in having length-fast character, lower birefringence, smaller extinction angle, and a generally larger 2V. The characteristic pleochroic colors distinguish it from other amphiboles.

GLAUCOPHANE

Composition. $Na_2(Mg,Fe^{2+})_3Al_2Si_8O_{22}(OH)_2$, in which $Mg > Fe^{2+}$. Crossite contains Fe^{3+} in substitution for Al. A series exists between low-iron actinolite (or tremolite) and glaucophane, in which intermediate types have been referred to as soda tremolite– or tremolite–glaucophane and differ from ordinary glaucophane through the replacement of Na and K by Ca and the replacement of Al by additional Mg, Fe^{2+}, and some Fe^{3+}. Richterite, $(Na,K)_2Ca(Mg_{5-4},Mn_{1-2})Si_8O_{22}(OH)_2$, a rare manganiferous amphibole, is closely allied to soda tremolite. There are also several very rare soda amphiboles [1] that resemble glaucophane and are characterized by $Mg > Fe^{2+}$ but contain Fe^{3+} and/or Fe^{2+} in place of Al. These may be considered as intermediate between glaucophane and riebeckite. In glaucophane K, Ca, Mn, Fe^{3+}, and Ti are minor constituents.

Indices.

	α	β	γ	$\gamma - \alpha$
glaucophane	1.606–1.637	1.615–1.650	1.627–1.655	0.018–0.021
crossite	1.640–1.659	1.645–1.666	1.652–1.670	0.004–0.015
soda tremolite–glaucophane series	1.605–1.651	1.616–1.661	1.623–1.670	0.016–0.022

In glaucophane the indices increase with increasing Fe^{2+} and the birefringence decreases slightly with increasing Fe^{2+}. In crossite the addition of Fe^{3+} also increases the refractive indices and lowers the birefringence. In the soda tremolite– (or actinolite) glaucophane series the indices of refraction increase with increase in percent of the glaucophane molecule and also with $Fe^{2+} + Fe^{3+}$ (Fig. 10.58).

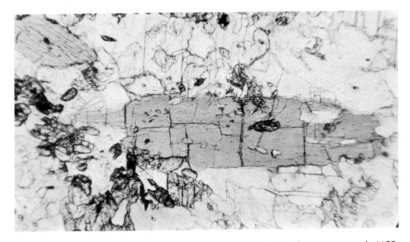

FIGURE 10.56 Glaucophane in glaucophane schist, Corsica. Polars not crossed, ×55.

[1] Examples are ternovskite, anophorite (also has Ti replacing Si), fluotaramite (also has F replacing OH), and eckermannite (lacks Fe^{2+} and has much K and some Li).

In glaucophane the maximum interference colors in thin section are in the upper first order. In crossite they range from lower to upper first order. In both, however, the color of the mineral tends to obscure interference tints. Both may show abnormal interference colors due to strong dispersion.

Color. Blue to violet. Strong pleochroism.

Glaucophane:

α = light yellowish violet, wine yellow, yellowish green, pale yellow, colorless, bluish green

β = light violet, violet, brownish green, bluish green, reddish violet, lavender blue

γ = light prussian blue, dark blue, greenish blue, deep bluish green, azure blue, bluish brown

$\gamma > \beta > \alpha$.

Crossite:

α = wine yellow, pale yellow, colorless, pale mauve

β = blue, prussian blue, purplish blue, deep blue

γ = gray violet, violet, blue, deep mauve

$\gamma > \beta > \alpha$ and $\beta > \gamma > \alpha$.

Soda tremolite–glaucophane:

α = colorless, pale apple green, pale yellow

β = pale greenish brown, smoky blue, pale green, dark olive green

γ = light green, blue green, dark violet brown, pale green

$\gamma > \beta > \alpha$. The depth of color increases toward the glaucophane end of the series.

Glaucophane–riebeckite:

α = sky blue, dark green, bluish green

β = red violet, bluish green, pale violet, light bluish green

γ = yellow brown, yellow, pale yellow, pale yellowish green

Absorption highly variable: $\alpha > \beta > \gamma$, $\gamma \geqq \alpha > \beta$, $\alpha \geqq \beta > \gamma$. Zonal variations in color commonly are present.

Form. Monoclinic. Elongated prismatic crystals, euhedral to subhedral in outline are common (Fig. 10.56). Columnar to nearly fibrous aggregates also occur; rarely granular. Cross sections are six-sided or rhombic and show (110) cleavage traces at 56 and 124°. Crushed grains tend to be

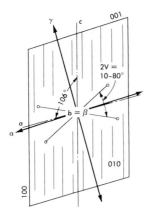

FIGURE 10.57 Orientation of glaucophane, section parallel with (010).

oriented on cleavage surfaces. Detrital pieces are generally elongated and striated parallel with c, and some are flattened tablets parallel with the cleavage. Inclusions of magnetite, sphene, zircon, and apatite are recorded. Overgrowths of glaucophane on hornblende are known.

Orientation. Glaucophane: Biaxial $(-)$. $\alpha \wedge a = 10$–$-8°$, $\beta = b$, $\gamma \wedge c = 3$–$21°$. Optic plane (010) (Fig. 10.57). $2V = 10$–$80°$, generally 40–$70°$, $r < v$ strong to moderate. Crossite: Biaxial $(-)$. $\alpha \wedge a = 5$–$-17°$, $\beta \wedge c = 2$–$30°$, $\gamma = b$. Optic plane is normal to (010). $2V = 12$–$65°$, usually $< 40°$, $r > v$ very strong. The strong dispersion causes incomplete extinction in white light. Less commonly, apparently, crossite may also have the orientation: Biaxial $(-)$. $\alpha \wedge a = 11$–$-10°$, $\beta = b$, $\gamma \wedge c = 2$–$3°$. Optic plane is (010), $2V = $ small. Possibly the shift in position of the optic plane from parallel with (010) in glaucophane and in some crossite to normal to (010) in most crossite accompanies an increase in Fe^{3+}, as in some ferrohastingsite and perhaps in riebeckite. Soda tremolite–glaucophane: Biaxial $(-)$. $\alpha \wedge a = -4$–$-44°$, $\beta = b$, $\gamma \wedge c = 17$–$60°$, generally $< 40°$. Optic plane is (010). $2V = 64$–$87°$, $r > v$ strong. The extinction angle and the dispersion increase toward the glaucophane end of the series. Normally in zoned members of the soda tremolite–glaucophane series, the cores have lower indices, smaller extinction angles, and paler colors. The extinction angle is particularly variable and is apparently very responsive to a small displacement of Al by Fe^{3+}. Irregular, "patchy" type of zoning occurs in some Fe^{3+}-rich glaucophanes. Wavy or flamboyant extinction may accompany marked zoning. Twins with (100) as the twin plane are rare. Cross sections show symmetrical extinction; longitudinal sections are length-slow.

Some of the zonal relations observed are:

Core	Margin
1. Crossite, $\beta \wedge c = 8°$, $2V = 30°$	1. Glaucophane, $\gamma \wedge c = 18°$, $2V = 60°$
2. Crossite, $2V = 12°$	2. Crossite, $2V = 65°$
3. Glaucophane, optic plane $=$ (010)	3. "Pseudoglaucophane" (crossite?), optic plane $\perp$ (010)
4. Glaucophane, $\gamma \wedge c = 17°$	4. Glaucophane higher in Fe^{3+}, $\gamma \wedge c = 12°$
5. Soda tremolite–glaucophane, $\gamma \wedge c = 48°$	5. Soda tremolite–glaucophane, $\gamma \wedge c = 60°$
6. Soda tremolite–glaucophane, $\gamma \wedge c = 55°$	6. Soda tremolite–glaucophane, $\gamma \wedge c = 30°$

Occurrence. Glaucophane is primarily a mineral of metamorphic rocks and is found in (1) schists containing various of the following: epidote, albite, sphene, garnet, quartz, sericite, chlorite, actinolite, calcite, lawsonite, chloritoid, pumpellyite, and less commonly muscovite, biotite, and aegirine–augite; (2) glaucophane eclogites with garnet, hornblende, omphacite, epidote, and pseudobrookite; and (3) garnet–glaucophane amphibolites ± hornblende and epidote. Crossite occurs in similar rocks associated with epidote, albite, chlorite, muscovite, sphene, and calcite. The members of the soda tremolite–glaucophane series occur in limestones pyrometasoma-

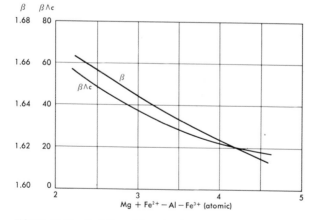

FIGURE 10.58 Variation of β and $\gamma \wedge c$ with composition in glaucophane–soda tremolite series (Larsen, 1942).

tized by alkalic intrusives. Glaucophane and crossite are relatively uncommon as detrital minerals. Glaucophane may form by the replacement of pyroxene. It can be replaced peripherally by actinolite and also alters to a green soda amphibole ("abnormal glaucophane"). The amphiboles related to glaucophane but rich in Fe^{3+} occur in various alkalic intrusive rocks.

Diagnostic features. Crossite differs from glaucophane in having higher indices of refraction and usually a different orientation. Riebeckite is normally length-fast and has higher indices. The rare amphibole, holmquistite, which contains Li, resembles glaucophane in its pleochroism: α = light greenish yellow, β = violet, γ = sky blue. However, it is orthorhombic and is restricted to the contact zones of some spodumene pegmatites. Tourmaline has parallel extinction, is uniaxial, and shows maximum absorption at right angles to c. Dumortierite is orthorhombic, has $c = \alpha$ (maximum absorption) and higher refractive indices.

CHAPTER ELEVEN

BIAXIAL MINERALS: SHEET SILICATES

MICA GROUP

GENERAL   The petrologically important micas include:

Muscovite, $K_2Al_4(Si_6Al_2)O_{20}(OH)_4$
Lepidolite, $K_2(Li,Al)_{2.5-3}(Si_{6-7},Al_{2-1})O_{20-21}(F,OH)_{3-4}$
Phlogopite, $K_2(Mg,Fe^{2+})_6(Si_6Al_2)O_{20}(OH)_4$
Biotite, $K_2(Fe^{2+},Mg)_{6-4}(Fe^{3+},Al,Ti)_{0-2}(Si_{6-5},Al_{2-3})\ O_{20-22}(OH,F)_{4-2}$

A series exists between muscovite and the highly variable lepidolites, likewise between phlogopite and biotite. Micas are structurally complex: muscovite occurs in three distinct structural types, lepidolite in four, phlogopite in three, and biotite in four, representing stacking variations of successive layers. Micas are characterized by perfect basal cleavage, strong birefringence, $(-)$ sign, and very small to nil extinction angles in sections normal to the cleavage.

MUSCOVITE

Composition. $K_2Al_4(Si_6Al_2)O_{20}(OH)_4$. Minor Na, Ba, Rb for K; some Mg, Fe^{2+}, Fe^{3+}, Mn for Al; minor Cr for Al (fuchsite); Li (as much as ca. 3.2% Li_2O); and variation in Si:tetrahedral Al from 6:2 to 7:1 (phengite); little F for OH.

Indices. $\alpha = 1.552-1.574$, $\beta = 1.582-1.610$, $\gamma = 1.587-1.616$, $\gamma - \alpha = 0.036-0.049$. The birefringence and the indices increase markedly with

Fe^{3+} and to a lesser extent with Fe^{2+} + Mn. In thin section the maximum interference tints fall into upper second and lower third orders. Fuchsite has $\alpha = 1.559$–1.571, $\beta = 1.593$–1.604, $\gamma = 1.595$–1.612, $\gamma - \alpha = 0.035$–0.042. Increasing Cr increases the indices as well as increasing Fe^{3+}. The birefringence is sufficiently high to cause a slight twinkle upon rotation of the stage. A few muscovites reveal marginal zonal structure through diminution of birefringence.

Color. Most muscovite is colorless. Some ferrian types are pleochroic in shades of light red and pale reddish brown. Fuchsite is colorless to green in thin section with $\alpha < \beta < \gamma$:

$\alpha =$ colorless, robin's-egg blue
$\beta =$ very pale green, yellow, pale yellowish green
$\gamma =$ pale green, blue green, bright chrome green

Form. Nearly all muscovites are monoclinic (two-layer) but a few rare, nearly uniaxial phengites are hexagonal (three-layer). Common as thin tablets, well outlined by (001) faces, but with poorly developed pinacoid faces, giving rise to basal plates with ragged, irregular outlines. Commonly in fine-grained, felted to shreddy aggregates (sericite) replacing many other minerals, particularly feldspar. Also in symplectitic intergrowths with quartz. Perfect (001) cleavage, which completely governs the shape and orientation of crushed flakes. Zircon crystals with pleochroic haloes are included, as well as grains of apatite, magnetite, and many other species in large books. Detrital muscovite forms relatively large, rounded, and thin basal plates (Fig. 11.2).

Orientation. Biaxial $(-)$. $\alpha \wedge c = 0$–$-2°$, $\beta \wedge a = 1$–$3°$, $\gamma = b$. Optic plane normal to (010) (Fig. 11.1). $2V = 30$–$47°$, although a few phengites have $2V = 0$–ca. $20°$. $r > v$ weak to distinct. Cleavage directions are

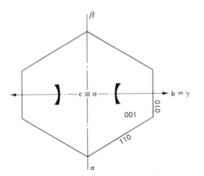

FIGURE 11.1 Orientation of muscovite, section parallel with (001).

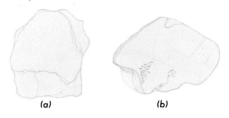

FIGURE 11.2   Detrital muscovite. (a) Yorkshire, England. (b) Dinard, Brittany.

length-slow. Near extinction positions a peculiar mottled or crinkly aspect appears, owing to minute buckling along the cleavages (birdseye maple structure). Wavy extinction characterizes bent plates. Twinning on (110) may be present. Two types of contrasting sections are encountered in thin section: (1) sections parallel or nearly parallel with (001), with irregular outlines, centered Bxa figures, γ and β, no cleavage traces, and first-order interference colors; and (2) sections normal or nearly normal to (001) with straight parallel sides, conspicuous cleavage traces, birdseye maple structure, and higher birefringent tints. Fuchsite has $2V = 32-46°$, $(-)$, $r > v$ weak to strong, and similar orientation.

Occurrence. In some granites, usually with biotite and microcline (binary or two-mica granites), where it may be formed by deuteric alteration of biotite. Primary muscovite does not occur in extrusive igneous rocks. Very common in granitic pegmatites. Also common and widespread in slates, phyllites, schists, quartzites, and a variety of gneisses. Detrital muscovite is widely distributed in sandy sediments. Secondary fine-grained muscovite, or sericite, replaces a variety of minerals, especially plagioclase, potash feldspars, nepheline, cordierite, kyanite, andalusite, topaz, corundum, and also tourmaline, beryl, spodumene, and scapolite. The replacement of feldspars by sericite may be highly selective, with calcic plagioclase normally replaced before sodic plagioclase, and plagioclase generally replaced before potash feldspar. In some cases only one set of plagioclase twin lamellae is preferentially altered. Some sericite is produced by weathering, but most sericitization appears to be a hydrothermal process. Wall rock sericitization is a common and extensive metasomatic transformation accompanying the formation of mesothermal ore deposits, particularly those in silicic igneous rocks. Fuchsite occurs principally in metamorphic rocks, including dolomitic marbles, micaceous schists and gneisses, and quartzites; also in some magmatic (?) corundum rocks and with some ore deposits.

Diagnostic features. Resembles colorless phlogopite, which is nearly uniaxial. Difficult to distinguish from talc, especially if the muscovite is fine-

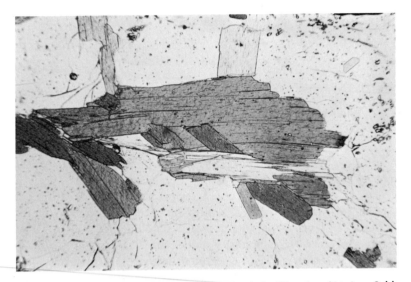

FIGURE 11.3 Biotite and muscovite in parallel growth, in sillimanite schist, Lues Gulch, Guffey, Colorado. Polars not crossed, $\times 85$ (J. E. Bever).

grained. Talc normally has a smaller 2V, but this may be difficult to determine. The association of talc with magnesium-rich rocks is helpful, however. It may be necessary to use chemical or x-ray methods to distinguish between sericite and fine-grained pyrophyllite; pyrophyllite, however, has a larger 2V than muscovite. Fuchsite has the characteristics of normal muscovite, but differs in color. Paragonite, $Na_2Al_4(Si_6Al_2)O_{20}(OH)_4$, with minor K for Na, is uncommon as a rock-forming mineral in some schists with quartz, kyanite, muscovite, and margarite. Its optical properties are within the muscovite range, and x-ray tests are required to differentiate it from muscovite.

LEPIDOLITE

Composition. $K_2(Li,Al)_{2.5-3}(Si_{6-7},Al_{2-1})O_{20-21}(F,OH)_{3-4}$. Rb and Cs replace K; Mn, Mg, Fe^{2+}, and Fe^{3+} are normally present in small amounts, although Mn may become important in some varieties. The F–OH ratio is variable, with F predominant normally. Polylithionite approaches the composition $K_2Li_4Al_2Si_8O_{20}(F,OH)_4$. The lithium content should be about 3.5% Li_2O or more for the mica to qualify structurally as a lepidolite.

Indices. $\alpha = 1.525-1.548$, $\beta = 1.551-1.580$, $\gamma = 1.554-1.586$, $\gamma - \alpha = 0.018-0.038$. Compositional-optical relations are complex, but the index

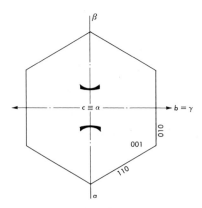

FIGURE 11.4   Orientation of lepidolite, section parallel with (001).

values depend not on the Li content, but rather on Mn, Fe^{2+}, and Fe^{3+}, with which they increase. The maximum interference colors fall in upper first to upper second order range in thin section.

Color. Colorless.

Form. Chiefly monoclinic, one-layer and six-layer; also pseudorhombohedral (three-layer) for the rare uniaxial types. In fine to coarse flakes usually with irregular pinacoidal outlines; also in stubby pseudohexagonal euhedra. May be finely felted to cryptocrystalline. Zircon (with pleochroic haloes) and cassiterite occur as inclusions.

Orientation. Biaxial $(-)$, rarely uniaxial $(-)$. Normally, $\alpha \wedge c = 0$--$3°$, $\beta \wedge a = 0$-$3°$, $\gamma = b$, optic plane parallel with (010) (Fig. 11.4). $2V = 25$-$58°$, $r > v$ weak. Also rarely essentially uniaxial. Cleavage traces are length-slow. Twinning on (110) may be present. Cleavage flakes show low birefringence, a centered Bxa figure and no cleavage traces.

Occurrence. Mainly in complex granitic pegmatites with albite, especially cleavelandite and such associates as alkali beryl and tourmaline, topaz, and spodumene. Rarely in high-temperature cassiterite veins.

Diagnostic features. Differs from muscovite optically mainly by lower indices and somewhat lower birefringence, although the ranges overlap. Chemical or x-ray data may be necessary in some cases for this distinction.

PHLOGOPITE

Composition. $K_2(Mg,Fe^{2+})_6(Si_6,Al_2)O_{20}(OH)_4$. Na can substitute for K in considerable amounts; also minor Ba for K. Fe^{2+} is almost always present, but Mg predominates. Small amounts of Mn, Fe^{3+}, and Ti may be present. The Si:tetrahedral Al ratio varies somewhat. Manganophyllites may be classed as phlogopites, normally with $Mg > Mn^{2+} > Fe^{2+}$, and in some types with some Fe^{3+} and Mn^{3+} as well.

Indices. $\alpha = 1.530–1.573$, $\beta = 1.557–1.617$, $\gamma = 1.558–1.618$, $\gamma - \alpha = 0.028–0.049$, with maximum interference colors in thin section ranging from second to third orders. Basal sections have very weak birefringence; for $\gamma - \beta = 0.000–0.001$. Indices and birefringence increase with Fe^{2+} and Mn, and even more rapidly with Ti and Fe^{3+}. Some phlogopites with much Ti, or Ti and Fe^{3+}, have indices as high as $\alpha = 1.599$, $\gamma = 1.643$. Manganophyllites have $\alpha = 1.548–1.612$, $\beta = 1.581–1.613$, $\gamma = 1.582–1.613$, $\gamma - \alpha = 0.024–0.040$.

Color. Colorless to pleochroic in brownish shades. Normally $\alpha < \beta \leqq \gamma$ with:

$\alpha = $ colorless, pale yellow, orange, salmon pink
$\beta = $ brownish yellow, buff, reddish orange, pale red brown
$\gamma = $ brownish yellow, buff, reddish orange, reddish brown, pale red brown, olive green

Also rarely with reversed pleochroism, $\alpha > \beta = \gamma$, with

$\alpha = $ deep red brown, deep chestnut brown
$\beta = \gamma = $ nearly colorless

May show color zoning with pale brown or, less commonly, green cores and outer zones either lighter or darker than cores. Outer zones may display the reverse pleochroism. Green zones around yellow-brown cores also are recorded. The depth of color and strength of pleochroism increase with Fe^{2+} and also with Fe^{3+} and Ti.

Form. Monoclinic (one- and two-layer types and pseudorhombohedral, three-layer type). Manganophyllites have principally the one-layer monoclinic structure. Typically in euhedral to subhedral, six-sided, thick tablets or stubby prismatic crystals. Fragments lie entirely on the perfect (001) cleavage planes, and detrital pieces are ragged flakes. Inclusions of rutile needles are common and may produce asterism. Hematite and tourmaline also may be included.

Orientation. Biaxial $(-)$, $\alpha \wedge c = 0–4°$, $\beta = b$, $\gamma \wedge a = 0–5°$, optic plane is (010) (Fig. 11.5). $2V = 0–12°$, $r < v$ weak to distinct. In rare types, 2V may be somewhat larger. In a few the optic plane is normal to (010). Cleavage traces are length-slow. Twinning (001) may be present. Manganophyllites are similar in orientation, with $2V = 4–9°$, but also rarely have the optic plane normal to (010).

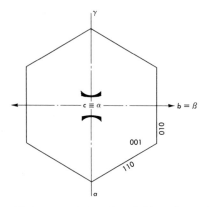

FIGURE 11.5 Orientation of phlogopite, section parallel with (001).

Occurrence. In ultramafic igneous rocks such as peridotites, kimberlites, leucite basalts, gabbros, and their serpentinized equivalents. In dolomitic marbles in association with tremolite, diopside, forsterite, talc, chlorite, chondrodite, sphene, spinel, and graphite. Large crystals occur in deposits with apatite and calcite in pyroxenite. Alters to vermiculite. Manganophyllites occur in contact manganese deposits.

Diagnostic features. It differs from muscovite in pleochroism and normally smaller 2V. Biotite has much stronger pleochroism and higher indices. Associated chlorites have much lower birefringence and possible abnormal interference colors.

BIOTITE

Composition. $K_2(Fe^{2+},Mg)_{6-4}(Fe^{3+},Al,Ti)_{0-2}(Si_{6-5},Al_{2-3})O_{20-22}(OH,F)_{4-2}$. Some Na, Ca, Ba, Rb, and Cs proxy for K; Mn for Fe^2; Mg may be almost absent in some types; others contain much Fe^{3+}; some Li may replace Al. Zinnwaldite is an Fe^{2+}–Li mica.

Indices. $\alpha = 1.565–1.625$, $\beta = 1.605–1.675$, $\gamma = 1.605–1.675$, $\gamma - \alpha = 0.040–0.060$. The indices increase with Fe^{2+} and more rapidly with Fe^{3+} and Ti (Fig. 11.8). In biotites in which total iron remains constant, 1% TiO_2 increases the refractive index by about 0.005. Mn for Mg effects a smaller increment than Fe^{2+}. Fe^{3+} and Ti also increase the birefringence. In thin section maximum birefringent colors are third and fourth orders but are difficult to detect because of the deep color. In some rare ferrian biotites γ may reach 1.73 and $\gamma - \alpha = 0.080$.

Color. Strongly pleochroic in shades of brown, less commonly in green.

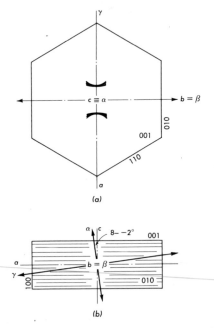

FIGURE 11.6   Orientation of biotite. (a) Section parallel with (001). (b) Section parallel with (010).

Red-brown biotites owe their color mainly to Ti and Fe^{3+}; greenish brown and brown biotites are colored mainly by Fe^{2+} and low Ti + Fe^{3+}, and blue-green biotites by Fe^{2+}. Biotites rich in Fe^{3+} and Ti also show stronger absorption. The absorption is normally $\alpha < \beta < \gamma$, but may be $\alpha < \gamma < \beta$. The direction parallel with the cleavage shows stronger absorption than that normal to the cleavage.

α = yellow, buff, light brown, pale greenish
β = reddish brown, greenish brown, bluish green, deep brown, opaque
γ = reddish brown, golden yellow brown, dark brown, green, opaque

Some biotites show color zoning of various combinations, with borders or rims either lighter or darker than central parts.

Form. Mainly monoclinic (one-layer and two-layer); less commonly triclinic (?) (six- and twenty-four-layer), and pseudorhombohedral (three-layer). In tabular, euhedral, six-sided crystals to anhedral ragged shreds and flaky aggregates. Perfect (001) cleavage which governs completely the orientation of flakes. Detrital pieces have jagged to rounded outlines.

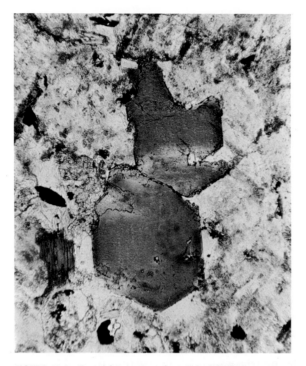

FIGURE 11.7   Zoned biotite in euhedral basal plates in minette, Steige, Vosges Mountains, France. Polars not crossed, ×94.

Inclusions are common, particularly zircon with its surrounding pleochroic halo, also apatite, magnetite, sphene, and allanite. Deformation produces bent flakes with curved cleavage traces.

Orientation. Biaxial $(-)$. $\alpha \wedge c = 8\text{--}2°$, $\beta = b$, $\gamma \wedge a = 0\text{--}9°$. Optic plane parallel with (010) (Fig. 11.6); less commonly normal to (010). Triclinic types depart only slightly from monoclinic orientation. $2V = 0\text{--}25°$, usually $< 10°$. Rare varieties have $2V > 25°$. Dispersion is either $r > v$ or $r < v$. Basal sections yield a centered Bxa figure and show little or no pleochroism. Cleavage traces are length-slow, but this may be difficult to determine owing to very strong absorption. Birdseye maple structure near the extinction position in sections showing cleavage is highly characteristic (Fig. 11.9). Wavy extinction occurs in bent plates. Twinning with (110) as the twin plane may be present.

Occurrence. In all types of intrusive igneous rocks from mafic to felsic: gabbros, norites, diorites, tonalites, syenites, foidal syenites, monzonites, granodiorites, granites, and various pegmatites. The Fe:Mg ratio increases

BIOTITE   **287**

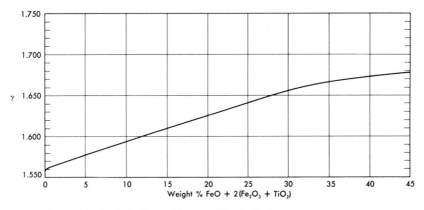

FIGURE 11.8 Variation in γ with composition in biotite.

progressively with increasing SiO_2 content of the host rock. Especially common and abundant in granites and granitic pegmatites. Also as phenocrysts in rhyolites, quartz latites, latites, trachytes, and andesites, but exceedingly uncommon in the matrix of such rocks. In intrusive rocks biotite may replace an amphibole. It may be replaced by deuteric muscovite in some granites. In extrusive rocks biotite phenocrysts show corrosion and replacement by matrix minerals: magnetite, pyroxene, and feldspar. In some cases only the biotite outline remains. Widespread in schists and gneisses of a variety of types, occurring with quartz, potash feldspar, albite, garnet, sillimanite, sericite, and chlorite. Biotite is altered to chlorite or vermiculite (Fig. 11.9). It may be changed to chlorite directly or pass first from a brown to a green biotite and thence to chlorite. By-product magnetite and sphene are commonly produced in this change. Calcite and epidote also replace biotite hydrothermally, and in high-grade metamorphic rocks biotite is replaced by sillimanite. The substance hydrobiotite represents interlayered biotite and vermiculite and is transitional in the alteration, biotite to vermiculite. Biotite, which is not uncommon as a detrital mineral, can show bleaching to pale brown or oxidation to reddish-brown colors. Biotite forms subparallel intergrowths with muscovite or chlorite (Fig. 11.3).

Diagnostic features. Distinguished by its darker color, stronger pleochroism, and higher indices from phlogopite, although the optical division between these two is not sharp. From brown amphiboles by its shape and cleavage, and from brown tourmaline by the position of maximum absorption and better cleavage. Green biotite differs from chlorite in having strong birefringence. Also resembles some stilpnomelane.

FIGURE 11.9 Biotite partly replaced by chlorite and magnetite and showing birdseye maple mottling near extinction, in granite, Eight Mile Park, Fremont County, Colorado. Polars crossed, ×55.

GLAUCONITE

Composition. $(K,Ca,Na)_{1-0.56}(Fe^{3+},Mg,Fe^{2+},Al)_2(Si,Al)_4O_{10}(OH)_2$. Celadonite is a variety that contains more Fe^{2+} and Mg and less Fe^{3+} and Al. The interlayer positions, which are occupied by the alkalies and Ca, are not always filled and are subject to base exchange. Glauconites that are low in K occur as mixed-layer structures mainly between expandable (montmorillonite) layers and nonexpandable layers (Burst, 1958).

Indices. $\alpha = 1.590-1.612$, $\beta = 1.609-1.643$, $\gamma = 1.610-1.644$, $\gamma - \alpha = 0.014-0.032$. In thin section the highest interference colors are of second order but are modified by the deep color of the mineral. The refractive indices and the birefringence increase with Fe^{3+} (Fig. 11.12), and indices increase with decreasing percentage of expandable layers (Fig. 11.11). Some types (with marked predominance of Al over Fe^{3+}), have indices as low as $\alpha = 1.557$, $\beta = 1.562$, $\gamma = 1.569$. Owing to very fine grain, exact indices are difficult to measure.

Color. Green in thin section and in grains green, olive green, and blackish green. Brown when altered. Pleochroism marked with:

α = lemon yellow, pale yellow green, bright green
β = γ = green, dark green, yellowish green, bluish green, olive green

$\gamma = \beta > \alpha$.

Form. Monoclinic. Usually in granules, pellets, or pseudomorphs (casts) after foraminifera (Fig. 11.10). These are commonly an aggregate of minute overlapping plates, but single-crystal grains, rudely hexagonal in outline, can also be found. Celadonite is of radial-fibrous habit and also forms vermicular aggregates composed of intergrown elongated plates. Distinct (001) cleavage.

Orientation. Biaxial $(-)$. $\alpha \wedge c$ = ca. $5°$, and from its close structural relations to biotite, the orientation probably is $\beta = b$, $\gamma \wedge a$ = small, optic plane is (010). $2V = 0\text{-}20°$, usually $> 10°$, $r > v$. Figures are difficult to secure because of the fine grain. Cleavage traces are length-slow.

Occurrence. Occurs in sedimentary rocks as a diagenetic mineral; a constituent of greensands, glauconitic sandstones, glauconitic limestones, and marls. Associated minerals are detrital quartz and feldspar, collophane, illite and other clay minerals, and calcite. Celadonite occurs as a secondary mineral filling vesicles and replacing olivine in basalt. Associated are calcite and saponite. Glauconite alters to limonite and goethite.

Diagnostic features. The occurrence, particle shape, color, and microcrystalline structure are distinctive for glauconite; for celadonite, the combination of occurrence, structure, color, and indices. The term "glauconite" has been loosely used for any green pellets found in sedimentary rocks. Burst (1958) has shown that these pellets are of four main types: (1) well-ordered single-layer glauconite; (2) interlayered glauconite–montmorillo-

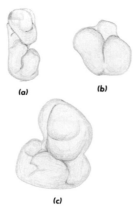

(a)　　　(b)

(c)

FIGURE 11.10　(a)–(c) Detrital glauconite, Chaldon Herring, Dorset, England (reflected light).

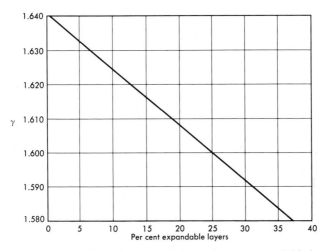

FIGURE 11.11 Relation between γ index and percent expandable layers in glauconite (Toler and Hower, 1959).

nite (if < two-thirds of all K positions are occupied); (3) interlayered clay minerals; and (4) mineral mixtures such as illite and montmorillonite or illite and chlorite.

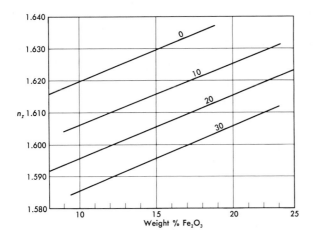

FIGURE 11.12 Relations between γ index, percent Fe_2O_3, and percent expandable layers in glauconite. The numbers on the lines of the graph indicate the percent of expandable layers (Toler and Hower, 1959).

STILPNOMELANE

Composition. $K(Fe^{2+},Fe^{3+},Al)_{10}Si_{12}O_{30}(O,OH)_{12}$ with $Al = $ ca. 2. K with some Ca or Na may be very low. K is not necessary to the structure and may be replaced by base exchange. Mg or Mn may be present in appreciable amounts, Ti in minor amounts. Some Al replaces Si. The major variation is between Fe^{2+} and Fe^{3+}.

Indices. $\alpha = 1.543–1.634$, $\beta = 1.576–1.745$, $\gamma = 1.576–1.745$, $\gamma - \alpha = 0.030–0.119$. The maximum interference colors in thin section vary all the way from middle second order to high-order white. Basal sections (cleavage pieces) are dark, since $\gamma - \beta = 0.000$. The indices and birefringence increase with Fe^{3+} (Fig. 11.14).

Color. Shades of green, brown, and yellow in thin section. Strongly pleochroic.

$\alpha = $ pale yellow, golden yellow, pale brown, colorless

$\beta = \gamma = $ deep brown, deep red brown, nearly black, chestnut brown, deep yellow brown, light greenish yellow

$\gamma = \beta > \alpha$. Ferroan stilpnomelane is generally pleochroic yellow to green; the ferrian variety, from brown to dark brown or black; and the manganoan type, from colorless or yellow to greenish yellow. With increasing oxidation of Fe^{2+} to Fe^{3+} the color becomes darker. Zoning has been observed with ferroan cores bordered by ferrian margins.

Form. Monoclinic. Thin micaceous plates, in some cases arranged in radial, plumose, sheaflike, or subparallel aggregates. Basal sections may

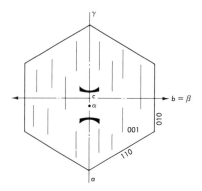

FIGURE 11.13 Orientation of stilpnomelane, section parallel with (001).

show pseudohexagonal outlines. Perfect (001) cleavage and an imperfect cleavage at right angles to (001). In veins it may show comb structure. Plates can be somewhat curved. Epidote and sphene are included, and parallel interlayerings with prochlorite have been observed.

Orientation. Biaxial ($-$). $\alpha \wedge c =$ ca. $7°$, $\beta = b$, $\gamma \wedge a = 0°$–small. Optic plane is (010) (Fig. 11.13). $2V = 0$–$40°$, generally nearly $0°$. Cross sections are length-slow. Cleavage pieces yield a centered, nearly uniaxial Bxa figure.

Occurrence. A relatively common mineral of certain low-grade schists and greenschists, in association with quartz, albite, sericite, chlorite, garnet, epidote, actinolite, magnetite, calcite, glaucophane, sphene, apatite, and pumpellyite. Also widespread in low-grade metamorphic iron-rich rocks

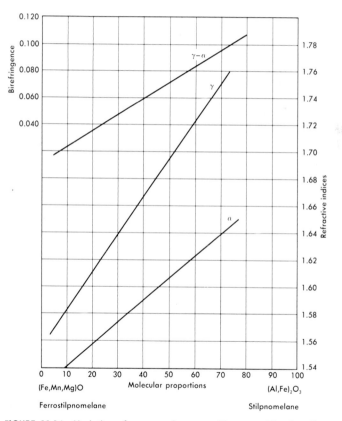

FIGURE 11.14 Variation of α, γ, and $\gamma - \alpha$ with composition in stilpnomelane (Hutton, 1956).

and ores (with minnesotaite), and in veins cutting them. In layered stilp-nomelane rocks the optical properties of the mineral may vary markedly from layer to layer.

Diagnostic features. From biotite, which it resembles very closely, it may be distinguished by:

1. Its less perfect basal cleavage
2. Its imperfect cleavage normal to (001)
3. Its strong yellow color in the α direction
4. Its lack of birdseye maple mottling near the extinction position
5. Its higher birefringence in varieties rich in Fe^{3+}
6. Its variable optic angle

MARGARITE

Composition. $CaAl_4Si_2O_{10}(OH)_2$. May contain some Fe^{3+} for Al, and Na, K for Ca; also small amounts of Li, Fe^{2+}, Mg, and Cr.

Indices. $\alpha = 1.595-1.638$, $\beta = 1.625-1.648$, $\gamma = 1.627-1.650$, $\gamma - \alpha = 0.010-0.032$. Indices decrease with substitution of Na for Ca, but birefringence increases markedly with this replacement. Maximum thin-section interference colors are normally about in middle first order but may be as high as lower second order.

Color. Colorless in thin section. May become pale brown with alteration. Rarely pale green (Cr).

Form. Monoclinic. Flakes or plates, usually in aggregates of thin laminae. May be euhedral with pseudohexagonal outline in basal section. Perfect (001) cleavage.

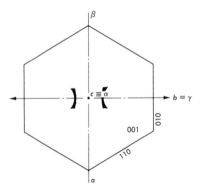

FIGURE 11.15   Orientation of margarite, section parallel with (001).

Orientation. Biaxial $(-)$. $\alpha \wedge c = 5°$ (varies), $\beta \wedge a = 6°$ (varies), $\gamma = b$. Optic plane normal to (010) and nearly normal to (001) (Fig. 11.15). $2V = 26$–$67°$, $r < v$. Elongate sections are length-slow. Multiple twinning with (001) as the composition plane may be present.

Occurrence. Usually associated with corundum in metamorphic rocks or in corundum veins. May replace corundum. Also in emery and rarely in mica schist with tourmaline and staurolite. Alters to vermiculite.

Diagnostic features. Differs from muscovite in having higher indices, lower birefringence, and variable 2V. Chlorites and chloritoid are colored or pleochroic, and the latter is optically $(+)$.

PREHNITE

Composition. $Ca_2(Al,Fe^{3+})_2Si_3O_{10}(OH)_2$. Al usually predominates greatly over Fe^{3+}.

Indices. $\alpha = 1.610$–1.637, $\beta = 1.615$–1.647, $\gamma = 1.632$–1.670, $\gamma - \alpha = 0.020$–0.035. In thin section maximum interference colors range from lower to upper second order. Abnormal interference tints appear in varieties that are complexly zoned. Indices and birefringence increase with increasing Fe^{3+}.

Color. Colorless to neutral in thin section. Parallel aggregates may display somewhat darker cores than margins.

Form. Orthorhombic. Common forms are sheaflike aggregates (Fig. 11.17c) with the characteristic bow-tie structure; spherulitic, fan-shaped, and radial groups; and columnar or lamellar masses. Euhedral crystals are tabular (001) or prismatic. (001) cleavage is distinct. Chlorite forms inclusions.

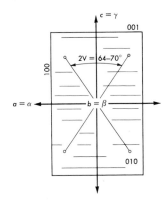

FIGURE 11.16   Orientation of prehnite, section parallel with (010).

Orientation. Biaxial ($+$). $\alpha = a$, $\beta = b$, $\gamma = c$. Optic plane is (010) (Fig. 11.16). 2V = 64–70°, but may appear much smaller, probably owing to submicroscopic optical anomalies. $r > v$ weak, but in anomalous types the dispersion may become $r < v$ strong, crossed. Anomalous types are relatively common. Some have thin twinning striae in two directions. In others basal sections show divisions into optical sectors or display variations of hourglass structure (Fig. 11.17a). Such crystals have marginal or terminal segments that are subparallel with (001) and are rotated 90° to each other about a central part that is optically normal. It is these lateral sectors that show abnormal interference colors, nonextinction, small and variable 2V, twinning, and abnormal dispersion.

Occurrence. Mainly a hydrothermal mineral in cavities, vugs, and veins in basalt, andesite, diabase, monzonite, peridotite, amphibolite, and rarely pegmatite. Commonly associated with some of the following: zoisite, clinozoisite–epidote, actinolite, chlorite, various zeolites, calcite, specular hematite, datolite, pectolite, and chalcedony. Also may be an important constituent of saussurite and replaces amphibole, plagioclase, and augite. Through saussuritization gabbros and other mafic rocks are transformed into zoisite-prehnite rocks. Occurs less commonly as a contact metamorphic mineral with calcite, diopside, grossularite, and wollastonite. Contact metamorphosed marl has been found altered to a rock consisting almost exclusively of spherulitic prehnite.

Diagnostic features. Wherever present, the bow-tie structure and optical anomalies are highly characteristic. Prehnite has much higher double refraction than topaz, lawsonite, andalusite, wollastonite, and datolite, which have about the same relief. Datolite is also ($-$).

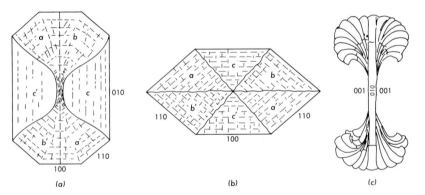

(a)   (b)   (c)

FIGURE 11.17 (a)–(c) Prehnite (Shannon, 1924). (a) Hourglass structure. (b) Twinned crystal. (c) Crystal aggregate, bow-tie structure.

CHLORITOID [1]

Composition. $(Fe^{2+},Mg)(Al,Fe^{3+})(Si,Al)O_5(OH)_2$. Mn also may be present in the variety ottrelite. Two-fifths of Fe^{2+} may be replaced by Mg, less than one-fifth by Mn (Halferdahl, 1961).

Indices. $\alpha = 1.713$–1.730, $\beta = 1.719$–1.734, $\gamma = 1.723$–1.740, $\gamma - \alpha = 0.006$–0.022. In thin section the interference colors range from lower to upper first order. The indices increase with increasing Fe^{3+} or with decreasing Mg.

Color. Colorless, green, or gray green in thin section. Usually somewhat pleochroic with:

α = greenish gray, olive green, colorless, greenish blue
β = blue gray, indigo, blue green
γ = colorless, greenish yellow, yellow, pale greenish brown

$\beta > \alpha > \gamma$. Some crystals show zoning with darker-colored cores and lighter margins; others show zoning of the hourglass type (Fig. 11.19).

Form. Both monoclinic and triclinic modifications occur. Platy or tabular (001) with pseudohexagonal or rhombic outline in basal sections. Subhedral to anhedral shapes are common. Cross sections are lath-shaped. In some rocks the plates are arranged parallel; in others radially curved or bent crystals occur. As a detrital mineral it forms laminated flakes parallel with (001), consisting usually of several superimposed plates. Perfect but difficult (001) cleavage, imperfect (110) cleavage, and (010) parting. Inclusions are very common, especially quartz but also magnetite, ilmenite, zircon, tourmaline, ankerite, and rutile. Inclusions may be arranged parallel with the rock structure.

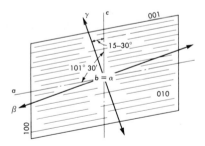

FIGURE 11.18   Orientation of chloritoid, section parallel with (010).

[1] Chloritoid structurally has atomic layers parallel with the basal plane, but the Si-O tetrahedra form individual units. Thus it is actually an orthosilicate.

Orientation. Biaxial $(+)$. $\alpha = b$, $\beta \wedge a = 7\text{-}-18°$, $\gamma \wedge c = 15\text{-}30°$. Optic plane is normal to (010) (Fig. 11.18). $2V = 36\text{-}89°$, usually $40\text{-}65°$, $r > v$. Elongate sections are length-fast. Multiple twinning is very common with (001) as the composition plane. Cleavage pieces yield a nearly centered Bxa figure. Some monoclinic types are reported to have $\beta = b$; others are $(-)$ with $2V = 89\text{-}55°$. 2V can be highly variable in material from one locality, and both monoclinic and triclinic forms occur together.

Occurrence. In mica and chlorite schists, phyllites, and quartzites. Quartz, albite, muscovite, biotite, garnet, chlorite, and kyanite are associated minerals. Also in quartz–ankerite veins as a hydrothermal mineral. Alters to pennine or kaolinite. Chloritoid of hydrothermal origin is commonly triclinic; both polymorphs occur in metamorphic rocks, but the monoclinic form is more commonly associated with almandite (Halferdahl, 1961).

Diagnostic features. Resembles chlorites, some micas, and margarite. Chlorites and biotite normally have smaller optic angles and small extinction angles. The combination of high indices, low birefringence, pleochroism, twinning, and, less commonly, zoning is very distinctive.

VERMICULITE

Composition. $(\mathrm{Mg,Fe^{2+}})_3(\mathrm{Si,Al,Fe^{3+}})_4\mathrm{O}_{10}(\mathrm{OH})_3\cdot 4\mathrm{H_2O}$. Some Ca may be present, and some types contain Ni and to a lesser extent Ti and Cr. Some of the Mg or Mg + Ca may be replaced by base exchange: substituted for them in laboratory are Li, Na, K, $\mathrm{NH_4}$, Rb, Cs, and Ba.

Indices. $\alpha = 1.525\text{-}1.561$, $\beta = 1.545\text{-}1.581$, $\gamma = 1.545\text{-}1.581$, $\gamma - \alpha = 0.020\text{-}0.031$. Varieties rich in Ni or in $\mathrm{Fe^{2+}} + \mathrm{Fe^{3+}}$ have indices at the upper end of the range. Maximum interference colors in thin section are in upper to lower second order but may be modified by the color of the mineral. Since $\gamma - \beta = 0.000$, basal sections are dark under crossed polars.

Color. Pale brown in thin section. Varieties containing Ni are green. May be pleochroic with:

α = colorless, pale green
$\beta = \gamma$ = pale brown, yellow green, brownish green

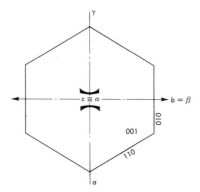

FIGURE 11.20 Orientation of vermiculite, section parallel with (001).

Form. Monoclinic. Perfect basal (001) cleavage. Usually subhedral lamellar or platy. Exfoliates with marked expansion upon heating. Resembles biotite greatly in form.

Orientation. Biaxial ($-$). $\alpha \wedge c =$ ca. 3°, $\beta = b$, $\gamma \wedge a =$ ca. 0°. Optic plane is (010) (Fig. 11.20). $2V = 0$–8°, $r > v$ weak. Cleavage traces are length-slow.

Occurrence. A hydrothermal mineral associated with altered peridotites and pyroxenites, commonly formed by the action of pegmatitic or syenitic intrusives on such rocks. Forms by the alteration of biotite or phlogopite in many types of rocks, both igneous and metamorphic. Submicroscopically interlayered biotite vermiculite is called hydrobiotite.

Diagnostic features. Resembles biotite but differs in having weaker pleochroism, less perfect cleavage, no birdseye mottling, and low birefringence. A test for exfoliation may be made.

CHLORITE GROUP

GENERAL The general formula for the chlorites is $(Mg,Fe^{2+},Fe^{3+},Al)_6$ $(Si,Al)_4O_{10}(OH)_8$. It is characteristic that some Al and less commonly some Fe^{3+} may enter into both the Mg and the Si positions. Other elements found in rarer varieties are Mn, Ni, and Cr. The group has been divided into two families (Berman, 1937): normal chlorites and leptochlorites. The latter are distinguished chemically by lower Mg and in many cases by much higher Fe^{2+} and Fe^{3+}. Chlorites are optically similar. Most are ($+$), with $r < v$ and absorption $\alpha = \beta > \gamma$ in shades of green and yellow. Negative chlorites have the dispersion and absorption formula reversed. 2V varies from 0 to 40° and is usually small. The acute bisectrix for both ($+$)

and $(-)$ types is nearly normal to the (001) cleavage. Birefringence is very low to low, varying from 0.000 to 0.015. An increase in iron, either Fe^{2+} or Fe^{3+}, brings about an increase in refractive indices and a deepening of the color. Chlorites differ from the micas in (1) being either $(+)$ or $(-)$, (2) having much lower birefringence, and (3) in their green colors.

CLINOCHLORE

Composition. $(Mg,Al,Fe^{2+})_6(Si,Al)_4O_{10}(OH)_8$, in which $Mg + Fe^{2+}$ equals about 5, Mg predominates greatly over Fe^{2+}, and Si equals approximately 3. Minor amounts of Fe^{3+}, Mn, and Cr may be present. Leuchtenbergite is the variety in which Fe^{2+} is very low or absent.

Indices. $\alpha = 1.571-1.588$, $\beta = 1.571-1.589$, $\gamma = 1.576-1.599$, $\gamma - \alpha = 0.005-0.015$, usually < 0.011. Maximum thin-section interference tints range from low first to low second order. Since $\beta - \alpha$ is very low, basal or near basal sections tend to be dark under crossed nicols. Indices rise with increasing Fe^{2+}.

Color. Colorless to green in thin section. Colorless, green, and olive green in detrital grains. $\alpha = \beta > \gamma$:

$\alpha =$ pale green
$\beta =$ pale green
$\gamma =$ pale yellow green, colorless

Manganiferous clinochlore:

$\alpha =$ colorless
$\beta =$ colorless
$\gamma =$ pale brownish yellow

Chromiferous clinochlore displays unusual pleochroic colors in shades of rose, lavender, and violet. Zircon inclusions in clinochlore cause brown pleochroic haloes. Margins of crystals may be somewhat darker colored than interior parts.

Form. Monoclinic. Perfect (001) cleavage. Thin to moderately thick plates which, if euhedral, display pseudohexagonal outlines. Fibrous, spherulitic, scaly, vermicular and cryptocrystalline structures also occur; plates may be curved. Crushed material lies on the (001) cleavage. Detrital pieces are plates with beveled margins.

Orientation. Biaxial $(+)$. $\alpha \wedge a = 0-9°$, $\beta = b$, $\gamma \wedge c = 0-9°$. Optic plane is (010) (Fig. 11.21). $2V = 0-40°$, $r < v$. Multiple twinning, with

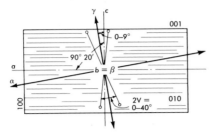

FIGURE 11.21 Orientation of clinochlore, section parallel with (010).

the twin plane normal to c, is common. Cleavage traces are length-fast. The pleochroic halo zone around zircon inclusions may show a reversal of sign from $(+)$ to $(-)$.

Occurrence. A low-grade metamorphic mineral occurring in schists, phyllites, and quartzites in combination with quartz, sericite, talc, actinolite, calcite, epidote, albite, chloritoid, and biotite. Also in some amphibolites with hornblende, anthophyllite, and cordierite and in marbles of contact-metamorphic origin with epidote, garnet, and diopside. Forms as a hydrothermal mineral in veins cutting serpentinites, commonly with talc, pennine, and antigorite. May replace hornblende and other silicates containing Mg.

Diagnostic features. Pennine is characterized by abnormal interference tints; prochlorite has higher refractive indices. Pennine and clinochlore both are characterized by a combination of relatively low refractive indices and low birefringence. Chlorites rich in Fe^{2+} have higher indices together with low birefringence; those with high Al contents are marked by low indices and relatively high birefringence; and types with both high indices and high birefringence contain much Fe^{3+}. Talc has much higher birefringence.

PENNINE

Composition. $(Mg,Fe^{2+},Al)_6(Si,Al)_4O_{10}(OH)_8$. Mg usually predominates greatly over Fe^{2+}, and the Mg:Fe ratio normally is similar to that in clinochlore. Pennine differs from clinochlore in having more Si. Minor Fe^{3+} may be present, and some varieties contain Cr.

Indices. $\alpha = 1.576-1.595$, $\beta = 1.576-1.600$, $\gamma = 1.579-1.600$, $\gamma - \alpha = 0.002-0.004$. Maximum interference colors are low first order and are commonly abnormal shades of blue, violet, or brown.

Color. Green. $(-)$ pennine has the absorption, $\gamma = \beta > \alpha$ and:

α = pale greenish yellow, colorless
$\beta = \gamma$ = pale green, green

(+) pennine has the absorption $\alpha = \beta > \gamma$ and the pleochroism:

$\alpha = \beta$ = green
γ = pale yellow green, colorless

Dark pleochroic haloes appear around included zircon and zonally arranged inclusions may be abundant (Fig. 11.23).

Form. Monoclinic. Crystals are relatively thick and pseudohexagonal in cross section. Pseudomorphs after biotite are common. Vermicular, radiating, and cryptocrystalline structures also occur. Perfect (001) cleavage, which governs the orientation of crushed material. Detrital grains are (001) plates or fibrous shreds.

Orientation. Biaxial (+) or (−) (Fig. 11.22). The sign may be reversed in the pleochroic haloes surrounding zircon inclusions. (+) pennine is apparently more common than (−) pennine.

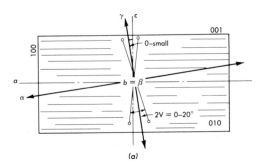

(a)

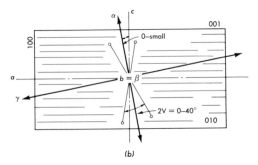

(b)

FIGURE 11.22 Orientation of pennine, sections parallel with (010). (a) (+) pennine. (b) (−) pennine.

$(+)$	$(-)$
$\alpha \wedge a =$ small to $0°$	$\alpha \wedge c =$ small to $0°$
$\beta = b$	$\beta = b$
$\gamma \wedge c =$ small to $0°$	$\gamma \wedge a =$ small to $0°$
Optic plane is (010).	Optic plane is (010).
$2V = 0\text{–}20°$	$2V \doteq 0\text{–}40°$
$r < v$	$r > v$
Length-fast cleavage traces.	Length-slow cleavage traces.

Size of 2V may vary considerably in a single crystal. In some cases a nearly uniaxial core is bordered by a biaxial rim. The sign may also alternate in adjacent laminae of the same crystal. Twinning with (001) as the composition and twin plane is characteristic; also less commonly with the twin plane normal to (001).

Occurrence. In chlorite schists and other low-grade metamorphic rocks associated with biotite, sericite, and talc. Also in veins in serpentinite with

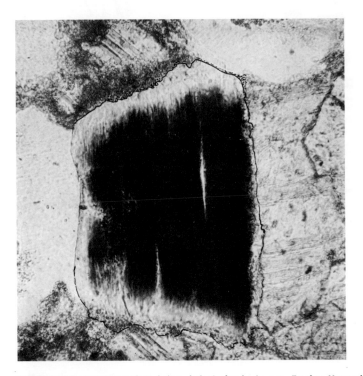

FIGURE 11.23 Pennine with inclusion-choked pleochroic core, Tendoy Mountains, Montana. Polars not crossed, ×148.

other chlorites, vermiculite, and serpentine. Forms in many metamorphic and igneous rocks as an alteration of biotite, which is replaced pseudomorphously with the separation of fine-grained by-product magnetite and sphene. May also replace garnet, staurolite, hornblende, and augite.

Diagnostic features. Distinguished from other chlorites by the combination of low indices and low birefringence; from clinochlore by lower extinction angles, variable sign, abnormal interference colors, and usually by small 2V.

PROCHLORITE

Composition. $(Mg,Fe^{2+},Al)_6(Si_{2.5}Al_{1.5})O_{10}(OH)_8$. $Mg + Fe = ca.$ 4.5 with the $Mg:Fe$ ratio, for many examples, varying from about 5.5 to 0.4.

FIGURE 11.24 Prochlorite with specular hematite, in quartz vein, Santa Maria del Oro, Durango, Mexico. Polars not crossed, $\times 70$ (R. I. Davis).

Minor Fe^{3+} and Mn are present in some varieties. Ripidolite is a synonym.

Indices. $\alpha = 1.605-1.648$, $\beta = 1.605-1.651$, $\gamma = 1.610-1.652$, $\gamma - \alpha = 0.001-0.006$. These figures apparently represent the range for most prochlorites, although values as low as $\alpha = 1.588$, $\beta = 1.589$, $\gamma = 1.599$ and as high as $\alpha = 1.658$, $\beta = 1.667$, $\gamma = 1.667$ have been recorded for material called prochlorite.

Color. Green. Weakly pleochroic. Absorption varies with sign:

$(+)$	$(-)$
$\alpha = \beta > \gamma$	$\alpha < \beta = \gamma$
$\alpha = \beta$ = green, brownish grass green	α = pale green
γ = pale green, pale greenish brown	$\beta = \gamma$ = olive green

Form. Monoclinic. Pseudohexagonal plates. In fine-grained scaly aggregates, in radial or fan-shaped groups (Fig. 11.24) and in spheroidal or vermicular masses. Perfect (001) cleavage on which crushed material is oriented.

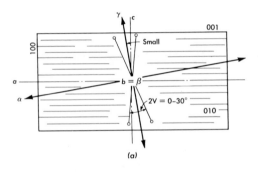

(a)

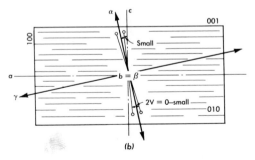

(b)

FIGURE 11.25 Orientation of prochlorite, sections parallel with (010). (a) (+) prochlorite. (b) (−) prochlorite.

Orientation. Biaxial $(+)$, less commonly $(-)$ (Fig. 11.25):

$(+)$	$(-)$
$\alpha \wedge a = $ small	$\alpha \wedge c = $ small
$\beta = b$	$\beta = b$
$\gamma \wedge c = $ small	$\gamma \wedge \alpha = $ small
Optic plane is (010).	Optic plane is (010).
$2V = 0\text{-}30°$, usually small	$2V = 0\text{-small}$
$r < v$	$r > v$
Length-fast cleavage traces.	Length-slow cleavage traces.

Occurrence. In chlorite schists and related rocks, with albite and magnetite. In serpentinites, and in veins cutting many types of igneous and metamorphic rocks, particularly with albite, adularia, rutile, and hematite (Fig. 11.24). Also in some Fe deposits with magnetite and hematite.

Diagnostic features. Has higher indices than clinochlore, pennine, and antigorite.

CHAMOSITE

Composition. $(Fe^{2+},Al,Mg)_6(Si,Al)_4O_{10}(OH)_8$. Minor Fe^{3+} occurs. Fe^{2+} usually predominates greatly over Mg, but magnesian types also are known. The ratio of Si to Al occupying Si positions is between 2.5 to 2.8.

Indices. Normally so fine-grained that only a median index of refraction can be measured: $n = 1.620\text{-}1.665$. The birefringence is low, near 0.007–0.008. Interference tints are masked by the mineral color. Indices increase with Fe^{2+}.

Color. Green. Brown when altered. May be slightly to markedly pleochroic from greenish yellow to grass green.

Form. Usually oolitic with concentric internal structure. The oolite may be spheroidal, flattened, or irregular (Fig. 11.26). Irregular tabular single crystals are not common. Occurs as fillings of microfossils. One good cleavage, probably (001).

Orientation. Biaxial $(-)$. 2V small. Orientation not known definitely.

FIGURE 11.26 Chamosite oolites, chamosite siltstone, Gloucestershire, England.

The cleavage traces are length-slow, and thus the orientation may be $\alpha \wedge c$ = small, $\beta = b$, $\gamma \wedge a$ = small, optic plane = (010).

Occurrence. An important constituent, generally as oolites, of some sedimentary iron ores. Also in associated beds of calcareous siltstones and limestones. Minerals found with it include calcite, siderite, collophane, pyrite, and clay minerals. May be replaced by pyrite or by another chlorite; alters to limonite. Some oolites are composed of alternating layers of hematite and chamosite.

Diagnostic features. Glauconite has higher birefringence. Greenalite is darker colored, nearly isotropic, and has a characteristic mineral association.

TALC

Composition. $Mg_3(Si_2O_5)_2(OH)_2$. Small amounts of Al, Fe^{2+}, and Ni may be present. Minnesotaite, which is essentially an iron talc, has the ideal formula $Fe_3(Si_2O_5)_2(OH)_2$ with some Mg for Fe^{2+} and some Fe^{3+} and Al for Si.

Indices. α = 1.538–1.550, β = 1.575–1.594, γ = 1.575–1.600, $\gamma - \alpha$ = 0.030–0.052. The maximum interference colors in thin section are in the upper third order. Basal sections show low first-order colors, since $\gamma - \beta$ = 0.000–0.006. Minnesotaite has α = 1.578–1.583, γ = 1.615–1.623, $\gamma - \alpha$ = 0.035–0.045.

Color. Colorless in thin section. Minnesotaite is colorless to slightly pleochroic:

$$\alpha = \text{colorless, pale yellow}$$
$$\beta = \gamma = \text{pale green}$$

Form. Monoclinic. Commonly in fine-grained shreddy aggregates or in foliated masses with random, subparallel, radial, or concentric arrangement. Curved plates are common. Perfect (001) cleavage. Minnesotaite occurs as minute plates or needles arranged in radiating aggregates.

Orientation. Biaxial $(-)$. $\alpha \wedge c$ = ca. 10°, $\beta \wedge a$ = ca. 0°, $\gamma = b$. Optic plane normal to (010) and nearly normal to (001) (Fig. 11.27). 2V = very small–30°, apparently 0° in some cases owing to rotated overlapping plates; $r > v$ noticeable. Minnesotaite has a similar orientation, with 2V = small.

Occurrence. In metamorphic rocks, such as talc schists with magnetite, and in talc–carbonate, talc–anthophyllite, talc–tremolite, and antigorite–talc schists. Also massive as steatite or soapstone. Common as a hydro-

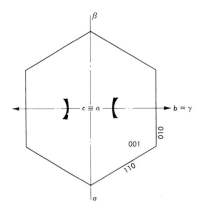

FIGURE 11.27 Orientation of talc, section parallel with (001).

thermal mineral in altered peridotites associated with magnesite, chlorite, actinolite, anthophyllite, vermiculite, and serpentine. Also in contact marbles in which it may replace tremolite pseudomorphously. May alter to magnesian clinochlore.

Diagnostic features. Pyrophyllite and muscovite have larger optic angles. Phlogopite has higher refractive indices and shows birdseye maple extinction. Brucite is biaxial ($+$) and may show abnormal interference colors. Sepiolite ($Mg_3Si_4O_{11} \cdot 5H_2O$), which occurs in similar deposits, has the properties $\alpha = 1.506-1.520$, $\gamma = 1.525-1.529$, $\gamma - \alpha = 0.009-0.020$, ($-$), $2V = 0-60°$, $\gamma \wedge c =$ small. It is difficult to distinguish optically between cryptocrystalline talc and sericite, and a chemical test for abundant Mg may be necessary to establish the identity of talc. The association of Mg minerals and ultramafic rocks assists in the recognition of talc.

SERPENTINE GROUP

GENERAL The serpentine group contains at least two well-defined, structurally distinct species, chrysotile and antigorite, which are dimorphous. Greenalite also has been shown to possess the serpentine structure. A few other minerals generally grouped here may deserve species status but are inadequately known. Most of the names listed under serpentine represent trivial varieties. Serpentines are natural mixtures in various proportions of the two end members, chrysotile and antigorite (Nagy and Faust, 1956). The electron microscope shows that antigorite forms flakes and laths, whereas chrysotile forms tubular fibers. Some serpentine is so fine-grained that it appears isotropic ("serpophite").

Composition. $Mg_3Si_2O_5(OH)_4$. Small amounts of Fe^{2+} and Fe^{3+} may be present, rarely Mn, Al, or Ni.

Indices. $\alpha = 1.529$–1.559, $\beta = 1.530$–1.564, $\gamma = 1.537$–1.567, $\gamma - \alpha = 0.004$–0.016, averaging 0.007–0.008. Indices as low as $\alpha = 1.493$, $\beta = 1.504$, $\gamma = 1.520$ and birefringence as high as $\gamma - \alpha = 0.024$ are recorded, but on poorly authenticated material. Maximum interference color in thin section rarely exceeds first-order yellow and usually is gray of the first order.

Color. Colorless, pale green, pale yellow, pale buff. May be faintly pleochroic:

$\alpha = $ colorless, greenish yellow
$\beta = $ colorless, yellow green
$\gamma = $ yellow, green

$\alpha \leqq \beta < \gamma$.

Form. Monoclinic. (110) cleavage indistinct. Usually fibrous either matted or parallel (Fig. 11.30); may be in stubby laths, in banded structures, and in crinkled microfolds. A wide variety of complex aggregate structures can be found.

Orientation. Biaxial $(-)$ or $(+)$. $(-)$ types may be more common than

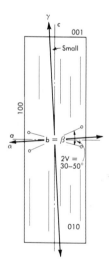

FIGURE 11.28 Orientation of chrysotile, section parallel with (010).

(+) ones. $\alpha \wedge a$ = small, $\beta = b$, $\gamma \wedge c$ = small. Optic plane is (010) (Fig. 11.28). $2V = 30$–$50°$ but may appear to be nearly $0°$, owing to overlapping randomly oriented lamellae. Figures may be difficult to obtain because of the fibrous nature. Fibers are length-slow.

Occurrence. Commonly associated with antigorite in serpentinite and in altered peridotite and pyroxenite. Forms by the alteration of ferromagnesian minerals such as olivine, augite, and less commonly hornblende. Dusty, fine-grained magnetite is a by-product in this replacement. Also occurs in veins, commonly as cross-fiber asbestos in ultramafic igneous rocks and in forsterite marbles; associated minerals are chlorite, talc, and vermiculite.

Diagnostic features. Other asbestiform minerals such as tremolite, actinolite, anthophyllite, and crocidolite have higher refractive indices and higher birefringence. Some types of antigorite (picrolite) also are asbestiform, and these two serpentine minerals are distinguished from each other most certainly by their x-ray diffraction patterns.

ANTIGORITE

Composition. $Mg_3Si_2O_5(OH)_4$. Minor Al, Fe^{3+}, Mn, and Ni may be present; some varieties contain considerable Fe^{2+}.

Indices. $\alpha = 1.555$–1.567, $\beta = 1.560$–1.573, $\gamma = 1.560$–1.573, $\gamma - \alpha = 0.004$–0.009, averaging 0.005. In thin section maximum interference colors may be as high as first-order yellow, but usually are first-order gray. Greenish-yellow abnormal interference tints may be observed. Lower indices than those listed above have been recorded on poorly authenticated material.

Color. In thin section colorless to light green. May be weakly pleochroic:

α = pale greenish yellow

$\beta = \gamma$ = light green

Form. Monoclinic. Commonly in fibrolamellar aggregates; also asbestiform, although apparently not as commonly as chrysotile. Complex radial, netlike (Fig. 11.30) and leaflike intergrowths are common and show undulatory or mottled extinction. Perfect (001) cleavage.

Orientation. Biaxial ($-$). $\alpha \wedge c$ = ca. $7°$, $\beta = b$, $\gamma \wedge a$ = ca. $0°$. Optic plane is (010) (Fig. 11.29). $2V$ =ca. 20–$50°$, larger optic angles have been reported, $r > v$ weak. Figures may be very difficult to obtain because of the fine-grained lamellar structure. Blades are length-slow.

Occurrence. A constituent of serpentinites, altered peridotites, and less commonly contact marbles and serpentine schists usually with chrysotile and in some cases associated with talc, vermiculite, chlorite, and chromite.

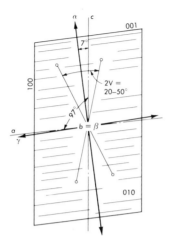

FIGURE 11.29 Orientation of antigorite, section parallel with (010).

FIGURE 11.30 Antigorite and cross-fiber chrysotile in serpentinite, Weistritz, Silesia. Polars crossed, ×28.

Forms pseudomorphously after olivine, augite, enstatite, and hornblende with separation of secondary fine-grained magnetite. May alter to magnesian clinochlore.

Diagnostic features. The lamellar, meshlike structure is more characteristic of antigorite, the fibrous asbestiform structure of chrysotile. X-ray powder patterns may be required to distinguish between the two. Antigorite usually has slightly higher indices and is invariably $(-)$. Brucite, which may occur with antigorite in some marbles, is uniaxial, has higher indices, whorl structure, and markedly abnormal interference colors.

GREENALITE

Composition. $(Fe^{2+},Fe^{3+},Mg)_6Si_4O_{10}(OH)_8$. The $Fe^{2+}:Fe^{3+}$ and $Fe^{2+}:$ Mg ratios vary.

Indices. Usually fine-grained, so that it appears isotropic or nearly so under the microscope. $n = 1.650–1.675$, increasing with Fe^{3+}. The birefringence is probably low. Some granules show both isotropic and mottled anisotropic patches.

Color. Dark olive green to very deep greenish brown. Normally without the accessory condenser it transmits relatively little light.

Form. Monoclinic. Typically in subrounded to subangular ellipsoidal granules in which it is very fine-grained and intimately intermixed with quartz, carbonates, minnesotaite, and stilpnomelane.

Orientation. Not known. Usually isotropic but distinctly crystalline to x-rays.

Occurrence. An important mineral in some sedimentary and metamorphic iron rocks and ores. Found in taconite with minnesotaite, stilpnomelane, siderite, fine-grained quartz, magnetite, and hematite. May be replaced by quartz.

Diagnostic features. The occurrence, associated minerals, color, and near-isotropic character.

PYROPHYLLITE

Composition. $Al_2Si_4O_{10}(OH)_2$. A small amount of Fe^{3+} may be present.

Indices. $\alpha = 1.534–1.556$, $\beta = 1.586–1.589$, $\gamma = 1.596–1.601$, $\gamma - \alpha = 0.046–0.062$. In thin section the maximum interference colors range from third to fourth order. Shows birdseye maple mottling at near-extinction positions.

Color. Colorless or neutral gray.

Form. Monoclinic. Tabular (010), subhedral, elongate crystals are

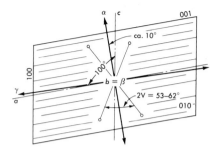

FIGURE 11.31   Orientation of pyrophyllite, section parallel with (010).

common. Aggregates are very fine-grained, matted, flaky, or coarsely radial. Single blades may show considerable curvature. The perfect (001) cleavage governs the orientation of crushed and detrital material. Rutile needles may be included.

Orientation. Biaxial $(-)$. $\alpha \wedge c = $ ca. $10°$, $\beta = b$, $\gamma \wedge a = $ ca. $0°$. Optic plane is (010) (Fig. 11.31). $2V = 53-62°$. Cleavage traces are length-slow. The elongate crystals are length-fast. Poorly defined twinning may be present.

Occurrence. In veins with quartz, diaspore, kyanite, and ankerite. Also in various altered wall rocks of some sulfide deposits with diaspore, siderite, and topaz. Common in hydrothermally altered silicic extrusive rocks, especially rhyolites and rhyolitic tuffs together with sericite, kaolinite, zoisite, diaspore, chloritoid, chlorite, rutile, alunite, quartz, and pyrite. Less common as constituent of metamorphic rocks, including quartz–sericite schist with carbonate and pyrophyllite, graphitic sericite schist, and in aluminous quartzites with diaspore, chloritoid, rutile, topaz, lazulite, sericite, apatite, and kyanite. Found in some soils. May be pseudomorphous after kyanite and replaces diaspore and andalusite.

Diagnostic features. Resembles closely talc or sericite, especially if fine-grained, but has a larger optic angle than either. Kaolinite has lower birefringence.

CLAY MINERALS GROUP

GENERAL   The members of the clay minerals group are difficult to identify by optical methods alone owing to (1) their fine grain size, (2) the impure nature of the samples—several clay species occurring together or with still other minerals, (3) the occurrence of mixed-layer combinations, (4) their optical similarity, (5) changes in refractive indices upon drying, and

(6) changes in refractive indices owing to absorption of some immersion liquids. Consequently, certain identification may require the combination of several additional techniques or devices including: (1) thermal analysis or dehydration curves, (2) electron photomicrographs, (3) x-ray powder photographs, (4) infrared absorption spectrograms, and (5) staining by means of organic dyes.

The group may be subdivided into several subgroups or series with their main species:

1. Kaolinite subgroup (kandites)
   Kaolinite, $Al_2Si_2O_5(OH)_4$
   Anauxite, $Al_2Si_3O_7(OH)_4$
   Dickite, $Al_2Si_2O_5(OH)_4$
   Nacrite, $Al_2Si_2O_5(OH)_4$
   Halloysite
      Hydrated halloysite, $Al_2Si_2O_5(OH)_4 \cdot 2H_2O$
      Meta-halloysite, $Al_2Si_2O_5(OH)_4$
   Allophane, $Al_2O_3 \cdot (1-2)SiO_2 \cdot nH_2O$
2. Montmorillonoids (also called smectites)
   Montmorillonite, $(Al,Mg,Fe^{3+})_2(Si_{4-3.5},Al_{0-0.5})O_{10}(OH)_2[Ca,Na,K]$
   Beidellite, $(Al,Mg,Fe^{3+})_2(Si_{3.5-3},Al_{0.5-1})O_{10}(OH)_2[Ca,Na,K]$

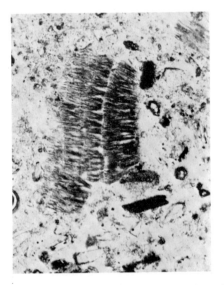

FIGURE 11.32 Illite with parting normal to the basal cleavage, from sediments of central Tyrrhenian Sea. Polars not crossed, ×125 (Erik Norin).

Nontronite, $(Fe^{3+},Al,Mg)_2(Si,Al)_4O_{10}(OH)_2[Ca,Na]$
Saponite, $(Mg,Al)_3(Si,Al)_4O_{10}(OH)_2$
3. Illite subgroup
Hydromuscovite, $(K,Na,H_3O)_2(Al,Mg,Fe^{3+},Fe^{2+})_4(Si_7,Al)O_{20}(OH)_4$

Another subgroup, the palygorskite clay minerals, includes the species attapulgite, $Al_2Si_5O_{13}\cdot6H_2O$; palygorskite, $Mg_2Si_3O_8\cdot4H_2O$; sepiolite, Mg_3-$Si_4O_{11}\cdot5H_2O$; and others. Palygorskite resembles montmorillonite optically but has higher refractive indices.

Clay minerals have a very wide range in composition, even within a single species. Ion substitution in the octahedral layer (sites of tri- and divalent cations) occurs among Al, Fe^{3+}, Fe^{2+}, Mg and less commonly Li, Zn, Cr, V; in the tetrahedral layers (sites of tri- and tetravalent cations) among Si, Al, $Fe^{3+}(?)$; in the intersheet positions (sites of exchangeable mono- and divalent cations) among Ca, Mg, K, Na, H_3O (Keller, 1956).

The 2:1 (Si:Al) clay minerals include not only clay minerals in the strict sense—illites and montmorillonoids, but also those genetically (and structurally) similar minerals—sedimentary chlorite and vermiculite. All of these form similar flake-shaped layers with different intersheet cations:

Illites: K^{1+},Na^{1+},H_3O^{1+}
Montmorillonoids: $Ca^{2+},Mg^{2+},Na^{1+},H_2O$
Chlorites: $(Mg\cdot Al)(OH)^{1+}$
Vermiculite: Mg^{2+}, Ca^{2+}

Nonmixed layer clay minerals have the same cation type in intersheet positions, but if, for example, 5 or 10 sheets are joined by K^{1+} (illite), whereas another 10 or 15 are linked by Ca^{2+} (montmorillonite), then a mixed-layer clay mineral is the product (Weaver, 1956). Apparently mixed-layer clay minerals can be formed with all possible combinations of different sheets:

Illite–montmorillonoid
Illite–chlorite
Different illites
Different montmorillonoids
Montmorillonoids (smectite)–chlorite
Chlorite–vermiculite (corrensite)
Illite–chlorite–montmorillonoids

When the different sheet types that form a mixed-layer clay mineral are randomly intercalated, the clay is referred to as a random mixed-layer clay. Regular mixed-layer clay minerals are those in which two or more different types of sheets appear in a regular sequence such as IM, IM, IM, or IMM, IMM, IMM.

Mixed-layer clay minerals are formed mainly by alteration of preexisting clay minerals during either weathering or diagenesis. Mixed-layer species are difficult to recognize optically. They look like single crystals and have intermediate refractive indices, but commonly show fuzzy interference figures. Their identification is best accomplished by means of powder x-ray diffraction methods (Weaver, 1956).

KAOLINITE

Composition. $Al_2Si_2O_5(OH)_4$. The Al–Si ratio varies from 2:2 to about 2:3. Minerals with the higher Si contents are called anauxite, $Al_2Si_3O_7(OH)_4$.

Indices. $\alpha = 1.553\text{–}1.563$, $\beta = 1.559\text{–}1.569$, $\gamma = 1.560\text{–}1.570$, $\gamma - \alpha = 0.005\text{–}0.010$. In thin section maximum interference colors vary from low to middle first order. Lower indices than those above have been reported but either on poorly identified material or on hydrated kaolinite. Basal sections may appear nearly isotropic, since $\gamma - \beta = 0.001\text{–}0.002$.

Color. Colorless or pale yellow. Rarely pleochroic, $\alpha < \beta = \gamma$:

$\alpha = $ colorless, pale yellow
$\beta = \gamma = $ buff, dark buff

Absorbs dyes readily and becomes markedly pleochroic.

Form. Triclinic, pseudomonoclinic. Typically in platy flakes or thin shreds, fine-grained and/or irregular outline; if euhedral, flakes are six- or three-sided. In elongated crystals with a twisted, vermicular form, angu-

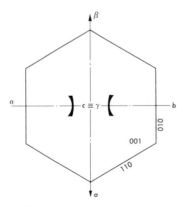

FIGURE 11.33 Orientation of kaolinite, section parallel with (001).

lar, shardlike pieces, plumose aggregates, spherulitic masses, and veinlets. Perfect (001) cleavage which gives the twisted crystals an "accordion-like" appearance (Fig. 11.34). Crushed material may or may not be oriented by the cleavage, depending on the relation of grain size to particle size.

Orientation. Biaxial $(-)$. Optic plane and γ nearly normal to (010) (Fig. 11.33). $\alpha \wedge \perp (001) = 1\text{-}4°$, β nearly coincident with a. $2V = 23\text{-}60°$, $r > v$ weak. Basal sections give a nearly centered Bxa figure, but may be very difficult to obtain owing to the fine grain size and low birefringence. In some clays and shales aggregates of flakes with parallel (001) orientation will yield a diffuse figure. Shardlike pieces may show a pseudo-interference figure under crossed nicols with convergent light (Fig. 11.35). This anomalous extinction effect is ascribed to strains produced during dessication. Cleavage traces are length-slow. Twinning is reported, but is uncommon; probably similar to the micas.

Occurrence. A weathering product of igneous and metamorphic rocks

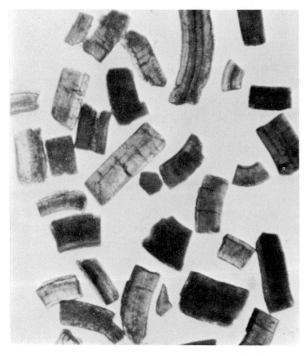

FIGURE 11.34 Vermiform crystals of kaolinite, Pontiac, South Carolina. Polars crossed, $\times 28$ (Ross and Kerr, 1930).

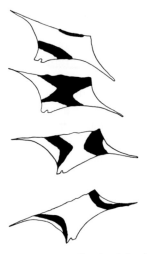

FIGURE 11.35 Shards of kaolinite with "polarization crosses," fire clay, Boksburg, South Africa (Bosazza, 1941).

mainly by alteration of feldspars. It also replaces feldspathoids, quartz, muscovite, beryl, and other aluminous silicates. It is the chief clay mineral in most residual or transported clays, in which it is associated with halloysite, quartz, orthoclase, sericite, limonite, leucoxene, and carbonaceous material. An important component of some shales, together with illite, chlorite, and quartz. In variable amounts in soils, bauxites, and some fire clays (with diaspore). May also be formed as a hydrothermal mineral in argillic wall rock alteration associated with sulfide veins and replacements, occurring with halloysite and dickite.

Diagnostic features. Cannot be distinguished from anauxite by optical tests, x-ray spacings, or thermal curves; thus a chemical analysis is here required. Sericite, illite, talc, and pyrophyllite all have stronger birefringence. Dickite is optically $(+)$ and has a larger extinction angle; nacrite also has a larger extinction angle. Montmorillonoids have small 2V's and much stronger birefringence. Montmorillonite itself also has considerably lower refractive indices.

DICKITE

Composition. $Al_2Si_2O_5(OH)_4$. Small amounts of Fe^{3+} may replace Al.

Indices. $\alpha = 1.558-1.562$, $\beta = 1.560-1.565$, $\gamma = 1.563-1.571$, $\gamma - \alpha = 0.003-0.008$. For a variety with some iron, $\alpha = 1.560$, $\gamma = 1.572$, $\gamma - \alpha =$

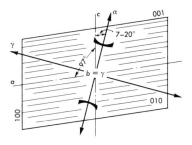

FIGURE 11.36   Orientation of dickite, section parallel with (010).

0.012. Maximum interference colors are middle first order in thin section.

Color. Colorless to very pale yellow. Does not absorb dyes readily.

Form. Monoclinic. Not uncommonly in minute, thin, six-sided tablets (Fig. 11.37). Also anhedral in veinlets and replacement masses. Groups in parallel or radial, fanlike arrangement also occur. Perfect (001) cleavage which orients crushed material.

Orientation. Biaxial $(+)$. $\alpha \wedge c = -7\text{--}-20°$, $\beta \wedge a = 14\text{--}27°$, $\gamma = b$. Optic plane normal to (010) (Fig. 11.36). $2V = 52\text{--}80°$, $r < v$. Ordinarily optic figures are difficult to obtain owing to fine grain size. Undulatory extinction may be present. Basal plates yield an eccentric Bxo figure.

Occurrence. Mainly a hydrothermal mineral, found with sulfide ores, or in their altered wall rocks. Replaces quartz, calcite, dolomite, and plagioclase. Also replaced by sericite and quartz and associated with kaolinite, nacrite, cherty quartz, sericite, and limonite.

Diagnostic features. Kaolinite has a smaller extinction angle. Nacrite, which is difficult to distinguish from dickite without x-ray photographs, is monoclinic, with $\alpha = 1.557\text{--}1.560$, $\beta = 1.561\text{--}1.563$, $\gamma = 1.563\text{--}1.566$, $\gamma - \alpha = 0.006$; $(-)$ $2V = 40\text{--}80°$, $r > v$; or $(+)$ $2V = $ ca. $90°$, $r < v$; $\alpha \wedge c = -7\text{--}-10°$, $\beta \wedge a = 10\text{--}13°$, $\gamma = b$, optic plane normal to (010).

HALLOYSITE

Composition. $Al_2Si_2O_5(OH)_4 \cdot 2H_2O$, (hydrated halloysite or endellite). Metahalloysite, $Al_2Si_2O_5(OH)_4$, can be produced by dehydration of the hydrated variety. Mixtures of the two and partly dehydrated intermediate members are known.

Indices. Both minerals are anisotropic, but the birefringence is so low that the gypsum plate may be required to detect it. Therefore, and because of the fine grain, measurement of separate indices of refraction is normally impracticable, and usually only a mean index is determinable:

Hydrated halloysite: $n = 1.526-1.542$
Metahalloysite: $n = 1.549-1.561$

Values for mixtures of the two are intermediate to these ranges. The birefringence is in the range 0.000–0.004.

Color. Colorless.

Form. Pseudohexagonal. Possible monoclinic or triclinic from analogy with kaolinite. Extremely fine-grained masses as replacements, layers, and veinlets. May show irregularly oriented shatter cracks or colloform structure (Fig. 11.38).

Orientation. Presumably biaxial but the orientation is not known. Reported to have a small extinction angle against one cleavage and to be length-slow.

Occurrence. A hydrothermal clay mineral in ore deposits and their altered walls. Associated with kaolinite, allophane, alunite, diaspore,

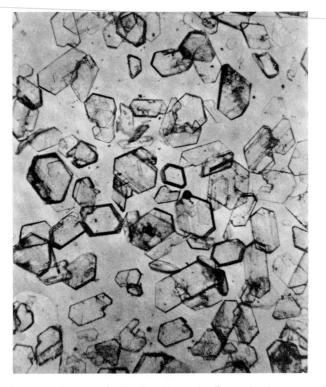

FIGURE 11.37 Crystals of dickite, National Belle Mine, Red Mountain, Colorado. Polars not crossed, ×260 (Ross and Kerr, 1930).

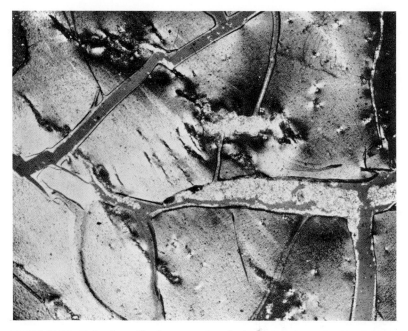

FIGURE 11.38 Halloysite with shatter cracks of alunite, Lawrence County, Indiana. Polars crossed, ×27 (Ross and Kerr, 1930).

gibbsite, sericite, quartz, pyrite, and limonite. Replaces feldspar and pyrite and is replaced by sericite. Also in some residual clay deposits with kaolinite and in some shales. Forms by hot springs action on perlite.

Diagnostic features. The low birefringence and shatter cracks are helpful, but for exact determination techniques other than optical may be needed. Allophane is an essentially amorphous material with the composition $Al_2O_3 \cdot (1-2)SiO_2 \cdot nH_2O$ and notable amounts of P_2O_5 in some types; $n = 1.470-1.496$.

MONTMORILLONITE

Composition. $(Al,Mg,Fe^{3+})_2(Si,Al)_4O_{10}(OH)_2[Ca,Na,K]$. Normally $Al > Mg > Fe^{3+}$; tetrahedral Al is usually < 0.5; Ca, Na in small amounts, and rarely K are present as extraneous interlayer ions and can replace one another through cation base exchange or even be replaced by large organic cations (amines, alcohols, glycols, nitriles, proteins, etc.). The term "beidellite" is generally applied to a mineral with less Si. Apparently a

MONTMORILLONITE   321 series exists in which the tetrahedral ratio varies from (Si_4,Al_0) to nearly (Si_3,Al_1).

Indices. $\alpha = 1.485\text{--}1.535$, $\beta = 1.504\text{--}1.550$, $\gamma = 1.505\text{--}1.550$, $\gamma - \alpha = 0.020\text{--}0.035$. The indices increase with increasing substitution of Mg for Al; acid treatment resulting in loss of Mg decreases indices to as low as $\alpha = 1.467$, $\gamma = 1.482$. Dehydration raises indices and birefringence. Indices also rise markedly with increasing Fe^{3+} for Al (Fig. 11.40). In thin section maximum interference colors usually are not above second order owing to extreme thinness of many flakes. Absorbs oils, especially those with NH_2 groups, and changes indices.

Color. Normally colorless in thin section, but iron-bearing types are pleochroic in shades of yellow, brown, and green, with $\alpha < \beta < \gamma$. Characteristically absorbs various dyes.

Form. Monoclinic. Usually in micro- to cryptocrystalline aggregates of scales and plates. Aggregates in the shape of shards pseudomorphous after volcanic glass are common; other inherited structures also occur. Vermicular crystals show the perfect (001) cleavage best.

Orientation. Biaxial $(-)$. α nearly normal to (001), thus $\alpha \wedge c$ is probably small, $\beta = b$, $\gamma \wedge a = $ small, probably generally $< 10°$. Optic plane parallel with (010) (Fig. 11.39). $2V = 5\text{--}30°$.

Occurrence. The chief clay mineral of bentonite, formed by the alteration of volcanic ash, occurring with unaltered glass, pyroclastic minerals, cristobalite, hydromica, kaolinite, zeolites, and pyrite. Also forms in soils of temperate or arid regions as hydrothermal veinlets and replacements in pegmatites, as a constituent of various hydrothermal ore deposits and their

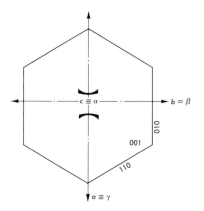

FIGURE 11.39   Orientation of montmorillonite, section parallel with (001).

altered wall rocks, and to a lesser extent as a mineral of some marine clays. Replaces a variety of other substances, especially volcanic glass, quartz, potash feldspar, and plagioclase, and is altered to kaolinite. Beidellite, whose occurrences are similar, also replaces ferromagnesian silicates.

Diagnostic features. Nontronite has higher refractive indices. Kaolinite has lower birefringence and higher indices. Montmorillonite characteristically expands or swells upon wetting.

NONTRONITE

Composition. $(Fe^{3+},Al,Mg)_2(Si,Al)_4O_{10}(OH)_2[Ca,Na]$. In most nontronites Fe^{3+} is in considerable excess over combined Al and Mg. $Si:Al = 3.8:0.2-3.3:0.7$. A series occurs between nontronite and montmorillonite-beidellite. Ca is normally $>$ Na.

Indices. $\alpha = 1.530-1.580$, $\beta = 1.555-1.612$, $\gamma = 1.560-1.615$, $\gamma - \alpha = 0.026-0.040$. Indices and birefringence increase with Fe^{3+} (Fig. 11.40). Air drying increases indices and birefringence somewhat. Indices also

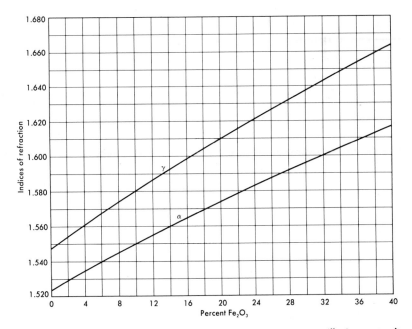

FIGURE 11.40 Variation of α and γ with composition in montmorillonite–nontronite series (Ross and Hendricks, 1945).

NONTRONITE

change with absorption of immersion oils, particularly liquids containing NH_2 groups. In thin section the maximum interference colors lie in second to third orders.

Color. Pleochroic in shades of green and yellow, with $\alpha < \beta = \gamma$:

α = yellow green

$\beta = \gamma$ = olive green, brownish green

Absorbs dyes and changes color.

Form. Monoclinic. Platy and fibrous in very small particles with matted to plumose texture, commonly showing structures inherited from other minerals. Also in veinlets. Perfect (001) cleavage.

Orientation. Biaxial ($-$). α nearly normal to (001), thus making a small angle with c, $\beta = b$, $\gamma \wedge \alpha =$ small; optic plane parallel with (010). 2V = small–66°, $r > v$ strong. Elongate sections or fibers are length-slow. Indistinct extinction because of the finely fibrous structure.

Occurrence. As an alteration product, either hydrothermal or less commonly supergene, in many iron-rich rocks, impure marbles, schists, gneisses, ferruginous quartzites and basalts, and may be pseudomorphous after iron minerals such as pyrite, hedenbergite, and olivine and associated with kaolinite, sericite, pyrite, and limonite.

Diagnostic features. Differs from montmorillonite in having higher indices, from kaolinite in higher birefringence.

HYDROMUSCOVITE

Composition. $(K,Na,H_3O)_2(Al,Mg,Fe^{3+})_4(Si_7,Al)O_{20}(OH)_4$ (Brown and Norrish, 1952). The oxonium ion (H_3O^{1+}) probably substitutes for K. Hydromica may be considered synonymous with hydromuscovite, but illite is a broader name including not only hydromuscovite but also brammalite (Na > K) and illitic minerals related to biotite.

Indices. $\alpha = 1.535$–1.570, $\beta = 1.555$–1.600, $\gamma = 1.565$–1.605, $\gamma - \alpha = 0.025$–0.037. Because of the minute character of the flakes, interference colors are usually in lower to middle second order. Indices of mixed-layer minerals vary with the proportions of montmorillonite to hydromuscovite.

Color. Colorless to neutral.

Form. Monoclinic. In minute shreds and flakes, in matted and plumose aggregates. May form subparallel intergrowths with kaolinite. Perfect (001) cleavage; parting normal to (001) may be present (Fig. 11.32).

Orientation. Biaxial ($-$). α is nearly coincident with c, $\beta \wedge a =$ small, $\gamma = b$. Optic plane normal to (010). 2V = small–25°, generally < 5° Interlayer mixtures are difficult to recognize on the basis of optical properties alone. They look like single minerals but commonly show fuzzy interference figures.

Occurrence. The common clay mineral of many transported clays and shales with kaolinite, quartz, feldspars, chloritic clay minerals, variable montmorillonite, limonite, calcite, and carbonaceous matter. In variable amounts in till and loess also with kaolinite and montmorillonite, and less commonly in some soils. Also of hydrothermal origin in ore deposits and particularly their altered wall rocks. An important constituent of some slates.

Diagnostic features. Differs from muscovite in having generally lower indices and smaller 2V. Has higher birefringence and higher indices than kaolinite.

CHAPTER TWELVE

BIAXIAL MINERALS: FRAMEWORK SILICATES

ZEOLITE GROUP

GENERAL The general composition of the common zeolites can be expressed by the formula

$$(Na,K)_m(Ca,Ba,Sr)_n(Al_{2n+m},Si_r)O_{2(2n+m+r)} \cdot H_2O$$

i.e., the sum of Al+Si $= \frac{1}{2}O$; $r > 2n+m$ and is variable. Some Mg and Mn also may be present. Although analcite is generally classed with the zeolites, because of its optical nature it has been described earlier. The group is large, embracing some 30 species or series under some half-dozen subgroups. The selection of the most common zeolites or those of greatest petrological significance is somewhat arbitrary. The following species are here described, and of these natrolite is perhaps most widely distributed:

Heulandite, $(Na,Ca)_5CaAl_6(Al,Si)_4Si_{26}O_{72} \cdot 24H_2O$
Stilbite, $(Na,Ca)_2Ca_4Al_{10}(Al,Si)_2Si_{24}O_{72} \cdot 28H_2O$
Chabazite, $(Na,K,Ca)_4Ca_3Al_{10}(Al,Si)_4Si_{26}O_{80} \cdot 40H_2O$
Thomsonite, $Na_3(Na,Ca)_4Ca_5Al_{17}(Al,Si)_4Si_{19}O_{80} \cdot 24H_2O$
Natrolite, $Na_{12}(Na,Ca)_4Al_{16}Si_{24}O_{80} \cdot 16H_2O$
Laumontite, $(Na,K)_{3-0}Ca_{2.5-4}Al_7(Al,Si)Si_{15}O_{48} \cdot 16H_2O$

Although water occupies definite structural positions, it is given off continuously owing to the open network, with preservation of the structural framework. Other liquids or gases may be inserted in place of water, e.g., ammonia, alcohol, or iodine. Zeolites also possess the property of base exchange, which permits the replacement of the alkali elements by each other (Na_2

or K_2 for Ca) and by such metals as Ag and Cu. The optical properties of zeolites not only vary with isomorphous substitution but also depend in part upon the degree of hydration, which may be incomplete. Although the properties vary considerably among the species, zeolites in general are characterized optically by low refractive indices and weak birefringence.

HEULANDITE

Composition. $(Na,Ca)_5CaAl_6(Al,Si)_4Si_{26}O_{72} \cdot 24H_2O$. Some K, Sr, and Ba may replace Na.

Indices. $\alpha = 1.487-1.499$, $\beta = 1.487-1.501$, $\gamma = 1.488-1.505$, $\gamma - \alpha = 0.001-0.007$. First-order white is the maximum thin-section interference color. Indices increase with Ca over Na and in types rich in Si. The latter have lower birefringence.

Color. Colorless.

Form. Monoclinic. Tabular (010) to nearly equant crystals, of orthorhombic appearance. Also in coarse-grained to fine-grained anhedral masses. Perfect (010) cleavage, upon which crushed material will lie. Such sections show extremely weak birefringence, for $\beta - \alpha = $ ca. 0.001.

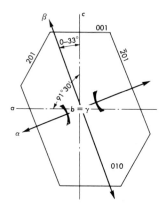

FIGURE 12.1 Orientation of heulandite, section parallel with (010).

Orientation. Biaxial (+). $\alpha \wedge a = 0\text{--}35°$, $\gamma = b$, $\beta \wedge c = 0\text{--}33°$. Optic plane normal to (010) (Fig. 12.1). $2V = 0\text{--}50°$, commonly ca. $30°$, $r > v$. Cleavage pieces are length-fast, yield a Bxo figure, and have inclined extinction. In sections showing cleavage traces, these traces extinguish parallel with the fast ray. Upon heating there is a rotation of the optic plane, with first a decrease and then an increase in 2V.

Occurrence. A low-intensity hydrothermal mineral in cavities and veins in basalt and in some schists and gneisses. Commonly accompanied by stilbite and chabazite, as well as by other zeolites. Also a low-grade replacement of glass in vitric tuffs. In volcanic graywackes, lightly altered (zeolite facies).

Diagnostic features. Differs from stilbite in having a (+) sign, better cleavage, and a different orientation.

STILBITE

Composition. $(Na,Ca)_2Ca_4Al_{10}(Al,Si)_2Si_{24}O_{72} \cdot 28H_2O$. Some K may proxy for Na.

Indices. $\alpha = 1.482\text{--}1.500$, $\beta = 1.491\text{--}1.504$, $\gamma = 1.493\text{--}1.508$, $\gamma - \alpha = 0.006\text{--}0.013$. Ca for Na increases the indices. Maximum interference colors in thin section are in first order.

Color. Colorless.

Form. Monoclinic. Uniformly twinned on (001) to composite, cruciform, pseudorhombohedral groups. Usually thinly tabular in subparallel sheafs of crystals, also radially arranged, and spherulitic. Cleavage (010) good.

Orientation. Biaxial (−). $\alpha \wedge a = 3\text{--}12°$, $\beta = b$, $\gamma \wedge c = 26\text{--}35°$. Optic

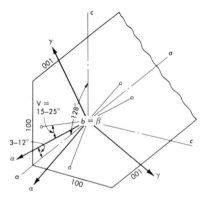

FIGURE 12.2 Orientation of stilbite, section parallel with (010), crystal twinned on (001).

plane is parallel with (010) (Fig. 12.2). $2V = 30{-}50°$, $r < v$. The extinction commonly is wavy.

Occurrence. In cavities and fractures of basalts and other rocks, including gabbro, granite, and gneisses, as a low-temperature hydrothermal mineral. Also a hot springs deposit and rarely as a cementing mineral in conglomerate.

Diagnostic features. The twinned form is distinctive. Heulandite is ($+$) and has a different orientation.

CHABAZITE

Composition. $(Na,K,Ca)_4Ca_3Al_{10}(Al,Si)_4Si_{26}O_{80} \cdot 40H_2O$. May be alkali-free, or much Na or K or both may be present.

Indices. $\alpha = 1.478{-}1.485$, $\gamma = 1.480{-}1.490$, $\gamma - \alpha = 0.002{-}0.010$. Maximum thin-section interference colors are first-order gray to white. Indices rise with Ca.

Color. Colorless.

Form. Probably monoclinic, pseudorhombohedral. Euhedral crystals may be cuboid in shape; also anhedral-granular. Imperfect pseudorhombohedral ($\cong 10\bar{1}1$) cleavage.

Orientation. Usually biaxial and usually ($+$), but may be ($-$). Basal sections consist of six sectors with different optical orientation. Some crystals are uniaxial, ($-$) or ($+$), and others are uniaxial in part, biaxial in part. $2V = 0{-}32°$. Extinction is symmetrical. The optical anomalies have been ascribed to Na–Ca variations, but may reflect degree of hydration.

Occurrence. In cavities and fractures in mafic volcanic rocks, schists, gneisses, and marbles. Also less commonly in hot springs deposits. Associated with stilbite, laumontite, other zeolites, calcite, and prehnite.

Diagnostic features. Has slightly higher birefringence than analcite. Gmelinite has lower refractive indices. The variable 2V and sign are helpful in a determination.

NATROLITE

Composition. $Na_{12}(Na,Ca)_4Al_{16}Si_{24}O_{80} \cdot 16H_2O$. A small amount of K may be present, and Ca may be entirely absent.

Indices. $\alpha = 1.473{-}1.489$, $\beta = 1.476{-}1.491$, $\gamma = 1.485{-}1.502$, $\gamma - \alpha = 0.011{-}0.014$. Yellow of the first order is the maximum thin-section interference tint.

Color. Colorless.

Form. Orthorhombic. Typically in elongated prisms and fibers with parallel or radial arrangement (Fig. 12.3). Cross sections show a cuboid outline. Fine-grained material replacing nepheline (hydronephelite) is scaly or fibrous. Cleavage (110) perfect, also (010) parting.

Orientation. Biaxial (+). $\alpha = a$, $\beta = b$, $\gamma = c$. Optic plane is (010) (Fig. 12.3). 2V = 60–63°, $r < v$ weak. In fine-grained material a figure may be difficult to secure. Length-slow with parallel extinction in elongate sections, symmetrical in cross sections. Cleavage pieces yield a flash figure. Some material may appear to be nearly uniaxial, an aggregate effect from subparallel fibers.

Occurrence. With calcite and other zeolites in cavities and veinlets in basalt, diabase, and serpentinite. Widespread as an alteration of nepheline, sodalite, analcite, leucite, and plagioclase in foidal rocks.

Diagnostic features. Differs from scolecite in being length-slow with parallel extinction; from thomsonite in being invariably length-slow.

THOMSONITE

Composition. $Na_3(Na,Ca)_4Ca_5Al_{17}(Al,Si)_4Si_{19}O_{80} \cdot 24H_2O$. Some may contain essentially no Na, other varieties carry appreciable K.

Indices. $\alpha = 1.511$–1.530, $\beta = 1.513$–1.532, $\gamma = 1.516$–1.545, $\gamma - \alpha = 0.006$–0.028. Indices increase in a general way with increase in Ca for Na, also with increasing H_2O (Fig. 12.6). Lower indices than the above range are reported for partly dehydrated types. The birefringence increases

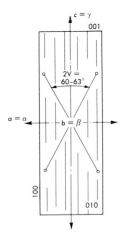

FIGURE 12.3 Orientation of natrolite, section parallel with (010).

FIGURE 12.4 Radial cluster of natrolite, tinguaite porphyry, Luster Peak Dome, Judith Mountains, Montana. Polars crossed, ×19 (S. B. Wallace).

markedly with Ca and slightly with H_2O. Maximum interference colors in thin section vary from white to blue of the first order.

Color. Colorless.

Form. Orthorhombic. Rarely euhedral; fibrous to columnar in parallel to radial arrangement. Perfect (010) and poor (100) cleavages. Crushed fragments usually lie on (010).

Orientation. Biaxial (+). $\alpha = a$, $\beta = c$, $\gamma = b$. Optic plane is (001) (Fig. 12.5). $2V = 44$–$75°$, $r > v$ distinct to strong. 2V decreases with increasing Ca. Cleavage pieces yield a centered Bxa figure with the optic plane normal to the elongation. Parallel extinction and either length-slow or length-fast (elongation parallel with β).

Occurrence. In cavities and cracks in basalts, diabases, phonolites, andes-

THOMSONITE

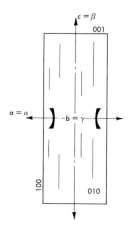

FIGURE 12.5 Orientation of thomsonite, section parallel with (010).

ites, schists, and contact rocks, usually with other zeolites. As authigenic cement in some sandstones.

Diagnostic features. Like cancrinite in indices and birefringence, but is biaxial. Has the strongest birefringence of the common zeolites and is the only common one in which the fiber orientation is either length-slow or -fast. Mesolite, which in this respect is similar, shows a small extinction angle and a larger 2V.

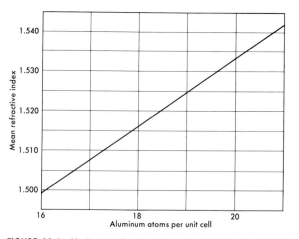

FIGURE 12.6 Variation of mean index of refraction, $(\alpha + \beta + \gamma)/3$, with composition in thomsonite (Hey, 1932).

Composition. $(Na,K)_{3-0}Ca_{2.5-4}Al_7(Al,Si)Si_{15}O_{48}\cdot16H_2O$. Some Fe^{3+} for Al. Varieties rich in Be and V have been reported. Loses water easily.

Indices. $\alpha = 1.502-1.519$, $\beta = 1.512-1.525$, $\gamma = {}^1.513-1.526$, $\gamma - \alpha = 0.008-0.016$. The indices appear to decrease with increasing $(Na+K)+Si$, but the effect may be obscured by dehydration, which also decreases the indices. In thin section interference colors maxima are in first and second orders.

Color. Colorless.

Form. Monoclinic. Euhedral crystals are of stout prismatic habit, commonly in masses of interlocking grains and in bladed interlacing aggregates, also with parallel or radial arrangement. Perfect (110) and (010) cleavages, which govern the orientation of crushed grains.

Orientation. Biaxial $(-)$, $\alpha \wedge a = 30-62°$, $\beta = b$, $\gamma \wedge c = -8--40°$. Optic plane is (010) (Fig. 12.7). $2V = 25-47°$, $r < v$ marked to extreme. With dehydration the mineral changes rapidly to leonhardite, which has lower indices, usually a smaller 2V and a larger extinction angle, $\gamma \wedge c$. This change may be produced by air drying, mild heating, thin-section preparation, and immersion in dehydrating agents, such as glycerol. Water immersion restores laumontite properties. Larger grains may show irregular extinction, owing to local H_2O variation. Cleavage pieces on (010) yield a flash figure; those on (110), an eccentric figure.

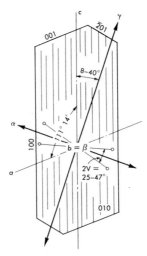

FIGURE 12.7 Orientation of laumontite, section parallel with (010).

Occurrence. As euhedra in cavities and veins in basalts, andesites, serpentinites, gabbro, granite, marble, gneisses, and schists. Also as a large-scale replacement mineral in vitric tuffs, spilites, graywackes and feldspathic sandstone, and conglomerate, forming cement or replacing plagioclase and glass, with associated prehnite, calcite, sericite, pumpellyite, and epidote.

Diagnostic features. The combination of three directions of cleavage and alteration to leonhardite is distinctive. Has higher birefringence than phillipsite.

ALKALI FELDSPAR GROUP

GENERAL Under alkali feldspars the main types recognized are:

1. Sanidine, $(K,Na)AlSi_3O_8$, monoclinic
   a. High sanidine
   b. Sanidine
2. Orthoclase, $(K,Na)AlSi_3O_8$, monoclinic
3. Microcline, $(K,Na)AlSi_3O_8$, triclinic
4. Anorthoclase, $(Na,K)AlSi_3O_8$, triclinic
5. Adularia, $KAlSi_3O_8$, monoclinic and triclinic

In addition, two Ba–feldspars may be noted: celsian, which is structurally related to sanidine, and hyalophane, structurally related to orthoclase or monoclinic adularia. All members of the group are characterized by the cleavages (001) perfect and (010) good, making an angle of nearly 90°.

The indices, except for some Na-rich types of anorthoclase and those of the Ba-rich or Fe^{3+}-rich feldspars, are below that of Canada balsam and those of quartz. The birefringence is uniformly low, ca. 0.007, resulting in first-order interference colors in thin section. Indices generally rise with Na for K, but the optical properties also are in part a function of the thermal history. Small amounts of Rb, Pb, and rare earths also may be present. Crystals are elongated usually parallel with *a,* or less commonly with *c,* resulting in nearly rectangular or square outlines for euhedral crystals in thin sections. Except for some high sanidine, the optic planes are parallel or nearly parallel (in triclinic types) with (010). Again, except sanidine, 2V is generally large. The sign normally is $(-)$ for all types, but very rare $(+)$ varieties have been reported: isosanidine, isoorthoclase, and isomicrocline. Zoning may be present in sanidine (Fig. 12.10), but is rare in the other members. The common types of twinning are:

TYPE	COMPOSITION PLANE	SPECIES
Carlsbad	(010)	orthoclase (common) sanidine (common)
Manebach	(001)	orthoclase (uncommon) sanidine (uncommon)
Baveno	(021)	orthoclase (uncommon) sanidine (uncommon)
combined albite and pericline	(010) rhombic section	microcline, anorthoclase, triclinic adularia
combination of albite + pericline, and Carlsbad		microcline, anorthoclase

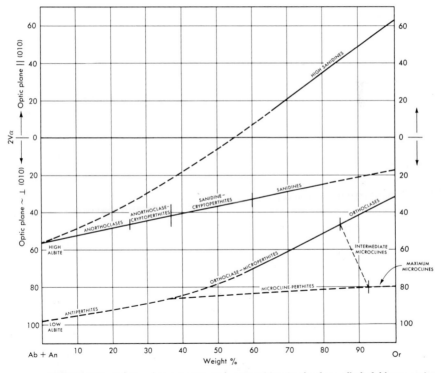

FIGURE 12.8 Relation between 2V and composition in the four alkali feldspar series (Tuttle, 1952a; MacKenzie and Smith, 1956).

The twinning of microcline, anorthoclase, and triclinic adularia is on a small and uniform scale and results in the crosshatch (quadrille, gridiron) structure under crossed polars (Fig. 12.16).

Regular to semiregular intergrowths with sodic plagioclase, called perthites, consist of single-crystal hosts of K–Na–feldspar and thin plates, spindles, blebs, or crystals of guest albite or oligoclase, oriented or partly oriented according to the host structure (Fig. 12.14). The scale of the intergrowths varies from types undetectable microscopically to megascopically visible varieties. Submicroscopic perthitic albite may give rise to optical anomalies in the microscopically (apparently) homogeneous feldspar. Most perthites are the result of exsolution of Na upon cooling, but others, less regular and with more abundant albite, are formed by replacement:

TYPE	HOST	GUEST	ORIGIN
x-ray perthite	sanidine	high albite	exsolution
cryptoperthite	sanidine anorthoclase	albite	exsolution
microperthite and perthite	microcline orthoclase anorthoclase	albite and oligoclase (rare)	exsolution and replacement

Alkali feldspars combined with members of the plagioclase series are the most abundant igneous minerals, being essential constituents of all igneous rocks, save peridotites and some rare feldspathoidal types. The presence or absence of various feldspars is one of the main bases for igneous rock classification.

Although the above division into the species listed is traditional and based primarily on optical properties, this separation is somewhat arbitrary on the basis of x-ray structural data. Goldsmith and Laves (1954) have demonstrated that there exists a structural series between sanidine (monoclinic) and microcline (with maximum triclinicity). This continuous range of triclinicity is believed to be a function of Al–Si order-disorder. Microcline with maximum triclinicity is considered to be ordered with respect to Al–Si, whereas sanidine is considered to be disordered. This relationship is similar to that between cordierite and indialite (p. 177). Most microclines contain the characteristic crosshatch, albite–pericline twinning resulting from inversion from an originally monoclinic crystal. K–feldspars are to be constructed of triclinic units of varying size, as shown by twinning units ranging from macro- to submicroscopic in size. Thus, according to Goldsmith and Laves (1954), K–feldspars may be variable continuously in two directions: (1) in degree of triclinicity and (2) in size of the units, which may show both variable size and triclinicity within a "single" crystal. It is suggested that as these triclinic units become very small the material may become optically monoclinic and that the optically monoclinic character of orthoclase may be the result of this diminution, in contrast to truly monoclinic sanidine.

SANIDINE

Composition. $(K,Na)AlSi_3O_8$.[1] Ba may also replace K to a considerable extent; minor Ca and Rb for K; minor Fe^{3+} for Al. Celsian has the ideal composition $BaAl(Si_2Al)_3O_8$, usually with some K and rarely with some Ca. There does not appear to be a complete series between barian sanidine and celsian, although some intermediate types have been reported.

Indices. $\alpha = 1.518-1.525$, $\beta = 1.523-1.530$, $\gamma = 1.525-1.531$, $\gamma - \alpha = 0.005-0.008$. In thin section maximum interference tints are in the first order. Indices increase with Na (and Ca) and also markedly with Ba (Fig. 12.20). Barian sanidines have $\alpha = 1.525-1.536$, $\beta = 1.531-1.542$, $\gamma = 1.533-1.546$, $\gamma - \alpha = 0.008-0.010$. Celsian has $\alpha = 1.580-1.584$, $\beta = 1.585-1.587$, $\gamma = 1.594-1.596$, $\gamma - \alpha = 0.010-0.012$.

Color. Colorless and commonly clear.

Form. Monoclinic. Typically in euhedral phenocrysts (Fig. 12.10), tabular parallel with (010) or somewhat elongated parallel with a. Sections are six-sided or square in outline. Also as subhedral microlites. Phenocrysts may show considerable resorption and contain abundant inclusions of glass, aegirine (may be oriented), sodic plagioclase, and calcite. Perfect (001) and good (010) cleavages, on which crushed material is oriented. Celsian may be triclinic, pseudomonoclinic.

Orientation. Biaxial $(-)$, very rarely $(+)$ in isosanidine. Two orientations occur (Fig. 12.9).

1. Optic plane parallel with (010). $\alpha \wedge a = 5-11°$ (as much as 17° reported) $\beta = b$, $\gamma \wedge c = 15-21°$, $2V = 60-0°$, $r < v$ distinct.
2. Optic plane normal to (010). $\alpha \wedge a = 5-8°$, $\beta \wedge c = 18-21°$, $\gamma = b$, $2V = 0-25°$, $r > v$.

The optical orientation and in part the Na content of sanidine are functions of its thermal history. With higher temperature of crystallization the optic plane is parallel with (010), and at lower crystallization temperatures

[1] Synthetic alkali feldspars are all part of the series high sanidine–high albite, of which only the potassian end member occurs naturally ($> Or_{67}$).

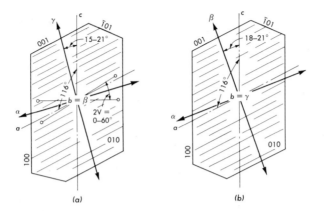

FIGURE 12.9 Orientation of sanidine, sections parallel with (010). (*a*) Optic plane parallel with (010). (*b*) Optic plane normal to (010).

it is normal to (010). Moreover, higher-temperature sanidines can contain more Na than lower-temperature types, provided exsolution of albite has not taken place, which is controlled by the rate of cooling. Different crystals of sanidine in the same rock may show considerable variation in 2V.

Although much sanidine is microscopically apparently homogeneous, x-ray studies have shown that many such crystals contain submicroscopic perthite, and some crystals may show iridescence owing to the presence of the very thin albite lamellae. The following sequence is recognized (Tuttle, 1952):

1. Homogeneous sanidine.
2. Sub-x-ray perthite. Recognized by comparing analyzed composition with that determined by x-rays.
3. x-ray perthite.
4. Cryptoperthite.

Most sanidines are x-ray-perthitic or cryptoperthitic. The optically monoclinic character of these intergrowths of a monoclinic host and a triclinic guest is ascribed to submicroscopic twinning of the albite.

Cleavage traces are nearly parallel with the fast ray. Zoning may occur, being distinguished by small index or birefringence variations (Fig. 12.10). In some crystals zoning is ascribed to Na variations. Zoning may be oscillatory. In others Ba-rich cores have Ba-poor mantles. Matrix sanidine is unzoned. Sanidine may be untwinned or have Carlsbad twinning.

CLEAVAGE	ORIENTATION	FIGURE	INDICES
(001)	optic plane ‖ (010)	Bxo	α,β
(001)	optic plane ⊥ (010)	near flash	α,γ
(010)	optic plane ‖ (010)	flash	α,γ
(010)	optic plane ⊥ (010)	Bxo	α,β

For celsian: optic plane parallel (010), $\alpha \wedge c = -3°$, $\beta = b$, $\gamma \wedge a = 28°$, (+) $2V = 86$–$90°$. $\wedge$ calcian celsian is (−) with $2V = 76°$. Some celsian may be triclinic.

Occurrence. As phenocrysts, matrix crystals, and microlites in extrusive (flow and tuff) and hypabyssal felsic rocks including rhyolites, quartz latites, phonolites, and trachytes. Rarely xenocrystic in andesites and basalts. In rhyolites and obsidians in spherulites as radiating fibers intergrown with cristobalite. Resorbed phenocrysts are known with a rapakivi-like overgrowth of sodic plagioclase. Also of metasomatic origin in sanidinites. Alters to kaolinite or sericite with by-product quartz possible, although more resistant to alteration than orthoclase.

Diagnostic features. From orthoclase it differs in optical orientation, pos-

FIGURE 12.10 Sanidine phenocryst, zoned and with Carlsbad twinning, in rhyolite, Apati, Hungary. Polars crossed, ×26.

sibly smaller 2V, and restricted occurrence. Microcline and plagioclase characteristically display distinctly different types of twinning.

ORTHOCLASE

Composition. $(K,Na)AlSi_3O_8$ with minor Fe^{3+}, Ba, and Ca. Ba-rich types are called hyalophane.

Indices. $\alpha = 1.518$–1.526, $\beta = 1.523$–1.530, $\gamma = 1.524$–1.533, $\gamma - \alpha = 0.005$–0.008, with maximum first-order interference colors in thin section. Indices and birefringence increase generally with Na (Fig. 12.20); also with Fe^{3+} and Ba. Hyalophane has $\alpha = 1.527$–1.539, $\beta = 1.530$–1.543, $\gamma = 1.532$–1.547, $\gamma - \alpha = 0.005$–0.008.

Color. Colorless but may be grayish because of kaolinization.

Form. Monoclinic (see p. 336). The form varies considerably from euhedral to anhedral. Euhedral phenocrysts show six-sided, square, rectangular, or diamond-shaped sections. Crystals are elongate parallel with a. In equant anhedral pieces, interstitial grains, and radiating fibers as in spherulites. Branching skeletal forms also are known. Detrital grains are commonly flattened parallel with (001) with irregular boundaries, but rectangular outlines also are encountered. Perfect (001) and good (010) cleavages. Most grains of crushed material lie on (001). Inclusions are common: particularly albite, quartz, hematite, biotite, and muscovite; also feldspathoids. Inclusions may be oriented or zonally arranged.

Orientation. Biaxial $(-)$. $\alpha \wedge a = 5$–$12°$, $\beta \wedge c = 20$–$13°$, $\gamma = b$, optic plane normal to (010) (Fig. 12.11). $2V = 35$–$85°$, $r > v$. The extinction

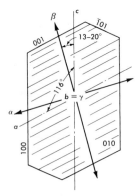

FIGURE 12.11 Orientation of orthoclase, section parallel with (010).

angle increases with Na (Fig. 12.21). Isoorthoclase, ($+$) with $2V = $ small–86°, is very rare. Fe^{3+} for Al lowers $2V$ about 5° for each 0.25% Fe_2O_3, and with more than about 1.8% Fe_2O_3 the optic plane becomes parallel with (010). Hyalophane is ($-$), $2V = 74$–78°, $\alpha \wedge a = 2$––25°, the angle apparently increasing with Ba content. Zoned orthoclase is not common, but oscillatory zoning with $2V$ between 59–78° and slight extinction angle variation is known. Carlsbad twinning with two individuals is common; Manebach and Baveno types are less common. Cleavage traces are parallel or nearly parallel with the fast ray. (001) cleavage pieces yield a near-flash figure; (010) cleavage pieces give a Bxo figure.

Occurrence. In plutonic igneous rocks including granites, quartz monzonites, granodiorites, syenites, monzonites, and to a lesser extent some granitic pegmatites, tonalites, diorites, and a few rare gabbros. Also in various lamprophyric dike rocks. In micaceous gneisses, quartzites, schists, granulites, and contact-metamorphic rocks; in sandstones, arkoses, and some graywackes and shales. Intergrowths with quartz and sodic plagioclase are common:

Micropegmatite (micrographic)	Rods to irregular blebs of quartz apparently unconnected, but commonly showing simultaneous extinction in single-crystal hosts of orthoclase.
Granophyre	Plumose, radially fibrous, and vermicular quartz individuals in single-crystal hosts of orthoclase (Fig. 12.12).
Armoring	Shell of oriented orthoclase on core of sodic plagioclase.
Rapakivi	Rim of unoriented granular sodic plagioclase around orthoclase phenocrysts.
Perthite	Plates, lenses, veins, irregular patches and single crystals of albite, oriented to unoriented in single-crystal orthoclase hosts. The size, shape, and abundance of the guest albite vary considerably, leading to various classifications: types based on size of the albite are cryptoperthite, microperthite, perthite; on shape, braid, string, film, vein or band, patch, bleb, and platy; on genesis, exsolution, replacement-deuteric or hydrothermal, and composite.
Antiperthite	Blebs of orthoclase in single-crystal plagioclase hosts.

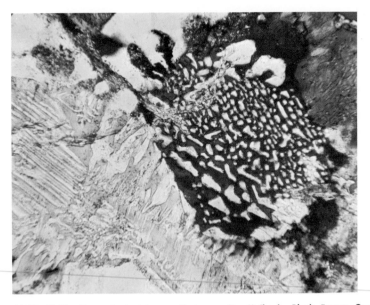

FIGURE 12.12 Granophyric intergrowth in granite, Haibach, Black Forest, Germany. Polars crossed, ×94.

Orthoclase alters readily to sericite and kaolinite and is replaced by many minerals including chlorite, quartz, calcite, glauconite, tourmaline, and albite. Hyalophane occurs in mineralized zones, veins, and replacement ore bodies. It may also be perthitic.

Diagnostic features. Sanidine has either a smaller 2V or a different optical orientation. Microcline normally shows the characteristic gridiron twinning pattern, but if untwinned, the extinction angle on (001) = 15° separates it from orthoclase. Orthoclase has indices below quartz and albite; nepheline is uniaxial.

MICROCLINE

Composition. (K,Na)AlSi$_3$O$_8$. Minor Ca. Some pegmatitic microcline contains considerable Rb. Na is normally less than in orthoclase.

Indices. $\alpha = 1.517$–1.522, $\beta = 1.522$–1.526, $\gamma = 1.524$–1.530, $\gamma - \alpha = 0.007$. Maximum thin-section interference colors fall in the first order. Indices increase with Na (Fig. 12.20) and with Rb.

Color. Colorless; cloudy owing to alteration.

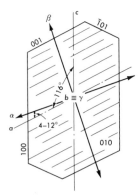

FIGURE 12.13 Orientation of microcline, section parallel with (010).

Form. Triclinic, pseudomonoclinic (see p. 336). Commonly anhedral, but also as subhedral to euhedral phenocrysts and metacrysts. Perfect (001) and good (010) cleavages, which govern orientation of crushed material. Detrital grains are platy (001); (010) plates are less common; irregular outlines characterize both.

Orientation. Biaxial ($-$). Very rarely ($+$) in isomicrocline. The axial plane and γ are nearly normal to (010) (Fig. 12.13). The angle on (010) between the (001):(010) edge and the trace of the optic plane = 3–7°. The extinction angle on (001) = 15–20°. 2V = 77–84°, $r > v$. Good figures are difficult to obtain on account of the twinning. Most microcline shows multiple fine-scale twinning of combined albite–pericline type (Fig. 12.16) (gridiron, scotch plaid, quadrille, crosshatch structure), which rarely may be entirely absent. Zonal growths with albite occur in some sodic rocks: microcline core, albite rim; albite core, microcline rim; albite core, microcline intermediate layer, albite rim. In some sediments detrital cores of microcline have adularia rims.

Occurrence. Not found in extrusive igneous rocks. Common in some granites, particularly muscovite granites; the common potash feldspar of granitic pegmatites; much less common in some syenites. Abundant in feldspathic micaceous gneisses and schists. Microcline, like orthoclase, forms graphic intergrowths with quartz and a variety of perthites with sodic plagioclase (Fig. 12.14). Much microcline is perthitic. Widespread as a detrital species in sandstone and arkose.

Diagnostic features. The gridiron twinning is highly distinctive. If it is absent, microcline can be distinguished from orthoclase by the large basal extinction angle.

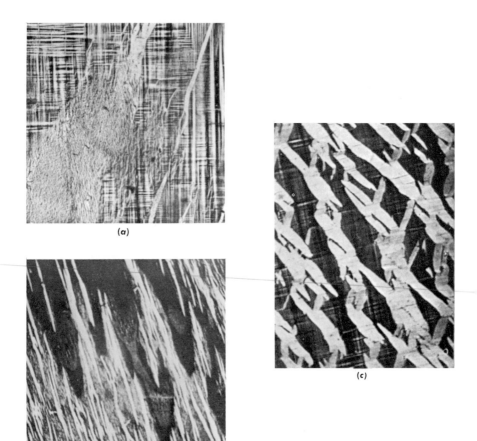

(a)

(c)

(b)

FIGURE 12.14 (a) Microcline perthite (vein, string, film), Stenne, Sannidal, Norway. Polars crossed, ×52 (Andersen, 1928). (b) Microcline perthite (vein), Stenne, Sannidal, Norway. Polars crossed, ×18 (Andersen, 1928). (c) Microcline perthite (braid), Tory Hill, Ontario. Polars crossed, ×66 (Goldich and Kinser, 1939).

ANORTHOCLASE

Composition. $(Na,K)AlSi_3O_8$. Minor Ca. Na is normally in considerable excess over K. Also called soda–microcline.

Anorthoclase is a feldspar that has been variously and imperfectly defined. MacKenzie and Smith (1956, p. 407), use the term to denote a

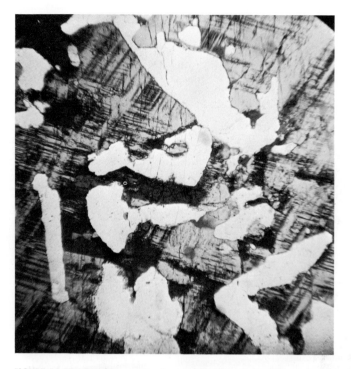

FIGURE 12.15 Graphic granite from pegmatite, Eight Mile Park, Colorado. Polars crossed, ×28.

high-temperature alkali feldspar, more sodium-rich than $Or_{37}(Ab-An)_{63}$, which when heated inverts from triclinic to monoclinic symmetry and reinverts to the triclinic form even on rapid quenching. In effect by this definition, anorthoclase forms a series with low sanidine (Fig. 12.20). Optically anorthoclase may be considered to form a series with microcline.

Indices. $\alpha = 1.519$–1.529, $\beta = 1.524$–1.534, $\gamma = 1.527$–1.536, $\gamma - \alpha = 0.005$–0.008. In thin section first-order gray whites are the maximum interference colors.

Color. Colorless.

Form. Triclinic, pseudomonoclinic. Euhedral phenocrysts to anhedral grains, also matrix microlites. Cleavages: (001) perfect, (010) good. Crushed material tends to lie mainly on (001). Detrital pieces resemble those of microcline. May be intergrown variously with albite (Fig. 12.18).

Orientation. Biaxial $(-)$. Optic plane is nearly normal to (010) (Fig. 12.17). Extinction on (001) = 1–6°, on (010) 4–12°. 2V = 34–60°, $r > v$. Gridiron twinning typical of microcline commonly is present, but

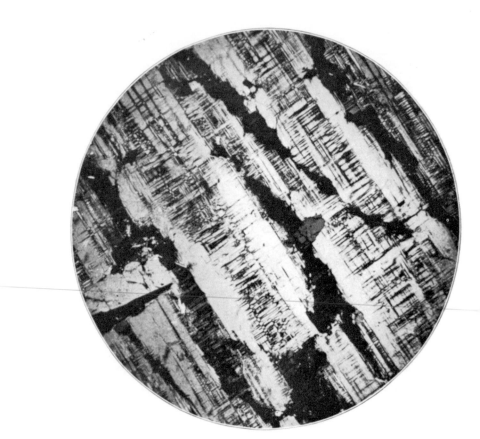

FIGURE 12.16 Microcline microperthite from pegmatite, Georgetown, Maine. Polars crossed, ×40 (Bastin, 1911).

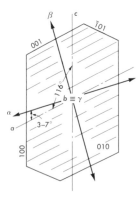

FIGURE 12.17 Orientation of anorthoclase, section parallel with (010).

BIAXIAL MINERALS: FRAMEWORK SILICATES

FIGURE 12.18 Overgrowths of albite on anorthoclase on albite, in nepheline syenite, Norra Kärr, Sweden. Polars crossed, ×35 (Adamson, 1944).

the lamellae may be very fine, and the result is an uneven extinction (moiré effect). Highest magnification may be necessary to resolve the twinning. Carlsbad twinning also may be present.

Occurrence. In sodic igneous rocks, both extrusive and intrusive, including soda trachytes, alkali rhyolites, nepheline syenites, alkali syenites, and some of their hypabyssal equivalents. Both as phenocrysts and as a matrix mineral. Rare as a detrital species. May be intergrown with sanidine. Uncommonly forms rim (armoring) on plagioclase. The "anorthoclase" of rhomb porphyries and allied types has now been shown to be plagioclase.

Diagnostic features. From microcline it differs in the finer twinning, indices, lower 2V, occurrence, and smaller basal extinction angle. Orthoclase lacks the gridiron twinning, as does sanidine.

ADULARIA

Composition. $KAlSi_3O_8$. Na very minor.

Indices. $\alpha = 1.518$–1.520, $\beta = 1.522$–1.524, $\gamma = 1.524$–1.526, $\gamma - \alpha = 0.006$–0.007. First-order maximum interference tints in thin section.

Color. Colorless.

Form. Monoclinic. Euhedral with a pseudoorthorhombic habit, with a rhombic cross section, owing to dominant development of (110) and $(1\bar{1}0)$; (010) is absent or narrow.

Orientation. Biaxial $(-)$. Adularia has commonly been considered as a habit modification, low-temperature variety of orthoclase. Many adularia crystals are a structural complex of a monoclinic and a triclinic feldspar, the latter commonly characterized by twinning of the microcline type and by the distinctive basal extinction angles. Some crystals are optically all monoclinic; others completely triclinic. Crystals are not uncommonly structurally zoned, and some show two pairs of alternatingly oriented

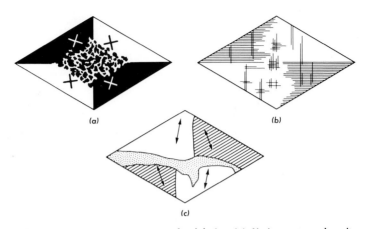

FIGURE 12.19 (a)–(c). Structures of adularia. (a) Native copper deposits, northern Michigan (Klein, 1939). (b) Alpine vein, St. Gotthard, Switzerland (Mallard, 1876). (c) Schwartzenstein, Austria (Köhler, 1948).

sectors around a core (Fig. 12.19). Triclinic adularia is now considered as an intermediate microcline with a considerable departure from maximum triclinicity (Goldsmith and Laves, 1954). Zoning with a core rich in hematite inclusions and an inclusion-free margin also is known. The orientation of monoclinic adularia is that of orthoclase, but the 2V may be small, like that of sanidine, or large, like that of orthoclase. As adularia becomes triclinic, the optic plane departs from parallel with (010) and trends toward normal to (010).

Occurrence. A low-temperature hydrothermal feldspar of ore deposits, both veins and replacements. Found in native copper ores, bonanza gold-silver ores, and in Alpine cleft veins. Also as authigenic overgrowths on detrital microcline in clastic rocks.

Diagnostic features. The combination of optical complexity and occurrence is distinctive.

PLAGIOCLASE SERIES

GENERAL The composition of the plagioclase series is conventionally expressed in molecular percentages of the two end members: albite, $NaAlSi_3O_8$, and anorthite, $CaAl_2Si_2O_8$. Generally adopted subdivisions are:

Albite	$Ab_{100}–Ab_{90}$
Oligoclase	$Ab_{90}–Ab_{70}$
Andesine	$Ab_{70}–Ab_{50}$

Labradorite	$Ab_{50}-Ab_{30}$
Bytownite	$Ab_{30}-Ab_{10}$
Anorthite	$Ab_{10}-Ab_0$

Many plagioclases also contain small amounts of K.

The plagioclases are triclinic (pseudomonoclinic) with two structural modifications: high-temperature ($Ab_{100}-Ab_{10}$) and low-temperature types ($Ab_{100}-Ab_0$), whose different optical properties are the result of their different thermal histories. High-temperature plagioclase occurs mainly as phenocrysts in extrusive rocks; low-temperature plagioclase in plutonic

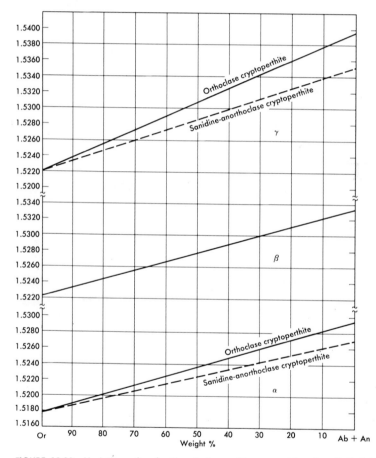

FIGURE 12.20 Variation of refractive indices with composition in alkali feldspars (Tuttle, 1952a).

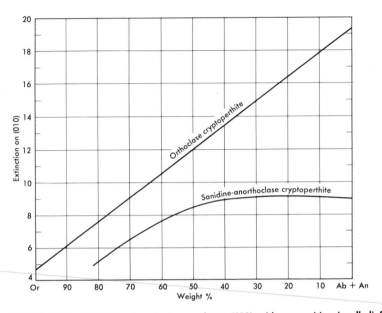

FIGURE 12.21 Variation of extinction angle on (010) with composition in alkali feldspars (Tuttle, 1952a).

igneous and metamorphic rocks. The low-temperature modifications can be changed to the high by heating to temperatures near those of the beginning of melting. Very rarely, both modifications occur in a single grain, and then usually the high-temperature type is an outer zone. Many natural plagioclases are structurally intermediate between the "maximum" high- and low-temperature states. "Maximum" high-temperature plagioclases appear to be rare in nature.

The optical changes that accompany the structural change from low- to high-temperature plagioclase have been summarized by J. R. Smith (1958):

1. The change in α in the range An_{0-20} is slight; in the range An_{20-100}, negligible (Fig. 12.33).
2. β and γ change measurably in the range An_{0-20}, negligibly for An_{20-100} (Fig. 12.33).

Therefore, measurements of the principal refractive indices can yield a reliable estimate of the An content of a plagioclase regardless of its structural state.

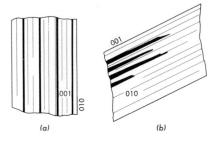

FIGURE 12.22 Cleavage fragments of plagioclase feldspars. Polars crossed. (a) (001) cleavage showing (010) cleavage and albite twinning. (b) (010) cleavage showing (001) cleavage and pericline twinning.

3. In the ranges An_{0-40} and An_{60-90} changes in 2V are such that they may be used to distinguish low- from high-temperature plagioclases of known composition (Fig. 12.32).

Plagioclases have (001) and (010) cleavages, perfect and very good. Crystals commonly are tabular parallel with (010) and may be elongate parallel with a, rarely with c. Crushed material tends to lie mainly on (001) but also on (010).

Twinning is nearly always present. Albite twinning is characteristic and is rarely absent. The more common twinning types are:

1. Albite, (010) = composition plane (Fig. 12.23)
2. Albite + Carlsbad
3. Pericline, rhombic section = composition face (Fig. 12.24)
4. Albite + Carlsbad + pericline (Fig. 12.26)

Rarer types include Carlsbad alone, Manebach, Baveno, ala, albite–ala, and acline.

In albite twinning the lamellae are narrowest and most numerous in oligoclase, and the width increases while the number of lamellae decrease both from oligoclase to albite and from oligoclase to anorthite. In the more calcic range combinations of Carlsbad and albite are more common and may be well developed. The twinning of plagioclase in metamorphic rocks is usually simple, with the albite type predominating and Carlsbad subordinate; complex twin combinations are absent, and in some schists and hornfelses plagioclase may be untwinned. J. V. Smith (1958) has shown that the obliquity (ϕ), i.e., the angular misfit of the twin components, which governs the ease (and thus frequency) of twinning, varies not only with the An content and the structural state of the plagioclase but

FIGURE 12.23   Albite twinning and normal zoning in labradorite in gabbro, Cuyamaca area, California. Polars crossed, ×55.

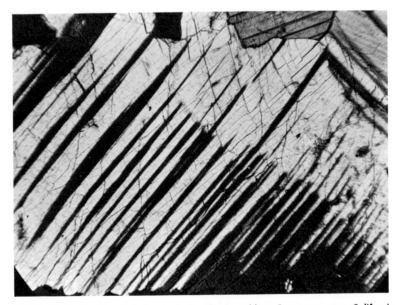

FIGURE 12.24   Pericline twinning in bytownite in gabbro, Cuyamaca area, California. Polars crossed, ×94.

FIGURE 12.25 Albite and pericline twinning in anorthite in gabbro, Cuyamaca area, California. Polars crossed, ×55.

FIGURE 12.26 Carlsbad, pericline, and minor albite twinning in labradorite in gabbro, Cuyamaca area, California. Polars crossed, ×55.

PLAGIOCLASE

also with the temperature and the amount of K (or Or content) proxying for Na.

Zoning is widespread, particularly in phenocrysts of volcanic rocks, and manifests itself most conspicuously by differences in extinction angles. Most types of zoning may be classified under (Figs. 12.27, 12.28):

1. *Normal.* Center more calcic, becoming more sodic toward margin; changes either gradual or in zones separated by sharp contacts. Some types have a ragged sodic core with a considerable compositional hiatus between it and marginal zones or zone.
2. *Reverse.* Center more sodic, becoming more calcic toward margin; change either gradual or in zones separated by sharp contacts. Sodic-core types show an irregularly outlined sodic core with a marked compositional gap between it and the outer part.
3. *Oscillatory.* Normally steplike progression from more calcic interior to more sodic margins with local reversals in adjoining zones.

Regular, semiregular, and irregular intergrowths with other minerals are common. The perthitic intergrowths are described under the potash feld-

FIGURE 12.27 Oscillatory zoning in andesine in quartz monzonite porphyry, Elk Peak, Judith Mountains, Montana. Polars crossed, ×104 (S. B. Wallace).

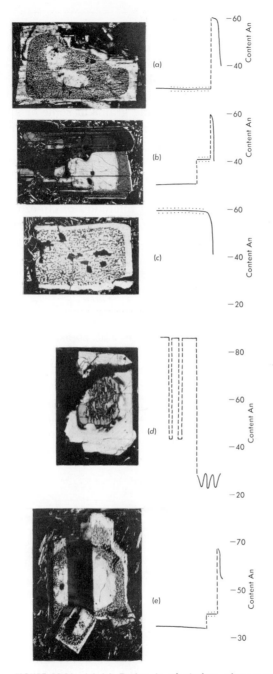

FIGURE 12.28 (a)–(e) Zoning in plagioclase phenocrysts, volcanic rocks, San Juan region, Colorado. (Larsen et al., 1938.)

spars. Antiperthite consists of single-crystal plagioclase host, commonly oligoclase or andesine, and bleblike inclusions of potash feldspar, usually orthoclase. Myrmekite is made up of single-crystal plagioclase hosts with bleblike to vermicular quartz patches. Plagioclase phenocrysts may carry numerous and abundant inclusions, including grains of some or all of the other minerals present in the rock, patches of fine-grained matrix material, bubbles, and glass. Such inclusions may be zonally arranged. Biotite inclusions in dusty plagioclase may be enveloped in a dustfree halo. Plagioclase porphyroblasts also may be rich in inclusions of other minerals, being even poikiloblastic in structure. "Stuffed" (gefüllte) plagioclase crystals contain thick cores of very abundant microlites of clinozoisite and sericite. Clouded plagioclases owe their turbid appearance to exceedingly numerous, oriented submicroscopic rods of iron minerals, commonly magnetite.

Plagioclases alter readily to sericite, kaolinite, and other clay minerals, saussurite, gibbsite, calcite, and zeolites. The alteration may be preferential: in zoned crystals the more calcic units are replaced first (Fig. 12.30); in some twinned crystals lamellae of one orientation are altered prior to the other set. In rocks with both plagioclase and potash feldspar, the plagioclase is normally altered first, even to sericite.

The indices of refraction increase from albite to anorthite. The birefringence also increases slightly from albite to anorthite. In thin section maximum interference colors are first-order grays to white. All members of the series are colorless to neutral or turbid, in many cases ascribable to partial kaolinization. The optic angle (for low-temperature plagioclase) is large, ca. 75–90°; albite is (+), oligoclase (+) and (−), andesine (−)

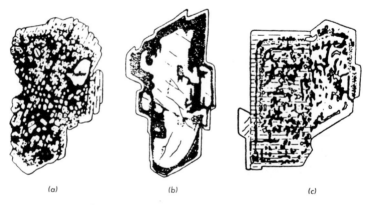

(a)   (b)   (c)

FIGURE 12.29 Inclusion zoning in plagioclase (Kuno, 1950). (a) Groundmass inclusions. (b) Dust inclusions. (c) Honeycomb structure, mainly glass and pyroxene.

FIGURE 12.30   Selective sericitization of calcic core of intermediate plagioclase in granodiorite, Hunza, Pakistan.   Polars crossed, ×33.

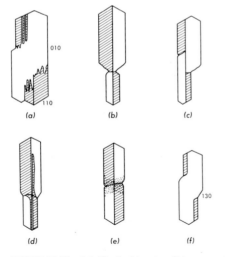

FIGURE 12.31   (a)–(f) Authigenic albite crystals, twinned, from sedimentary rocks, Zweisimmen, Switzerland (Füchtbauer, 1948).

PLAGIOCLASE   357

and (+), labradorite (+), bytownite (+) and (−), and anorthite (−) (Fig. 12.32). The dispersion is variable. The optic plane shifts its position about 90° with changes in composition, from nearly normal to (010) and approaching parallelism with (001) in albite and oligoclase, to approaching parallelism with (100) in anorthite. Thus in the sodic members γ emerges nearly normal to (010) and α to (100), whereas in the calcic types γ emerges nearly normal to (100).

Determination of approximate composition. Numerous techniques have been employed for determining the approximate composition of members of the series without a chemical analysis: specific gravity determination, x-ray methods, and measurement of optical constants. The last includes the greatest number of methods, which fall generally into three categories:

1. Methods employing the universal stage.
2. Methods applicable to plagioclases in thin sections, without the universal stage.
   a. Michel-Lévy statistical method
   b. Combined Carlsbad–albite twin method
   c. Fouqué method
   d. Microlite method

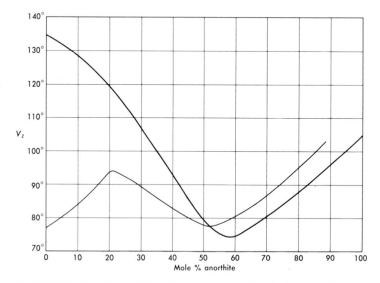

FIGURE 12.32 Variations of 2V with composition in the plagioclase series (J. R. Smith, 1958). Light-line curve is for low-temperature plagioclase; heavy-line curve is for high-temperature plagioclase.

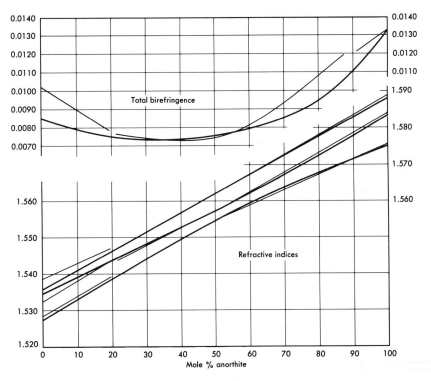

FIGURE 12.33 Variation of α, β, γ, and $\gamma - \alpha$ with composition in the plagioclase series (J. R. Smith, 1958). Light-line curves are for low-temperature plagioclase; heavy-line curves for high-temperature plagioclase.

3. Methods applicable to cleavage fragments immersed in index liquids, without the universal stage.
   a. Schuster method
   b. Tsuboi method
   c. Rhombic section method

For some determination techniques it is important to recognize readily certain sections (Fig. 12.22). The criteria for (010) are:

1. Albite twin lamellae absent
2. Pericline lamellae may be present
3. (001) cleavage nearly normal to section
4. Zoning may be conspicuous
5. γ normal to the section only in the more sodic members
6. Outline is a parallelogram with a 64° angle

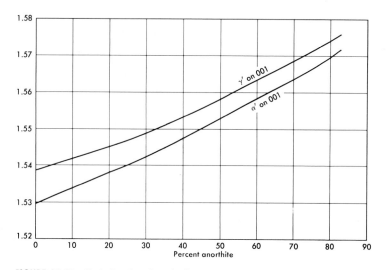

FIGURE 12.34 Variation in α' and γ' on (001) with composition in the plagioclase series (Emmons, 1953).

The criteria for (001) are:

1. Albite lamellae, Carlsbad individuals, and less commonly pericline lamellae all may be visible
2. Albite twins show symmetrical extinction
3. (010) cleavage nearly normal to section
4. Outline is rectangular

The content of K in plagioclase influences the size of the extinction angles and other optical properties, but normally the K content can be neglected; for the differences lie within the range of the experimental error of the measuring technique.

Michel-Lévy statistical method (Fig. 12.39). This, perhaps the single most useful method in thin section, requires the determination of the maximum extinction angles of albite twins in sections normal to (010):

1. Determine that the section is normal or nearly normal to (010) by noting: (*a*) equality of illumination of the two sets of twin lamellae in positions parallel with the vibration directions of the polars and (*b*) equality or near equality of the extinction angles of the two sets of lamellae.
2. Measure the extinction angles of the two sets against the fast ray and average them. The values should agree within 3–4° at least.

3. Repeat for a large number of similarly oriented sections. A minimum of 10 sections is recommended, but the precision increases if more are measured.
4. Choose the maximum value obtained and apply it to the curve of Fig. 12.39. Values of 0–20° appear in both + and − values and, if the sign of the angle cannot be determined, which commonly is the case (see p. 29), other optical data (preferably indices of cleavage flakes or relief against Canada balsam) must be obtained for a unique determination.

Combined Carlsbad-albite twin method (*Fig. 12.40*). In sections normal to (010) in which both Carlsbad and albite twins occur together, a determination on a single section is sufficient:

1. Check the orientation as above, noting equality of illumination and equality or near-equality of twin extinction angles.
2. Measure the four extinction angles, two for the albite lamellae of one Carlsbad individual and two for the other individual. Within each Carlsbad unit the two values should agree within 3–5°, but the averages for the different Carlsbad twins may differ considerably.
3. Apply the smaller extinction angle to the ordinate values of Fig. 12.40 and the larger angle to the curved lines on the Fig. 12.40. The intersection indicates the percentage of albite. Again, if the sign of the extinction angle is indeterminate, several combinations are repeated and a unique determination may not be possible without additional optical data.

Fouqué method (*Fig. 12.37*). This method requires sections normal to a bisectrix, either α or γ:

1. Secure a centered bisectrix figure and determine its optical sign. The use of figures even 10° off-center leads to relatively large errors in determination.
2. In (+) figures (normal to γ) measure the extinction angle of the faster ray either against the trace of (010) i.e., cleavage lines or albite twinning lamellae (Fig. 12.37) or against the trace of (001), usually cleavage lines (Fig. 12.37).
3. In (−) figures (normal to α) measure the extinction angle of the faster ray against the trace of (001), i.e., cleavage lines (Fig. 12.37), or, for the more calcic end of the series, against the trace of (010), either cleavage or albite twin lamellae (Fig. 12.37).
4. For some values of extinction angles (ca. 0–20°) unique determinations require the (+) or (−) sign of the angle.

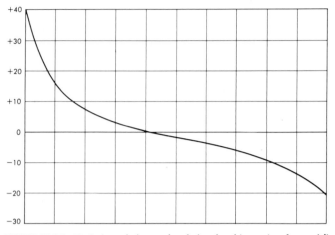

FIGURE 12.35 Variation of the angle of the rhombic section for pericline twins with composition in the plagioclase series (Schmidt, 1919).

Microlite method (Fig. 12.38). Plagioclase microlites commonly are elongate parallel with *a*. Sections of such crystals cut parallel with *a* have (010) cleavage traces and albite twin boundaries parallel with the long edges of the microlite, except in sections parallel with (010), on which the

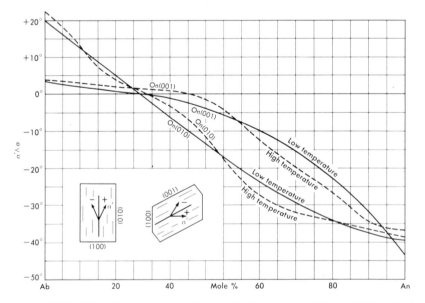

FIGURE 12.36 Variation of extinction angles of $\alpha' \wedge a$ on (001) and (010) with composition in the plagioclase series (Schuster, 1880; Köhler, 1942).

albite twinning does not appear. Measure the extinction angle from the direction of elongation, repeat for 10–12 microlites, or more for greater precision, choose the maximum value, and apply to microlite curve of Fig. 12.38. Note that it is not possible to distinguish (+) and (−) extinction angles, so that for values 0–20°, an index comparison with Canada balsam is required.

Schuster method (Fig. 12.36). The Schuster method requires the measurement of the extinction angle between the faster ray and the trace of (100) either on (001) sections or cleavage flakes (Fig. 12.36) or on (010) sections or cleavage flakes (Fig. 12.36). On (010) the values 0–20° are

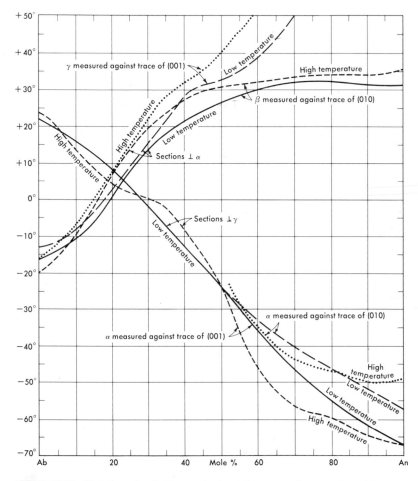

FIGURE 12.37 Variation of extinction angles in sections normal to α and γ with composition in the plagioclase series (Fouqué, 1894; Köhler, 1942).

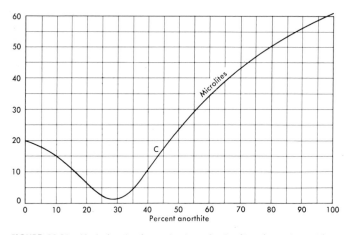

FIGURE 12.38 Variation in the extinction of microlite elongation with composition in the plagioclase series.

repeated, and on (001) 0–4° are repeated; thus the sign of the extinction is required for these ranges. From Ab_{100}–Ab_{60} the (001) curve is very flat and small errors in angle determination may cause relatively large errors in estimating the composition.

Tsuboi method. This method is based on the measurement of the maxi-

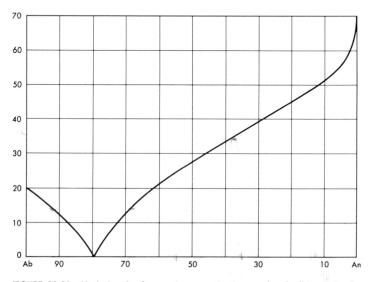

FIGURE 12.39 Variation in the maximum extinction angle of albite twins in sections normal to (010) with composition in the plagioclase series (Michel-Lévy, 1877).

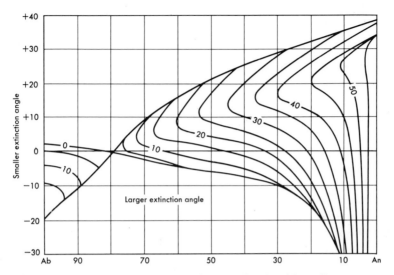

FIGURE 12.40 Variation in extinction angles of combined Carlsbad–albite twins in sections normal to (010) with composition in the plagioclase series.

mum and minimum refractive indices of cleavage flakes, n_1, n_2 on (001) and n_1, n_2 on (010) (Fig. 12.34). The method offers some difficulty if the crystals have numerous zones of different composition or contain inclusions. For most determinations of the lower index of refraction, it is unnecessary to distinguish between (001) and (010) cleavage flakes, for the values of n_2 are nearly the same for both types of cleavage pieces.

Rhombic section method. The position of the composition plane of pericline twins varies with the composition of the plagioclase. Thus it is possible to estimate the Ab–An ratio by measuring on (010) the angle between the trace of pericline twin lamellae and the trace of (001) face or (001) cleavage. The values are referred to Fig. 12.35. Angles between 0 and 20° appear in both (+) and (−) values.

ALBITE

Composition. Ab_{100}–Ab_{90}, K also may be present in minor amounts. The high-temperature form can contain considerably more K than the low modification.

Indices. For the low-temperature modification: $\alpha = 1.528–1.533$, $\beta = 1.532–1.537$, $\gamma = 1.538–1.542$, $\gamma - \alpha = 0.009–0.010$. The indices of both temperature modifications increase with An content (Fig. 12.33). In thin section the maximum interference color is gray white of the first order.

Color. Colorless.

Form. Triclinic, pseudomonoclinic. In thin plates or laths flattened parallel with (010), in subparallel or radial groups (cleavelandite); in anhedral granular aggregates ("sugary" albite); less commonly as tabular euhedral crystals; also as minute wisps, lenses, veinlets, and blebs in potash feldspar to form perthite. Cleavages, (001) perfect, (010) excellent. The larger number of crushed grains will lie on (001). Inclusions of other minerals, particularly quartz, microcline, and muscovite, are common.

Orientation. Biaxial (+). Optic plane nearly normal to (010) and approaching parallelism with (001). For low albite $2V = 76\text{--}83°$, increasing with An, $r < v$ weak; for high albite, $(-)$ $2V = 54\text{--}55°$, increasing with An (Fig. 12.32). Extinction angles for significant sections are given in Figs. 12.36 to 12.40. Albite twinning rarely absent. Combinations of albite and Carlsbad or albite and pericline are much less common. Zoning may be present but it is not common.

Occurrence. In igneous rocks, mainly in complex granitic pegmatites, also in sodic granites and rhyolites, alkali syenites and trachytes, foidal syenites, and phonolites. In perthites and myrmekites; also in saussuritized gabbros and albitized basalts and diabases (spilites), replacing calcic plagioclase, and in their wall rocks (adinoles). Widespread in various metamorphic rocks, particularly those of low grade, including some hornfelses, chloritoid schists, chlorite–sericite schists, muscovite–biotite schists ± garnet and epidote, calc schists, glaucophane schists, feldspathic quartzites and schistose grits, and hornblende–epidote amphibolites, also as a hydrothermal mineral in the wall rocks of some metallic mineral deposits. In sediments both as a detrital and as an authigenic mineral (Fig. 12.31), the former in sandstones and arkoses, the latter particularly in limestones and dolomites. Alters to sericite and clay minerals. Albite formed by replacement of other mineral may show a distinctive mosaic structure which combined with twinning gives rise to a "checkerboard" albite.

Diagnostic features. From the other plagioclases it differs mainly in indices and extinction angles; from the potash feldspars, in its distinctive albite twinning.

OLIGOCLASE

Composition. $Ab_{90}\text{--}Ab_{70}$, some K may be present.

Indices. For the low-temperature modification: $\alpha = 1.533\text{--}1.543$, $\beta = 1.537\text{--}1.547$, $\gamma = 1.542\text{--}1.552$, $\gamma - \alpha = 0.009$. In thin section the maximum interference colors are first-order gray white. The indices increase with An content (Fig. 12.33).

Color. Colorless.

Form. Triclinic, pseudomonoclinic. Euhedral to anhedral. Euhedral crystals are elongate parallel with a, with rectangular or six-sided sections. Subhedral and anhedral grains are common; some may be interstitial. Porphyroblasts may be poikiloblastic with crenulated margins. Cleavages, (001) perfect, (010) excellent. Most crushed grains will lie on (001). Inclusions of other minerals, especially quartz, potash feldspar, muscovite, and biotite, are common. In phenocrysts of extrusive rocks glass and matrix material may also be included. Matrix inclusions, irregularly or zonally arranged, may appear in porphyroblasts.

Orientation. Biaxial $(+)$ or $(-)$. For low oligoclase 2V varies from 82° $(+)$ (Ab_{90}) to 83° $(-)$ (Ab_{70}), $r > v$ weak; $2V = 90°$ for Ab_{82}. High oligoclase is $(-)$ with $2V = 52\text{--}73°$, increasing with An content (Fig. 12.32). For some uncommon matrix oligoclase of glassy andesites 2V $(+)$ ranges from large to very small, with variations of as much as 20° in different grains in the same thin section, and is reportedly due to abnormal K content. Extinction angles are given in Figs. 12.36 to 12.40. Albite twinning may be absent, or the lamellae may be very fine. Pericline and other twin types are uncommon.

Occurrence. In granites, quartz monzonites, granodiorites, and to a lesser extent in syenites, monzonites, tonalites, and diorites. The high-temperature type occurs in their extrusive equivalents: rhyolites, quartz latites, and less commonly trachytes, latites, dacites, and andesites. Not uncommon as a primary constituent in some granitic pegmatites. Widely distributed in metamorphic rocks, particularly granitic gneisses, garnet–biotite gneisses, feldspathic quartzites, granulites, and some schists (quartz–mica, quartz–garnet–mica); may form porphyroblasts. It is the most common detrital plagioclase in sands and sandstones. Intergrown with blebs of quartz it forms myrmekite; with orthoclase blebs it forms antiperthite. Alters principally to sericite and kaolinite.

Diagnostic features. If twinning is absent, may be difficult to tell from quartz without determining the biaxial character. Has higher refractive indices than potash feldspars. Differs from other members of the plagioclase series in extinction angles and indices.

ANDESINE

Composition. $Ab_{70}\text{--}Ab_{50}$; very minor K may be present.

Indices. For the low-temperature modification: $\alpha = 1.543\text{--}1.554$, $\beta = 1.547\text{--}1.558$, $\gamma = 1.552\text{--}1.562$, $\gamma - \alpha = 0.008\text{--}0.009$. In thin section the maximum interference color is gray white of the first order. Indices increase with An content (Fig. 12.33).

Color. Colorless.

Form. Triclinic, pseudomonoclinic. Commonly euhedral or subhedral, in both intrusive and extrusive igneous rocks. Elongated parallel with a; sections are commonly rectangular in outline. Common as lath-shaped groundmass microlites. Phenocrysts may be clustered (glomeroporphyritic), embayed by reaction with matrix material, or composite by welding of smaller crystals. Zoning is common both in phenocrysts and in plutonic types and may be of various types (Fig. 12.27). Inclusions are abundant: hematite microlites, magnetite, rutile, hornblende, apatite, matrix material, glass, and liquid ($\pm$ bubbles). Microlites are usually oriented, either parallel with a or parallel with c. Other inclusions may be present in zones or outer jackets. Parallel overgrowths of orthoclase form armored crystals. Cleavages (001) perfect, (010) excellent. Crushed material tends to lie on (001).

Orientation. Biaxial ($-$) and ($+$). Low-temperature andesine has 2V = 83° ($-$) to 77° ($+$). $r < v$ weak. 2V = 90° for Ab_{63}. The high-temperature type has 2V = 73° ($-$) to 76° ($+$) (Fig. 12.32). 2V of both types varies with the An content. The variations of extinction angles are shown in Figs. 12.36 to 12.40. Albite twinning is almost always present; other types are much less common.

Occurrence. Occurs in tonalites, diorites, some monzonites, and some anorthosites; the high-temperature equivalent is characteristic of andesites, dacites, and less commonly latites. Occurs in some gneisses of intermediate composition: Hornblende gneiss, biotite–plagioclase gneiss, garnet–hornblende–plagioclase gneiss, and hypersthene–plagioclase gneiss; also in plagioclase amphibolites and a few schists. Not common as a detrital species.

Diagnostic features. Has higher indices than quartz and potash feldspars. Separated from other plagioclases by differences in indices and extinction angles.

LABRADORITE

Composition. An_{50}–An_{70}

Indices. For the low-temperature type: $\alpha = 1.554$–1.563, $\beta = 1.558$–1.568, $\gamma = 1.562$–1.573, $\gamma - \alpha = 0.007$–0.009. The maximum thin-section interference colors are first-order gray to white. An increase in An content increases the refractive indices (Fig. 12.33).

Color. Colorless.

Form. Triclinic, pseudomonoclinic. May be euhedral, as phenocrysts of extrusive rocks, subhedral or anhedral in intrusive types. In anorthosites may be in anhedral, equant grains. Microlites are lath-shaped. Commonly elongate parallel with a, and sections are typically nearly square or rectangular in shape. Cleavages (001) perfect, (010) excellent; (110) and (1$\bar{1}$0) also may be present. Crushed fragments normally lie on (001). Augite, pigeonite, magnetite, ilmenite, apatite, glass, and matrix material form inclusions. Oriented microlites of hematite, ilmenite, or rutile appear as minute plates. Zoning of several types is particularly common in phenocrysts, with zones ranging from labradorite to andesine. An unusual type has a checkerboard core of labradorite–andesine and a labradorite rim. Zonally arranged inclusions, such as a core of matrix material with a labradorite rim ("stuffed" crystals), or a core with oriented microlites and a clear rim are known.

Orientation. Biaxial (+). The low-temperature type has 2V = 77–76–86°, first decreasing then increasing with An content; $r < v$. The high-temperature modification shows a very similar 2V range, with slightly lower values (Fig. 12.32). Extinction angle ranges are listed in Figs. 12.36 to 12.40. Albite twinning is common, and the lamellae are typically broad; also common are albite–Carlsbad combinations and some pericline twinning.

Occurrence. An essential mineral of basic rocks including gabbro, norite, anorthosite, diabase, and basalt, which contains the high-temperature type. Also as phenocrysts in some andesites (high-temperature type) and in basic metamorphic rocks such as pyroxene granulites, some hornblende–pyroxene gneisses, as well as in some hornfels. Uncommon as a detrital mineral. Alters readily to sericite, kaolinite, calcite, or saussurite.

Diagnostic features. Distinguished by its refractive indices and extinction angles from the other plagioclases.

BYTOWNITE

Composition. An_{70}–An_{90}.
Indices. $\alpha = 1.563$–1.572, $\beta = 1.568$–1.578, $\gamma = 1.573$–1.583, $\gamma - \alpha = 0.008$–0.010. In thin section grays and whites of first order are the maximum interference colors. Indices increase with the An content (Fig. 12.33).
Color. Colorless.
Form. Triclinic, pseudomonoclinic. Subhedral to anhedral. Cleavages (001) perfect, (010) excellent. Crushed fragments lie mainly on (001). Zoning may be present but is rare. Inclusions of pyroxene, olivine, or microlites may be observed.
Orientation. Biaxial (+) and (−). For the low-temperature type 2V = 86°(+)–79°(−), $r > v$. 2V = 90° for An_{73} and decreases again with increasing An. The 2V values for the uncommon high-temperature types are

lower (Fig. 12.32). Extinction angle variations are shown in Figs. 12.36 to 12.40. Albite twinning common as broad lamellae, also in combinations with Carlsbad and pericline.

Occurrence. An uncommon member of the plagioclase series. Found in some gabbros, in some anorthosites, in troctolites, as an accessory in several types of peridotites, in a few meteorites, and rarely in basalts.

Diagnostic features. Distinguished from other plagioclases by means of extinction angles and indices of refraction.

ANORTHITE

Composition. An_{90}–An_{100}.

Indices. $\alpha = 1.572–1.576$, $\beta = 1.578–1.583$, $\gamma = 1.583–1.588$, $\gamma - \alpha = 0.011–0.012$. In thin section first-order white to yellow represents the maximum interference tints. Indices increase with An content (Fig. 12.33).

Color. Colorless.

Form. Triclinic, pseudomonoclinic. Commonly anhedral but may be euhedral. Cleavages (001) perfect, (010) excellent. Crushed material tends to lie on (001). Inclusions of other minerals may be present.

Orientation. Biaxial (−). $2V = 79–77°$, decreasing with An content, $r > v$. The high-temperature type has slightly lower 2V values (Fig. 12.32). Figures 12.36 to 12.40 show the variation in extinction angles. Twinning in broad albite lamellae or combinations of albite with pericline types occurs. Zoning is not common.

Occurrence. In a few igneous rocks: some anorthosites, gabbros, and peridotites. Not uncommon in contact-altered limestones with diopside, grossularite, wollastonite, melilite, and rarer calcium silicates. In some coal-ash slags and clinkers and some slagged firebricks.

Diagnostic features. Distinguished from the other plagioclases through indices and extinction angles.

PART THREE

IDENTIFICATION TABLES

1. Isotropic (including some species essentially isotropic owing to submicroscopic grain size); listed in order of increasing n
2. Uniaxial (including some biaxial species with $2V = 0°$ or nearly $0°$)
   a. Positive $(+)$ }
   b. Negative $(-)$ } listed in order of increasing ω
3. Biaxial
   a. Parallel extinction in principal sections: orthorhombic
      (1) Positive $(+)$ }
      (2) Negative $(-)$ } listed in order of increasing β
   b. Inclined extinction in one or more principal sections: monoclinic and triclinic
      (1) Positive $(+)$ }
      (2) Negative $(-)$ } listed in order of increasing β

Isotropic Minerals

n	SPECIES	COLOR (IN T.S.) (C = COLORLESS)	PAGE
1.406–1.460	opal	c, gray, p. brown	56
1.434	fluorite	c, purple, blue	58
1.436–1.457	rare-earth fluorite	c, purple, blue	58
1.44 –1.45	opal	c, gray, p. brown	56
1.459–1.462	lechatlierite	c, gray	58
1.470–1.496	allophane	c	321
1.483–1.487	sodalite	c, p. blue	66
1.485–1.487	hackmanite	pink	66
1.485–1.487	cristobalite	c	102
1.485–1.495	nosean	c, gray, blue	66
1.487–1.493	analcite	c	70
1.490	sylvite	c	59, 135
1.496–1.500	haüyne	blue, lilac blue, blue green	67
1.500–1.510	lazurite	blue	67
1.508–1.511	leucite	c	68
1.526–1.542	hydrated halloysite	c	320
1.53 –1.56	serpentine ("serpophite")	c, p. green	308
1.53 –1.72	allanite	yellow, brown	194
1.535	langbeinite	c	135
1.542	apophyllite	c	108
1.544	halite	c	59
1.549–1.561	metahalloysite	c	320
1.58 –1.63	collophane	brown	81
1.636	melilite	c	107
1.650–1.675	greenalite	olive green, d. brownish green	312
1.72 –1.74	spinel	c	54
1.720–1.774	pyrope	pink	60
1.735–1.770	grossularite	c	60

Isotropic Minerals (*Cont.*)

n	SPECIES	COLOR (IN T.S.) (C = COLORLESS)	PAGE
1.736–1.745	periclase	c	51
1.75 –1.79	pleonaste	green, blue green	54
1.770–1.820	almandite	c, pink	60
1.78 –1.80	hercynite	d. green	54
1.790–1.810	spessartite	c	60
1.82	zircon	gray, white	110
1.850–1.890	andradite	p. brown	60
1.94	melanite	d. brown	59
2.00	picotite	olive brown, brown	54
2.0 –2.4	limonite	red, brown	56
2.1 –2.2	chromite	opaque, d. brown on edges	54
2.30 –2.37	perovskite	yellow, orange, brownish red, gray, black	124
2.5 –2.8	leucoxene	gray, white	76
2.9 –3.2	hematite (earthy)	red	74

Minerals of Isometric Habit That May Show Weak Birefringence

analcite   spessartite   gahnite
sodalite   grossularite
leucite

Nonisometric Minerals That May Appear To Be Essentially Isotropic, Usually Owing to Submicroscopic Grain Size or Owing to Metamictization

allanite   hydrated halloysite   metahalloysite
collophane   kaolinite   serpentine ("serpophite")
greenalite   leucoxene   zircon
hematite   melilite

Uniaxial Minerals

Uniaxial Positive (+)

ω	ε	SPECIES	COLOR (IN T.S.) (C = COLORLESS)	PAGE
1.461	1.474	tincalconite	c	142
1.478–1.485	1.480–1.490	chabazite	c	329
1.488–1.500	1.487–1.495	heulandite	c	327
1.531–1.536	1.533–1.537	apophyllite	c	108
1.531–1.544	1.539–1.553	chalcedony	c, gray	98
1.544	1.553	quartz	c	94
1.545–1.547	1.549–1.551	osumilite	p. blue	183
1.559–1.590	1.580–1.600	brucite	c	78

Uniaxial Minerals (Cont.)

Uniaxial Positive (+) (Cont.)

ω	ε	SPECIES	COLOR (IN T.S.) (C = COLORLESS)	PAGE
1.569–1.578	1.590–1.601	alunite	c	80
1.575–1.585	1.580–1.596	clinochlore	green	300
1.576–1.595	1.579–1.600	pennine	green	301
1.585	1.595	natroalunite	c	80
1.605–1.648	1.610–1.652	prochlorite	green	304
1.626–1.633	1.632–1.639	åkermanite	c	107
1.715–1.719	1.720–1.724	vesuvianite	c	112
1.716–1.722	1.816–1.824	bastnäsite	yellow, red brown	86, 137
1.720–1.724	1.816–1.828	xenotime	c	85
1.920–1.960	1.967–2.015	zircon	c, gray	110
1.992–1.997	2.091–2.093	cassiterite	yellow, orange, red, brown	77
2.605–2.616	2.890–2.903	rutile	red brown, amber, yellow brown, red	74

Uniaxial Negative (−)

ω	ε	SPECIES	COLOR (IN T.S.) (C = COLORLESS)	PAGE
1.480–1.490	1.478–1.485	chabazite	c	329
1.486–1.488	1.482–1.484	cristobalite	c	102
1.502–1.528	1.491–1.503	cancrinite	c	106
1.53 –1.57	1.52 –1.56	indialite (high)	c	177
1.531–1.549	1.528–1.544	nepheline	c, gray	104
1.534–1.550	1.522–1.540	scapolite (marialite)	c	113
1.537–1.543	1.535–1.542	apophyllite	c	108
1.542	1.537	kalsilite	c	105
1.552–1.595	1.541–1.565	scapolite (mizzonite, dipyre, wernerite)	c	113
1.554–1.580	1.525–1.540	lepidolite	c	282
1.555	1.530	vermiculite	c, p. amber	298
1.557–1.604	1.530–1.560	phlogopite	c, p. amber	283
1.568–1.602	1.564–1.595	beryl	c	109
1.575–1.595	1.538–1.550	talc	c	307
1.579–1.600	1.576–1.595	pennine	green	301
1.583	1.550	stilpnomelane	yellow, brown	292
1.596–1.607	1.566–1.571	scapolite (meionite)	c	113
1.597–1.610	1.555–1.570	muscovite (phengite)	c	279

Uniaxial Negative (—) (*Cont.*)

ω	ε	SPECIES	COLOR (IN T.S.) (C = COLORLESS)	PAGE
1.605–1.675	1.565–1.625	biotite	brown, green	285
1.610–1.652	1.605–1.648	prochlorite	green	304
1.619	1.600	glauconite	d. green	289
1.620–1.635	1.610–1.631	carbonate apatite (incl. dahllite and francolite)	c, p. green	81
1.631–1.655	1.612–1.630	dravite	yellow, orange, yellow brown	119
1.632–1.649	1.630–1.645	fluorapatite	c	81
1.635–1.650	1.615–1.632	elbaite	c	120
1.638	1.634	strontian apatite	c	82
1.640–1.660	1.635–1.654	mangan–fluorapatite	c	81
1.640–1.698	1.622–1.675	schorl	gray, d. blue, d. green, black	115
1.642–1.669	1.636–1.658	gehlenite	c	107
1.645–1.651	1.640–1.645	hydroxylapatite	c	82
1.658	1.486	calcite	c, gray	86
1.660–1.739	1.490–1.545	manganoan calcite	c	86
1.680	1.500	dolomite	c, gray	90
1.685–1.743	1.505–1.540	ankeritic dolomite	c, gray, brownish	90
1.700	1.509	magnesite	c	92
1.702–1.742	1.698–1.736	vesuvianite	c, gray, green, greenish brown	112
1.705–1.785	1.515–1.565	ferroan magnesite	c, gray, brownish	92
1.740–1.810	1.546–1.595	calcian rhodochrosite	c	89
1.749	1.547	ankerite	c, gray, brownish	90
1.767–1.772	1.760–1.763	corundum	c, gray, blue, red	72
1.788–1.875	1.570–1.633	siderite	c, gray, brown	93
1.815–1.820	1.713–1.715	jarosite	p. yellow	80
1.816	1.597	rhodochrosite	c, gray, pink	88
1.830	1.750	natrojarosite	p. yellow	80
2.40	2.26	goethite	red	79
2.488	2.561	anatase	brown, d. blue, black	76
3.22	2.94	hematite	opaque, d. red	74

TABLE I   375

Tetragonal and Hexagonal Minerals That May Appear in Biaxial Form

anatase	calcite	quartz
apatite	cassiterite	rutile
apophyllite	corundum	vesuvianite
beryl	jarosite	zircon
brucite	natrojarosite	

Orthorhombic and Monoclinic Minerals That May Appear in Uniaxial (or Near-Uniaxial) Form

muscovite (phengite)	stilpnomelane	heulandite
lepidolite	talc	chabazite
phlogopite	prochlorite	zoisite
biotite	pennine	pigeonite
glauconite	clinochlore	sanidine
vermiculite		

Biaxial Minerals

Orthorhombic Parallel extinction in principal sections

POSITIVE (+)

β	α	γ	SPECIES	2V	COLOR (IN T.S.) (C = COLORLESS)	PAGE
1.469	1.469	1.473	tridymite (synthetic)	small	c	100
1.472-1.475	1.465-1.467	1.494-1.497	carnallite	70	c	125
1.474	1.464	1.485	thenardite	83	c	135
1.476-1.491	1.473-1.489	1.485-1.502	natrolite	60-63	c	329
1.478-1.483	1.477-1.482	1.481-1.488	tridymite	36-90	c	100
1.513-1.532	1.511-1.530	1.518-1.545	thomsonite	44-75	c	330
1.532-1.574	1.527-1.560	1.538-1.578	cordierite (high, anhydrous)	80-89	c	178
1.535-1.545	1.53 -1.54	1.54 -1.55	cordierite (low, hydrous)	75-89	c	178
1.566-1.573	1.561-1.563	1.587-1.590	norbergite	44-50	c	152
1.572-1.579	1.569-1.573	1.613-1.618	anhydrite	36-45	c	126
1.605-1.685	1.598-1.674	1.615-1.697	anthophyllite	89-59	c, tan, p. green	246
1.610-1.631	1.607-1.629	1.618-1.638	topaz	48-68	c	157
1.618-1.647	1.610-1.637	1.635-1.670	prehnite	64-70	c	295
1.619-1.653	1.607-1.643	1.639-1.675	humite	68-81	c, p. yellow	154
1.623-1.624	1.621-1.622	1.630-1.632	celestite	50-51	c	132
1.636-1.638	1.634-1.637	1.646-1.648	barite	36-40	c	131
1.644-1.655	1.642-1.653	1.654-1.679	mullite	20-50	c	164
1.652-1.673	1.650-1.668	1.659-1.679	enstatite	54-90	c	200
1.653	1.646	1.661	boehmite	80	c	123
1.654-1.670	1.653-1.661	1.669-1.684	sillimanite	20-30	c, gray	162
1.672-1.676	1.665	1.684-1.686	lawsonite	84-85	c	176
1.667-1.759	1.659-1.730	1.695-1.772	forsterite (Fo_{88-50})	89-74	c	144
1.688-1.711	1.685-1.707	1.698-1.725	thulite	60-30	pink	184

TABLE I 377

Biaxial Minerals (Cont.)

Orthorhombic Parallel extinction in principal sections (*Cont.*)

Positive (+) (*Cont.*)

β	α	γ	SPECIES	2V	COLOR (IN T.S.) (C = COLORLESS)	PAGE
1.695–1.75	1.67 –1.73	1.72 –1.77	iddingsite	large	red, brown	150
1.696–1.700	1.696–1.698	1.702–1.714	zoisite	30–0	c	184
1.698–1.702	1.698–1.700	1.705–1.718	ferrian zoisite	0 –60	c	184
1.715	1.714	1.720	vesuvianite	5 –30	c, gray, green	112
1.720–1.722	1.700–1.702	1.747–1.752	diaspore	80–88	c, p. blue, pink	121
1.740–1.754	1.736–1.747	1.745–1.762	staurolite	80–89	yellow, brown	167
1.752–1.763	1.744–1.755	1.761–1.773	ferroan hypersthene (orthoferrosilite)	90–80	c	203
2.33 –2.38			perovskite	90	yellow, orange brown, p. brown red	124
2.584	2.583	2.700	brookite	30	yellow, orange, p. brown	76

Negative (−)

β	α	γ	SPECIES	2V	COLOR (IN T.S.) (C = COLORLESS)	PAGE
1.455	1.433	1.461	epsomite	51	c	135
1.532–1.574	1.527–1.560	1.538–1.587	cordierite (high, anhydrous)	89–70	c	178
1.532–1.545	1.53 –1.54	1.54 –1.55	cordierite (low, hydrous)	89–40	c	178
1.605–1.660	1.598–1.650	1.615–1.672	anthophyllite	57–90	c, buff	246
1.633–1.644	1.629–1.640	1.639–1.647	andalusite	71–86	c, pink, p. green	159
1.645–1.674	1.639–1.663	1.653–1.680	monticellite	88–72	c	148
1.650–1.677	1.636–1.659	1.669–1.695	forsterite (Fo_{100-88})	87–89	c	144

1.664–1.667	1.516–1.520	1.666–1.668	strontianite	7	c	141
1.670–1.680	1.527–1.529	1.676–1.685	strontian aragonite	18–19	c	139
1.671	1.662	1.691	manganian andalusite	75–71	green, golden	159
1.674–1.752	1.669–1.744	1.680–1.761	hypersthene	89–46–89	c, pink, p. green	203
1.680–1.682	1.529–1.530	1.685–1.686	aragonite	18–19	c	139
1.683–1.695	1.531–1.542	1.687–1.699	plumbian aragonite	19–23	c	139
1.684–1.691	1.659–1.678	1.686–1.692	dumortierite	20–52	blue, green, pink, lilac	170
1.695–1.75	1.67 –1.73	1.72 –1.77	iddingsite	35–80	red, brown	150
1.710	1.694	1.722	ferroan anthophyllite	59	tan	246
1.716–1.723	1.679–1.686	1.729–1.736	glaucochroite	61	c	150
1.719–1.722	1.715–1.718	1.720–1.724	vesuvianite	30–60	c, gray, green	112
1.760–1.864	1.731–1.824	1.773–1.875	fayalite	74–47	c, p. yellow	147
1.765–1.770	1.760–1.763	1.767–1.772	corundum	small–58	c, gray, blue, red	72
1.786	1.759	1.797	tephroite	65	c	147
2.39	2.26	2.40	goethite	small–30	red, brown	79

Monoclinic and Triclinic   Inclined extinction in one or more principal sections

POSITIVE (+)

1.478–1.485	1.478–1.485	1.480–1.490	chabazite	0 –32	c	329
1.487–1.501	1.487–1.499	1.488–1.505	heulandite	0 –50	c	327
1.505	1.496	1.519	ulexite	73	c	135
1.506	1.492	1.519	bischofite	79	c	135
1.522–1.523	1.520–1.521	1.529–1.530	gypsum	58	c	129
1.530–1.564	1.529–1.559	1.537–1.567	chrysotile	30–50	c, p. green	309
1.532–1.537	1.528–1.533	1.538–1.542	albite	76–83 (low) / 54–55 (high)	c	365
1.535	1.523	1.586	kieserite	57	c	135
1.537–1.543	1.533–1.539	1.542–1.548	oligoclase	82–89	c	366
1.551–1.558	1.548–1.554	1.555–1.562	andesine	89–76	c	367

TABLE I

Biaxial Minerals (Cont.)

Monoclinic and Triclinic Inclined extinction in one or more principal sections (Cont.)

POSITIVE (+) (Cont.)

β	α	γ	SPECIES	2V	COLOR (IN T.S.) (C = COLORLESS)	PAGE
1.558–1.568	1.554–1.563	1.562–1.573	labradorite	77–76–86	c	368
1.560–1.565	1.558–1.562	1.563–1.571	dickite	52–80	c	318
1.568–1.570	1.563–1.565	1.573–1.575	bytownite	86–89	c	369
1.568–1.570	1.568–1.570	1.586–1.587	gibbsite	0 –small	c, gray	122
1.571–1.589	1.571–1.588	1.576–1.599	clinochlore	0 –40	c, green	300
1.576–1.600	1.576–1.595	1.579–1.600	pennine	0 –20	green	301
1.585–1.587	1.580–1.584	1.594–1.596	celsian	86–89	c	337
1.592	1.586	1.614	colemanite	56	c	135
1.595–1.596	1.586–1.592	1.598–1.606	cuspidine	63	c	149
1.602–1.655	1.592–1.643	1.619–1.675	chondrodite	64–89	yellow, brown	152
1.603–1.614	1.594–1.610	1.631–1.642	pectolite	50–54	c	244
1.605–1.651	1.605–1.648	1.610–1.652	prochlorite	0 –30	green	304
1.606	1.599	1.621	scawtite	74	c	149
1.611–1.615	1.610–1.614	1.648–1.651	pseudowollastonite	0 –6	c	243
1.617–1.660	1.606–1.649	1.631–1.672	edenite	52–83	c, p. green	257
1.632–1.635	1.612–1.617	1.652–1.654	tilleyite	89	c	149
1.636–1.709	1.623–1.702	1.651–1.728	clinohumite	52–89	c, yellow, orange	304
1.644	1.640–1.641	1.650	rankinite	64	c	155
1.647–1.689	1.639–1.671	1.664–1.708	cummingtonite	65–89	c, gray	249
1.652	1.651	1.661	clinoenstatite	50–54	c	227
1.655–1.670	1.648–1.661	1.662–1.679	spodumene	54–69	c	239
1.659–1.670	1.654–1.665	1.667–1.685	omphacite	60–67	p. green	238
1.659–1.679	1.654–1.673	1.667–1.693	jadeite	70–78	c, p. green	238
1.671–1.705	1.663–1.699	1.693–1.728	diopside	54–58	c, p. green	209
1.673–1.745	1.699–1.739	1.728–1.757	hedenbergite	58–63	c, p. green, p. brown	213

1.674–1.716	1.662–1.708	bustamite	1.676–1.724	44–85	c	244
1.678–1.754	1.677–1.748	pumpellyite	1.688–1.764	80–0	c, green	197
1.679–1.744	1.673–1.720	aegirine–augite	1.691–1.759	60–89	p. green	236
1.683–1.700	1.680–1.698	riebeckite	1.685–1.706	50–85	blue	271
1.684–1.711	1.680–1.703	augite	1.706–1.729	40–60	c, gray, p. green, lilac	215
1.684–1.722	1.683–1.722	pigeonite	1.704–1.752	0–32	c, p. green	224
1.706–1.718	1.699–1.712	ferroan augite	1.728–1.742	55–70	c, gray, green	215
1.708–1.729	1.706–1.724	clinozoisite	1.712–1.735	14–89	c	188
1.711–1.712	1.706–1.708	merwinite	1.718–1.724	67	c	149
1.714–1.741	1.711–1.738	rhodonite	1.724–1.751	61–76	c	240
1.715	1.707	larnite	1.730	large	c	149
1.717–1.721	1.708–1.713	johannsenite	1.735–1.740	70	c, gray	214
1.719–1.734	1.713–1.730	chloritoid	1.723–1.740	36–89	c, green, gray, green	297
1.730–1.789	1.725–1.756	piedmontite	1.750–1.832	50–89	pink, red, golden, violet	193
1.764	1.763	clinoferrosilite	1.794	40	c	227
1.787–1.801	1.785–1.800	monazite	1.837–1.849	6–19	c, gray	136
1.870–1.970	1.840–1.950	sphene	1.943–2.093	20–56	c, yellow, p. brown	173

NEGATIVE (−)

1.469–1.470	1.447	borax	1.42	39–40	c	142
1.472	1.454	kernite	1.488	80	c	142
1.479–1.486	1.478–1.485	chabazite	1.480–1.490	0–32	c	329
1.491–1.504	1.482–1.500	stilbite	1.493–1.508	30–50	c	328
1.492	1.412	trona	1.540	72	c	135
1.504–1.550	1.485–1.535	montmorillonite	1.505–1.550	5–30	c	321
1.505	1.375	nahcolite	1.583	75	c	135
1.505	1.494	kainite	1.516	85	c	135
1.512–1.525	1.502–1.519	laumontite	1.513–1.526	25–47	c	333
1.522–1.524	1.518–1.520	adularia	1.524–1.526	small–large	c	347

TABLE I

Biaxial Minerals (Cont.)

Monoclinic and Triclinic   Inclined extinction in one or more principal sections (Cont.)

NEGATIVE (−) (Cont.)

β	α	γ	SPECIES	2V	COLOR (IN T.S.) (C = COLORLESS)	PAGE
1.522–1.526	1.517–1.522	1.524–1.530	microcline	77–84	c	342
1.523–1.530	1.518–1.525	1.525–1.531	sanidine	0 –60	c	337
1.523–1.530	1.518–1.526	1.524–1.533	orthoclase	35–85	c, gray	340
1.524–1.534	1.519–1.529	1.527–1.536	anorthoclase	34–60	c	344
1.527–1.535	1.507–1.515	1.529–1.536	glauberite	0 –7	c	133
1.530–1.543	1.527–1.539	1.532–1.547	hyalophane	74–78	c	340
1.530–1.564	1.529–1.559	1.537–1.567	chrysotile	30–50	c, p. green	309
1.531–1.542	1.525–1.536	1.533–1.546	barian sanidine	30–60	c	337
1.543–1.547	1.539–1.543	1.548–1.552	oligoclase	89–83 (low) 52–73 (high)	c	366
1.545–1.581	1.525–1.561	1.545–1.581	vermiculite	0 –8	green, p. brown	298
1.547–1.551	1.543–1.548	1.552–1.555	andesine	83–89 (low) 73–89 (high)	c	367
1.551–1.580	1.525–1.548	1.554–1.586	lepidolite	25–58	c	282
1.555–1.600	1.535–1.570	1.565–1.605	hydromuscovite	5 –35	c	324
1.555–1.612	1.530–1.580	1.560–1.615	nontronite	5 –66	green, yellow	323
1.557–1.617	1.530–1.573	1.558–1.618	phlogopite	0 –12	c, buff, brown	283
1.558–1.562	1.546–1.548	1.567	polyhalite	60–62	c	134
1.559–1.569	1.553–1.563	1.560–1.570	kaolinite	23–60	c, gray	316
1.560–1.573	1.555–1.567	1.560–1.573	antigorite	20–50	c, p. green	310
1.562	1.557	1.569	aluminian glauconite	20	green	289
1.570–1.578	1.565–1.572	1.575–1.583	bytownite	89–79	c	369
1.575–1.594	1.538–1.550	1.575–1.600	talc	5 –30	c, p. green	307
1.576–1.600	1.576–1.595	1.579–1.600	pennine	0 –40	green	301

1.578–1.583	1.572–1.576	1.583–1.588	anorthite	79–77	green	292
1.581–1.613	1.548–1.612	1.582–1.613	manganophyllite	4–9	c	370
1.582–1.610	1.552–1.574	1.587–1.616	muscovite	30–47	p. brown	284
1.586–1.588	1.534–1.552	1.596–1.600	pyrophyllite	53–62	c	279
1.593–1.604	1.559–1.571	1.595–1.612	fuchsite	32–46	c	312
1.605–1.651	1.605–1.648	1.610–1.652	prochlorite	0–10	blue, green	280
1.605–1.675	1.565–1.625	1.605–1.675	biotite	0–25	green	304
1.606	1.592	1.617	chondrodite	80	brown, green	285
1.609–1.643	1.590–1.612	1.610–1.644	glauconite	0–20	c, yellow green, olive green	152
1.613–1.625	1.602–1.615	1.624–1.635	hornblende	36–60	green	289
1.613–1.626	1.599–1.612	1.625–1.637	tremolite	88–80	c	258
1.615–1.623	1.578–1.583	1.615–1.623	minnesotaite	small	p. green	253
1.615–1.650	1.606–1.637	1.627–1.655	glaucophane	10–80	blue, violet	307
1.616–1.661	1.605–1.651	1.623–1.670	soda tremolite	64–87	yellow, green, blue green	273
1.620–1.665	1.595–1.638	1.627–1.650	chamosite	5–10	green	274
1.625–1.648	1.616–1.645	1.636–1.660	margarite	26–67	c, p. green	306
1.626–1.655	1.613–1.628	1.638–1.655	hornblende	40–65	green	294
1.627–1.644	1.615–1.646	1.629–1.662	actinolite	84–73	p. green	258
1.627–1.659	1.604–1.620	1.642–1.659	wollastonite	35–63	gray	255
1.631–1.648	1.617	1.652	lazulite	65–70	blue	242
1.635	1.599	1.643	tilleyite	89	c	138
1.642			titanian phlogopite	5	red, brown	149
1.645–1.666	1.640–1.659	1.652–1.670	crossite	12–65	yellow, blue, violet	284
1.648–1.667	1.621–1.637	1.659–1.677	scorzalite	61–65	blue	274
1.656–1.675	1.646–1.665	1.661–1.685	hornblende	60–85	brown, green	138
1.660–1.705	1.652–1.699	1.666–1.708	arfvedsonite	30–70	d. brown, blue green	258
1.661–1.690	1.653–1.670	1.669–1.700	magnesiohastingsite	60–89	green greenish yellow, blue green	269 266

TABLE I

383

Biaxial Minerals (Cont.)

Monoclinic and Triclinic   Inclined extinction in one or more principal sections (*Cont.*)

NEGATIVE (−) (*Cont.*)

β	α	γ	SPECIES	2V	COLOR (IN T.S.) (C = COLORLESS)	PAGE
1.674-1.676	1.640-1.641	1.679-1.681	spurrite	40	c	149
1.674-1.716	1.662-1.708	1.676-1.724	bustamite	44-85	c	244
1.676-1.695	1.666-1.680	1.686-1.700	hornblende	65-85	brown, green	258
1.677-1.698	1.672-1.690	1.681-1.699	axinite	69-85	c, lavender	156
1.680-1.709	1.663-1.686	1.696-1.729	grunerite	89-70	c, gray	252
1.682-1.695	1.669-1.680	1.684-1.705	femaghastingsite	40-81	yellow, p. green, blue green	266
1.683-1.700	1.680-1.698	1.685-1.706	riebeckite	70-87	d. blue	271
1.683-1.769	1.650-1.702	1.689-1.796	oxyhornblende	56-88	d. brown	264
1.694-1.731	1.679-1.705	1.703-1.732	ferrohastingsite	12-76	blue green, yellow green, brown	266
1.696-1.702	1.681-1.690	1.701-1.705	hornblende	65-89	brown, d. green	258
1.700-1.707	1.687-1.694	1.701-1.712	barkevikite	40-75	d. brown	266
1.703-1.732	1.701-1.729	1.705-1.734	sapphirine	51-69	p. blue, gray	172
1.714-1.741	1.711-1.738	1.724-1.751	rhodonite	61-76	c	240
1.718-1.815	1.715-1.791	1.733-1.822	allanite	40-80	yellow, brown, red brown	194
1.720-1.725	1.712-1.718	1.728-1.734	kyanite	82	c	165
1.720-1.734	1.715-1.730	1.725-1.740	chloritoid	89-55	c, green	297
1.730-1.784	1.723-1.751	1.736-1.797	epidote	89-64	c, p. yellow	191
1.740-1.819	1.720-1.778	1.757-1.839	aegirine	60-89	green, brown	230
2.19	2.13	2.20	baddeleyite	30	c, brown, black	76

TABLE II   COLORED SPECIES ARRANGED BY (1) COLOR AND (2) OPTICAL GROUPS

		YELLOW	ORANGE, GOLDEN AMBER	RED, PINK	GREEN	BLUE	VIOLET, LILAC	BROWN, TAN
	ISOTROPIC	allanite perovskite	allanite	hackmanite pyrope almandite limonite hematite	serpentine greenalite pleonaste hercynite	fluorite sodalite nosean haüyne lazurlite	fluorite haüyne	opal allanite collophane andradite melanite picotite limonite chromite perovskite
UNIAXIAL	POSITIVE (+)	bastnäsite xenotime zircon cassiterite	cassiterite	zircon cassiterite rutile	clinochlore pennine prochlorite vesuvianite	osumilite	zircon	vesuvianite bastnäsite cassiterite rutile
UNIAXIAL	NEGATIVE (−)	stilpnomelane dravite jarosite natrojarosite	phlogopite dravite	corundum goethite hematite	talc pennine stilpnomelane biotite prochlorite glauconite schorl vesuvianite	schorl corundum anatase		vermiculite phlogopite stilpnomelane biotite carbonate–apatite vesuvianite ankerite siderite anatase

TABLE II   385

TABLE II (Cont.)

	YELLOW	ORANGE, GOLDEN AMBER	RED, PINK	GREEN	BLUE	VIOLET, LILAC	BROWN, TAN
ORTHORHOMBIC — POSITIVE (+)	humite staurolite perovskite brookite	staurolite perovskite brookite	thulite iddingsite diaspore hypersthene perovskite	anthophyllite vesuvianite hypersthene	diaspore		anthophyllite iddingsite vesuvianite staurolite perovskite brookite
ORTHORHOMBIC — NEGATIVE (−)	fayalite	andalusite	andalusite hypersthene dumortierite iddingsite corundum goethite	andalusite hypersthene dumortierite vesuvianite	dumortierite corundum	dumortierite	anthophyllite iddingsite vesuvianite
MONOCLINIC AND TRICLINIC — POSITIVE (+)	chondrodite clinohumite piedmontite monazite	chondrodite clinohumite piedmontite	piedmontite	chrysotile clinochlore pennine prochlorite edenite omphacite jadeite diopside hedenbergite pumpellyite aegirine–augite augite pigeonite chloritoid	riebeckite	augite piedmontite	hedenbergite sphene

BIAXIAL

	YELLOW	ORANGE, GOLDEN, AMBER	RED, PINK	GREEN	BLUE	VIOLET, LILAC	BROWN, TAN
BIAXIAL / **MONOCLINIC AND TRICLINIC** / **NEGATIVE (−)**	nontronite stilpnomelane chondrodite soda tremolite crossite hastingsite allanite epidote	phlogopite chondrodite	axinite	crysotile vermiculite nontronite antigorite glauconite talc pennine stilpnomelane fuchsite prochlorite hornblende minnesotaite soda tremolite chamosite margarite actinolite arfvedsonite hastingsite chloritoid aegirine	fuchsite glaucophane soda tremolite lazulite crossite scorzalite hastingsite riebeckite sapphirine	glaucophane crossite axinite holmquistite	vermiculite phlogopite stilpnomelane mangano-phyllite hornblende arfvedsonite oxyhorn-blende hastingsite barkevikite allanite acmite baddeleyite

TABLE II

TABLE II (Cont.)

Minerals That May Show Conspicuous Color Zoning

ORANGE	cassiterite, dravite, allanite
RED, PINK	corundum, piedmontite, thulite, andalusite
GREEN	hornblende, hastingsite, soda–tremolite, aegirine, aegirine–augite, chloritoid, pennine
BLUE	haüyne, corundum, anatase, schorl, glaucophane, riebeckite, crossite
VIOLET, LILAC	zircon, titanian augite
BROWN	melanite, grossularite, brookite, anatase, vesuvianite, dravite, ankerite, allanite, phlogopite, biotite, hornblende, arfvedsonite, oxyhornblende

Minerals That Show Sector ("Hourglass") Zoning

prehnite, chloritoid, titanian augite, hornblende (brown)

Minerals That May Show Irregular ("Blotchy") Color Variations

ORANGE	cassiterite, dravite
RED	corundum, andalusite
BLUE	haüyne, corundum, schorl, glaucophane
PURPLE	fluorite
BROWN	allanite

Minerals in Which Pleochroic Haloes Are Found Developed Around Radioactive Inclusions

muscovite, biotite, phlogopite, lepidolite, chlorites, hornblende, cordierite

Also listed is "relief in thin section," for which the following designations
apply:

INDEX	RELIEF	ABBREVIATION
< 1.45	negative, moderate	−M
1.54–1.45	negative, low	−L
1.54	neutral	N
1.54–1.60	positive, low	+L
1.60–1.80	positive, moderate	+M
1.80–2.00	positive, high	+H
> 2.00	positive, extreme	+E

For a thin section with a thickness, $t = 0.025$ mm, the following applies:

$n_z - n_1$ (BIREFRINGENCE)	ORDER OF INTERFERENCE COLORS
.001–.024	first
.025–.044	second
.045–.065	third
> .065	fourth and higher

Uniaxial Minerals

Positive (+)

BIREFRINGENCE	RELIEF IN T.S.	SPECIES
.000–.003	−L	apophyllite
.001–.005	−L	heulandite
.001–.006	+L, M	prochlorite
.002–.004	+L	pennine
.002–.010	−L	chabazite
.004	+L	osumilite
.005	+M	vesuvianite
.006	+M	åkermanite
.006–.010	+L	clinochlore
.008–.009	N	chalcedony
.009	N	quartz
.010	+L	natroalunite
.010–.021	+L	brucite
.013	−L	tincalconite
.021–.023	+L	alunite
.042–.062	+H	zircon
.095–.107	+M, H	xenotime

TABLE III   389

TABLE III (Cont.)

Uniaxial Minerals (Cont.)

Positive (+) (Cont.)

BIREFRINGENCE	RELIEF IN T.S.	SPECIES
.096–.099	+H, E	cassiterite
.100–.102	+M, H	bastnäsite
.285–.287	+E	rutile

NEGATIVE (−)

.000–.003	−L, N	apophyllite
.001–.006	+M	prochlorite
.001–.012	+M	vesuvianite
.002–.006	−L	cristobalite
.002–.010	−L	chabazite
.003–.005	+M	fluorapatite
.003–.005	+L	pennine
.003–.005	−L, +L	nepheline
.004–.007	+M	hydroxylapatite
.004–.008	+L	beryl
.005	N	kalsilite
.005–.006	+M	mangan–fluorapatite
.005–.029	−L	cancrinite
.006–.011	+M	gehlenite
.007–.010	+M	corundum
.007–.017	+M	carbonate apatite
.008–.012	−L, +L	scapolite (marialite)
.009–.034	+L	scapolite (mizzonite, dipyre, wernerite)
.012–.040	+M	schorl
.015–.024	+M	elbaite
.019	+M	glauconite
.019–.026	+M	dravite
.025	−L, +L	vermiculite
.027	+L	stilpnomelane
.028–.044	+L	phlogopite
.029–.040	+L	lepidolite
.033–.038	+L	scapolite (meionite)
.035–.050	+L	talc
.040–.042	+L, M	muscovite (phengite)
.040–.060	+L, M	biotite
.073	+E	anatase
.080	+M, H	natrojarosite
.101–.105	+M, H	jarosite
.140	+E	goethite
.172	−L, +M	calcite
.172–.194	−L, +M	manganoan calcite

TABLE III (*Cont.*)

Uniaxial Minerals (*Cont.*)

Negative (—) (*Cont.*)

BIREFRINGENCE	RELIEF IN T.S.	SPECIES
.180	−L, +M	dolomite
.180–.203	−L, +M	ankeritic dolomite
.191	−L, +M	magnesite
.191–.220	−L, +M	ferroan magnesite
.144–.215	+L, H	calcian rhodochrosite
.202	+L, M	ankerite
.218–.242	+L, H	siderite
.219	+L, H	rhodochrosite
.28	+E	hematite

Biaxial Minerals

POSITIVE (+)

.001–.006	+M	prochlorite
.001–.007	−L	heulandite
.002–.004	−L	tridymite
.002–.004	+L	pennine
.002–.010	−L	chabazite
.002–.020	+M	pumpellyite
.003–.008	+L	dickite
.003–.009	+M	riebeckite
.004–.016	−L, +L	chrysotile
.004–.023	+M	clinozoisite
.005–.015	+L	clinochlore
.005–.016	+M	zoisite
.006	+M	vesuvianite
.006	+E	perovskite
.006–.022	+M	chloritoid
.006–.028	−L	thomsonite
.007–.009	+L	labradorite
.007–.012	−L, +L	cordierite (low, hydrous)
.007–.015	−L, +L	cordierite (high, anhydrous)
.007–.022	+M	thulite
.008–.009	+M	celestite
.008–.009	+L	andesine
.008–.010	+L	bytownite
.009	−L, +L	oligoclase
.009–.010	−L	gypsum
.009–.010	−L	albite
.009–.011	+M	enstatite

TABLE III   391

TABLE III *(Cont.)*

Biaxial Minerals *(Cont.)*

Positive (+) *(Cont.)*

BIREFRINGENCE	RELIEF IN T.S.	SPECIES
.009–.011	+M	topaz
.009–.015	+M	staurolite
.009–.018	+M	ferrian zoisite
.010	+M	clinoenstatite
.010	+M	merwinite
.010–.012	+L	celsian
.010–.013	+M	barite
.011–.014	+M	rhodonite
.011–.014	−L	natrolite
.012–.027	+M	jadeite
.012–.031	+M	mullite
.013–.022	+M	bustamite
.013–.025	+M	anthophyllite
.014–.027	+M	spodumene
.015	+M	boehmite
.016–.019	+L	gibbsite
.016–.023	+M	edenite
.017–.018	+M	ferroan hypersthene
.018–.029	+M	hedenbergite
.018–.039	+M	aegirine–augite
.019–.021	+M	lawsonite
.020–.023	+M	sillimanite
.020–.035	+M	prehnite
.021	−L	thenardite
.021–.028	+M	pigeonite
.022	+M	scawtite
.023	−L	ulexite
.023	+M	larnite
.023–.030	+M	omphacite
.024–.028	+M	augite
.025–.037	+M	chondrodite
.025–.038	+M	cummingtonite
.026–.029	+M	johannsenite
.026–.030	+M	ferroan augite
.027	+L	norbergite
.027	−L	bischofite
.028	+L, M	colemanite
.028–.031	−L	carnallite
.028–.031	+M	diopside
.028–.036	+M	humite
.028–.045	+M	clinohumite
.028–.082	+M, H	piedmontite

TABLE III (Cont.)

Biaxial Minerals (Cont.)

Positive (+) (Cont.)

BIREFRINGENCE	RELIEF IN T.S.	SPECIES
.029-.031	+M	clinoferrosilite
.032-.038	+M	pectolite
.034-.041	+M	pseudowollastonite
.035	+M	tilleyite
.036-.042	+M	forsterite
.04 -.05	+M	iddingsite
.044-.047	+L, M	anhydrite
.045-.055	+M, H	monazite
.048-.052	+M	diaspore
.063	−L, +L	kieserite
.100-.182	+H, E	sphene
.117	+E	brookite

Negative (−)

BIREFRINGENCE	RELIEF IN T.S.	SPECIES
.001-.006	+M	prochlorite
.002-.004	+L	pennine
.002-.010	−L	chabazite
.003-.009	+M	riebeckite
.004-.009	+L	antigorite
.004-.015	+M	crossite
.004-.016	−L, +L	chrysotile
.005-.008	−L, +L	hyalophane
.005-.008	−L	anorthoclase
.005-.008	−L	orthoclase
.005-.008	−L	sanidine
.005-.009	+M	vesuvianite
.005-.010	+L	kaolinite
.005-.014	+M	arfvedsonite
.006-.007	−L	adularia
.006-.013	−L	stilbite
.007	−L	microcline
.007-.008	+M	chamosite
.007-.008	+M	corundum
.007-.012	−L, +L	cordierite (low, hydrous)
.007-.015	−L, +L	cordierite (high, anhydrous)
.008-.009	+L	andesine
.008-.010	−L, +L	barian sanidine
.008-.010	+M	bytownite
.008-.016	−L	laumontite
.009	+L	oligoclase

TABLE III   393

TABLE III (Cont.)

Biaxial Minerals (Cont.)

Negative (—) (Cont.)

BIREFRINGENCE	RELIEF IN T.S.	SPECIES
.009–.011	+M	andalusite
.009–.013	+M	axinite
.009–.020	+M	chloritoid
.010–.032	+M	margarite
.011–.012	+M	anorthite
.011–.014	+M	rhodonite
.011–.017	+M	hypersthene
.011–.027	+M	dumortierite
.012–.015	+M	kyanite
.013–.017	+M	wollastonite
.013–.020	+M	anthophyllite
.013–.022	+M	bustamite
.013–.046	+M	epidote
.014–.017	+M	monticellite
.014–.023	+M	barkevikite
.014–.029	+M	ferrohastingsite
.014–.032	+L, M	glauconite
.015–.030	+M	hornblende
.015–.031	+M, H	allanite
.015–.033	+M	femaghastingsite
.016–.022	+M	soda tremolite
.016–.030	+M	magnesiohastingsite
.018–.021	+M	glaucophane
.018–.038	+L	lepidolite
.019–.021	+L	polyhalite
.020–.031	−L, +L	vermiculite
.020–.035	−L, +L	montmorillonite
.020–.094	+M	oxyhornblende
.021–.023	−L, +L	glauberite
.021–.034	+M	hornblende
.022	−L	kainite
.022–.027	+M	tremolite
.024–.028	+M	actinolite
.024–.040	+L, M	manganophyllite
.025	−M, L	borax
.025	+M	chondrodite
.025–.037	−L, +L, M	hydromuscovite
.026–.040	−L, +L, M	nontronite
.028	−M, L	epsomite
.028–.049	−L, +L, M	phlogopite
.030–.052	+L	talc
.030–.119	+L, M	stilpnomelane

TABLE III (Cont.)

Biaxial Minerals (Cont.)

Negative (—) (Cont.)

BIREFRINGENCE	RELIEF IN T.S.	SPECIES
.032–.060	+M, H	aegirine
.033–.036	+M	forsterite
.034	−L	kernite
.034–.037	+M	lazulite
.035	+M	tilleyite
.035–.042	+L, M	fuchsite
.035–.045	+L, M	minnesotaite
.036–.049	+L, M	muscovite
.037–.040	+M	scorzalite
.038	+M	tephroite
.038–.045	+M	grunerite
.039	+M	spurrite
.04 −.05	+M	iddingsite
.042–.051	+M, H	fayalite
.044	+M	titanian phlogopite
.045–.051	+M	glaucochroite
.046–.062	−L, +L	pyrophyllite
.07	+E	baddeleyite
.128	−M, N	trona
.14	+E	goethite
.147–.150	−L, +L, M	strontianite
.155–.156	−L, +M	aragonite
.208	−M, L, +L	nahcolite

Minerals That May Display Abnormal Interference Colors

anatase	clinohumite (titanian)	pennine
antigorite	clinozoisite	prehnite
apophyllite	crossite	sanidine
augite (titanian)	epidote	sphene
borax	glaucophane	vesuvianite
brookite	melilite	zoisite
brucite		

The *interference color chart* is a graphical representation of the equation $\Delta = t(n_2 - n_1)$ in terms of the interference colors that result under white light, assuming no dispersion of the birefringence. For each mineral of specific chemical composition, the greatest double refraction (characteristic birefringence), where $n_2 - n_1$ is at its maximum, is a constant. If such values are substituted, the equation becomes (e.g.):

$$\Delta = tK_{quartz} \quad \text{or} \quad \Delta = t \times 0.009$$
$$\Delta = tK_{calcite} \quad \text{or} \quad \Delta = t \times 0.172$$

Thus for each species there can be plotted a straight line that relates birefringence to thickness of the plate or grain.

The chart is divided into groups of interference color, or *orders*. For the first order $\Delta = 0$–550 mμ; the second order includes $\Delta = 550$–1128 mμ; the third order comprises $\Delta = 1128$–1652 mμ. With increasing order number brilliance of the colors diminishes, fourth- and higher-order colors becoming pastel shades which tend to intergrade, the net result approximating a creamy white or gray ("high-order grays").

The chart is useful in several ways. The thickness of the section can be estimated, a determination usually made by noting the maximum interference color displayed by a known species. For example, since most thin sections should be 0.02–0.03 mm thick, the maximum interference color for quartz (max $n_2 - n_1 = 0.009$) should be first-order white, $\Delta = $ ca. 200 mμ. If some quartz grains display first-order yellow ($\Delta = 300$ mμ), the section is thicker than normal ($t = 0.034$ mm). Thus, by knowing the section thickness it is possible to determine the birefringence for an unknown species. With $t = 0.025$ mm and the maximum interference color shown by the unknown being a second-order green ($\Delta = 800$ mμ), the birefringence can be read off at the top of the chart by following the diagonal birefringence line that crosses the intersection of $t = 0.025$ and $\Delta = 800$ mμ, which in this case is $n_2 - n_1 = 0.032$, the same as that for an epidote. If the species has been identified, the thickness of a grain may be read from the chart after noting the maximum interference color displayed. In crushed material interference color bands distribution reveals the topography of grains (Fig. 3.2).

TABLE IV   MINERALS THAT COMMONLY SHOW TWINNING

Grouped (1) by types of twins: paired; multilamellar (polysynthetic), either in one direction or in two or more directions; and cyclic or sector and (2) by optical group.

| | | MULTILAMELLAR (POLYSYNTHETIC) | | |
	PAIRED	ONE DIRECTION	TWO OR MORE DIRECTIONS	CYCLIC, SECTOR
ISOMETRIC (VERY WEAK BIREFRINGENCE)		perovskite	leucite	grossularite
UNIAXIAL — POSITIVE (+)		rutile cassiterite		apophyllite rutile
UNIAXIAL — NEGATIVE (−)		corundum calcite dolomite	cristobalite	jarosite apatite
BIAXIAL — POSITIVE (+)	forsterite chondrodite diopside augite aegirine–augite cummingtonite edenite staurolite cordierite sphene	carnallite gypsum barite clinohumite zoisite pumpellyite diopside hedenbergite augite pigeonite aegirine–augite jadeite rhodonite pseudowollas- tonite larnite cummingtonite lawsonite cordierite sphene clinochlore	gibbsite anhydrite tridymite chondrodite lawsonite cordierite labradorite bytownite	tridymite vesuvianite staurolite cordierite prehnite chabazite

TABLE IV   397

TABLE IV   MINERALS THAT COMMONLY SHOW TWINNING (*Cont.*)

| | | MULTILAMELLAR (POLYSYNTHETIC) | | |
	PAIRED	ONE DIRECTION	TWO OR MORE DIRECTIONS	CYCLIC, SECTOR
BIAXIAL POSITIVE (+)		albite oligoclase andesine labradorite bytownite		
BIAXIAL NEGATIVE (−)	fayalite spurrite cordierite kyanite actinolite hornblende hastingsite arfvedsonite orthoclase sanidine	baddeleyite polyhalite strontianite lazulite epidote aegirine rhodonite spurrite kyanite cordierite grunerite tremolite actinolite hornblende margarite oligoclase andesine bytownite anorthite	baddeleyite polyhalite spurrite cordierite microcline anorthoclase adularia bytownite anorthite	aragonite strontianite monticellite vesuvianite cordierite dumortierite stilbite adularia

REFERENCES

Adamson, Olge J. (1944)   The petrology of the Norra Kärr district. *Geol. Fören. Förh.*, **66**, 113–255.

Allen, Roy M. (1954)   "Practical Refractometry by Means of the Microscope." R. P. Cargille Lab., Inc.

Andersen, Olaf (1928)   The genesis of some types of feldspar from granite pegmatites. *Norsk. Geol. Tidssk.*, **10**, 113–205.

Bailey, Edgar M., and Rollin E. Stevens (1960)   Selective staining of plagioclase and K feldspar on rock slabs and thin sections (abs.). *Bull. Geol. Soc. Am.*, **71**, 2047.

Bastin, E. S. (1911)   Geology of the pegmatites and associated rocks of Maine, including feldspar, quartz, mica, and gem deposits. *U.S. Geol. Surv. Bull.* **445.**

Baumann, Henry N., Jr. (1957)   The preparation of petrographic sections with bonded diamond wheels. *Am. Mineral.*, **42**, 416–421.

Berman, Harry (1937)   Constitution and classification of the natural silicates. *Am. Mineral.*, **22**, 342–408.

Billings, M. P. (1928)   The chemistry, optics, and genesis of the hastingsite group of amphiboles. *Am. Mineral.*, **13**, 287–296.

Bosazza, V. L. (1941)   Occurrence of a coarsely crystalline kaolin mineral in some South African fire-clays. *Am. Mineral.*, **26**, 290–292.

Bowen, N. L., J. F. Schairer, and E. Posnjak (1933)   The system $CaO-FeO-SiO_2$. *Am. Jour. Sci.* (5th Ser.), **26**, 193–284.

Braitsch, Otto (1957)   Über die natürlichen Faser- und Aggregationstypen beim SiO_2, ihre Verwachsungsformen, Richtungsstatistik and Doppelbrechung. *Heidelb. Beitr. Mineral. Petrog.*, **5**, 331–372.

Brammall, Alfred (1928)   Dartmoor detritals: A study in provenance. *Proc. Geol. Assoc.*, **39**, 31–34.

Brown, George, and Keith Norrish (1952) Hydrous micas. *Mineral. Mag.*, **29**, 929–932.

———— and I. Stephen (1959) A structural study of iddingsite from New South Wales, Australia. *Am. Mineral.*, **44**, 251–260.

Brown, M. C., and P, Gay (1959) Identification of oriented inclusions in pyroxene crystals. *Am. Mineral.*, **44**, 592–602.

Burri, C. (1950) "Das polarisations Mikroskop." E. Birkhäuser und Cie.

———— (1941) Zur optischen Bestimmung der orthorhombischen Pyroxene. *Schweiz. Mineral. Petrog. Mitt.*, **21**, 177–182.

Burst, J. F. (1958) Mineral heterogeneity in "glauconite" pellets. *Am. Mineral.*, **43**, 481–497.

Butler, Robert D. (1933) Immersion liquids of intermediate refraction (1.450–1.630). *Am. Mineral.*, **18**, 386–401.

Chayes, F. (1952) Relations between composition and indices of refraction in natural plagioclases. *Am. Jour. Sci., Bowen Vol.*, 85–105.

Clavan, Walter, Wallace M. McNabb, and Edward H. Watson (1954) Some hypersthenes from southeastern Pennsylvania and Delaware. *Am. Mineral.*, **39**, 566–580.

Coombs, D. S. (1953) The pumpellyite mineral series. *Mineral. Mag.*, **30**, 113–135.

Deer, W. A. (1937) The composition and paragenesis of the hornblendes of the Glen Tilt complex, Perthshire. *Mineral. Mag.*, **25**, 56–74.

————, R. A. Howie, and J. Zussman (1962, 1963) "Rock-forming Minerals." Vols. I–IV. John Wiley & Sons, Inc., New York.

———— and L. R. Wager (1938) Two new pyroxenes included in the system clinoenstatite, clinoferrosilite, diopside, and hedenbergite. *Mineral. Mag.*, **25**, 15–22.

Dodge, Nelson B. (1948) The dark-field color immersion method. *Am. Mineral.*, **33**, 541–549.

Donnay, J. D. H. (1945) Form birefringence of nemalite. *Univ. Toronto Studies, Geol. Ser.*, **49**, 5–15.

Emmons, R. C. (1943) The universal stage. *Geol. Soc. Am. Mem.*, **8**.

———— (ed.) and others (1953) Selected petrogenetic relationships of plagioclase. *Geol. Soc. Am. Mem.*, **52**.

———— and R. M. Gates (1948) The use of Becke line colors in refractive index determination. *Am. Mineral.*, **33**, 612–618.

Fairbairn, H. W. (1943) Gelatin-coated slides for refractive index immersion mounts. *Am. Mineral.*, **28**, 396–397.

Ferguson, J. B., and A. F. Buddington (1920) The binary system åkermanite-gehlenite. *Am. Jour. Sci.*, **50**, 131–140.

Foslie, Steinar (1945) Hastingsite and amphiboles from the epidote-amphibolite facies. *Norsk. Geol. Tidssk.*, **25**, 74–98.

Fouqué, F. (1894) Contribution a l'étude des feldspaths des roches volcaniques. *Bull. Soc. franc. Mineral.*, **17** (7, 8).

Friedman, Gerald M. (1959) Identification of carbonate minerals by staining methods. *Jour. Sed. Petrol.*, **29**, 87–97.

Füchtbauer, Hans (1948) Einige Beobachtungen an authigenen Albiten. *Schweiz. Mineral. Petrog. Mitt.*, **28**, 709–716.

Goldich, Samuel S., and James H. Kinser (1939) Perthite from Tory Hill, Ontario. *Am. Mineral.*, **24**, 407–427.

Goldsmith, Julian R., Donald L. Graf, and Oiva I. Joensuu (1955) The occurrence of magnesian calcites in nature. *Geochim. Cosmochim. Acta, 7,* 212–230.

———— and Fritz Laves (1954) Potassium feldspars structurally intermediate between microcline and sanidine. *Geochim. Cosmochim. Acta, 6,* 100–118.

Gravenor, C. P. (1951) A graphical simplification of the relationship between 2V and N_x, N_y, and N_z. *Am. Mineral.*, **36**, 162–164.

Grout, Frank F. (1946) Acmite occurrences on the Cuyuna Range, Minnesota. *Am. Mineral.*, **31**, 125–130.

Halferdahl, L. B. (1961) Chloritoid, its composition, x-ray and optical properties, stability and occurrence. *Jour. Petrology, 2,* 49–135.

Hallimond, Arthur F. (1953) "Manual of the Polarizing Microscope," 2d ed. Cooke, Troughton and Simms, York, England.

Hardy, Arthur C., and Fred H. Perrin (1932) "The Principles of Optics." McGraw-Hill Book Company, New York.

Harrington, V. F., and M. J. Buerger (1931) Immersion liquids of low refraction. *Am. Mineral.*, **16**, 45–54.

Hartshorne, N. H., and A. Stuart (1950) "Crystals and the Polarizing Microscope," 2d ed. Edward Arnold (Publishers) Ltd., London.

Hauswaldt, Hans (1902) "Interferenzerscheinungen an doppeltbrechenden Krystallplatten im konvergenten polarisirten Licht." Magdeburg, 33 plates.

———— (1904) "Interferenzerscheinungen im polarisirten Licht," Neue Folge. Magdeburg, 80 plates.

———— (1907) "Interferenzerscheinungen im polarisirten Licht," Dritte Reihe. Magdeburg, 72 plates.

Heinrich, E. Wm. (1956) "Microscopic Petrography." McGraw-Hill Book Company, New York.

Hess, H. H. (1949) Chemical composition and optical properties of common clinopyroxenes. *Am. Mineral, 34,* 621–666.

———— (1952) Orthopyroxenes of the Bushweld type, ion substitutions and changes in unit cell dimensions. *Am. Jour. Sci., Bowen Vol.,* 173–187.

——— (1941) Pyroxenes of common mafic magmas. Part 1. *Am. Mineral*, **26**, 515–535. Part 2. *Am. Mineral*, **26**, 573–594.

Hey, Max H. (1932) Studies on the zeolites. Part II. Thomsonite (including faroelite) and gonnardite. *Mineral. Mag.*, **23**, 51–125.

Howie, R. A. (1963) Cell parameters of orthopyroxenes. *Mineral. Soc. Am. Spec. Paper 1*, 213–222.

Hunt, Walter F., and George T. Faust (1937) Pencatite from the Organ Mountains, New Mexico. *Am. Mineral*, **22**, 1151–1160.

Hurlbut, C. S., Jr. (1947) An improved heating and circulating system to use in double-variation procedure. *Am. Mineral*, **32**, 487–492.

——— and John L. Rosenfeld (1952) Monochromator utilizing the rotary power of quartz. *Am. Mineral*, **37**, 158–165.

Hutton, C. Osborne (1956) Further data on the stilpnomelane mineral group. *Am. Mineral*, **41**, 608–615.

——— (1950) Studies of heavy detrital minerals. *Bull. Geol. Soc. Am.*, **61**, 635–716.

Iiyama, Toshimichi (1956) Optical properties and unit cell dimensions of cordierite and indialite. *Mineral. Jour. Jap.*, **1**, 372–394.

Johannsen, Albert (1918) "Manual of Petrographic Methods," 2d ed. McGraw-Hill Book Company, New York.

——— (1922) "Essentials for Microscopical Determination of Rock-forming Minerals and Rocks in Thin Section." The University of Chicago Press, Chicago.

Johnston, R. W. (1949) Clinozoisite from Camaderry Mountain, Wicklow. *Mineral. Mag.*, **28**, 505–515.

Juurinen, Aarno (1956) Composition and properties of staurolite. *Ann. Acad. Sci. Fenn. Ser. A III, Geol.-Geogr.*, **47**.

Kaaden, Gerrit van der (1951) Optical studies on natural plagioclase feldspars with high- and low-temperature optics. Diss. Utrecht Ryksuniv.

Kaiser, E. P., and William Parrish (1939) Preparation of immersion liquids for the range $n_D = 1.411–1.785$. *Ind. Eng. Chem.*, **11**, 560–562.

Kamb, W. Barclay (1958) Isogyres in interference figures. *Am. Mineral*, **43**, 1029–1067.

Keller, W. D. (1956) Clay minerals as influenced by environments of their formation. *Bull. Am. Assoc. Petrol. Geol.*, **40**, 2689–2710.

Kennedy, George C. (1947) Charts for correlation of optical properties with chemical composition of some common rock-forming minerals. *Am. Mineral*, **32**, 561–573.

Kerr, Paul F. (1959) "Optical Mineralogy," 3d ed. McGraw-Hill Book Company, New York.

Klein, Ira (1939)   Microcline in the native copper deposits of Michigan. *Am. Mineral*, **24**, 643–650.

Köhler, Alexander (1941)   Die Abhangigkeit der Plagioklasoptik vom vorange-gangen Wärmeverhalten. *Min. Petr. Mitt.*, **53**, 24–66.

———— (1942)   Drehtischmessungen an Plagioklaszwillingen von Tief und Hochtemperaturoptik. *Min. Petr. Mitt.*, **53**, 159–221.

———— (1948)   Zur Optik des Adulars. *Neues Jahrb. Min.*, A (5–8), 49–55.

———— (1949)   Recent results of investigations of the feldspars. *Jour. Geol.*, **57**, 592–599.

Krumbein, W. C., and F. J. Pettijohn (1938)   "Manual of Sedimentary Petrography." Appleton-Century-Crofts, Inc., New York.

Kuno, H. (1936)   Petrological notes on some pyroxene–andesites from Hakone volcano, with special reference to some types of pigeonite phenocrysts. *Jap. Jour. Geol. Geogr.*, **13**, 107–140.

———— (1950)   Petrology of Hakone volcano and the adjacent areas, Japan. *Bull. Geol. Soc. Am.*, **61**, 957–1020.

Larsen, E. S. (1921)   The microscopic determination of the nonopaque minerals. *U.S. Geol. Surv. Bull.*, **679**.

———— (1942)   Alkalic rocks of Iron Hill, Gunnison County, Colo. *U.S. Geol. Surv. Prof. Paper*, **197-A**.

———— and Harry Berman (1934)   The microscopic determination of the non-opaque minerals. *U.S. Geol. Surv. Bull.*, **848**.

————, C. S. Hurlbut, Jr., B. F. Buie, and C. H. Burgess (1941)   Igneous rocks of the Highwood Mountains, Montana. Part VI. Mineralogy. *Bull. Geol. Soc. Am.*, **52**, 1841–1856.

————, John Irving, F. A. Gonyer, and Esper S. Larsen, 3d (1938)   Petrologic results of a study of the minerals from the Tertiary volcanic rocks of the San Juan Region, Colorado. *Am. Mineral*, **23**, 227–257.

Leake, Bernard E. (1960)   Compilation of chemical analyses and physical constants of natural cordierites. *Am. Mineral.*, **45**, 282–298.

Loupekine, I. S. (1947)   Graphical derivation of refractive index ϵ for the trigonal carbonates. *Am. Mineral*, **32**, 502–507.

Lowell, W. R. (1952)   Phosphatic rocks in the Deer Creek–Wells Canyon area, Idaho. *U.S. Geol. Surv. Bull.*, **982-A**.

MacKenzie, W. S., and J. V. Smith (1956)   The alkali feldspars. III. An optical and x-ray study of high-temperature feldspars. *Am. Mineral*, **41**, 405–427.

Mallard, F. (1876)   Explication des phénomènes optiques anomaux, qui présentent un grand nombre de substances cristallisées. *Ann. Mines. Mem. X.*

Marshall, C. E., and C. D. Jeffries (1945)   Gelatin slide method for mounting soil separates for microscopic study. *Soil Sci. Soc. Am., Proc.*, **10**, 397–405.

Maschke, O. (1872) Über Abscheidung krystallisirter Kieselsäure aus wässerigen Lösungen. *Pogg. Ann.*, *5th Ser.*, **25**, 549–578.

McConnell, Duncan, and John W. Gruner (1940) The problem of the carbonate-apatites. III. Carbonate-apatite from Magnet Cove, Arkansas. *Am. Mineral*, **25**, 157–167.

Mertie, John B., Jr. (1942) Nomograms of optic angle formulae. *Am. Mineral*, **27**, 538–551.

Meyer, C. F. (1948) "Electromagnetic Waves and Light." J. W. Edwards, Publisher, Inc., Ann Arbor, Mich.

Meyer, Charles (1946) Notes on the cutting and polishing of thin sections. *Econ. Geol.*, **41**, 166–172.

Meyrowitz, Robert (1952) A new series of immersion liquids. *Am. Mineral*, **37**, 853–856.

——— (1955) A compilation and classification of immersion media of high index of refraction. *Am. Mineral*, **40**, 398–409.

——— (1956) Solvents and solutes for the preparation of immersion liquids of high index of refraction. *Am. Mineral*, **41**, 49–59.

——— and Esper S. Larsen, Jr. (1951) Immersion liquids of high refractive index. *Am. Mineral*, **36**, 746–750.

Michel-Lévy, A. (1877) De l'emploi du microscope polarisant a lumière parallèle. *Ann. Mines*, **12**, 392–471.

——— (1904) "Étude sur la détermination des feldspathes." Trois. fasc., Paris.

Milner, Henry B. (1952) "Sedimentary Petrography," 3d ed. Thomas Murby and Co., London.

Miyashiro, Akiho (1957*a*) Cordierite-indialite relations. *Am. Jour. Sci.*, **255**, 43–62.

——— (1957*b*) The chemistry, optics, and genesis of the alkali amphiboles. *Jour. Fac. Sci. Univ. Tokyo Ser. II*, **11** (1), 57–83.

——— (1956) Osumilite, a new silicate mineral and its crystal structure. *Am. Mineral*, **41**, 104–116.

——— and Toshimichi Iiyama (1954) A preliminary note on a new mineral, indialite, polymorphic with cordierite. *Japan Acad. Proc.*, **30**, 746–751.

———, Toshimichi Iiyama, Masao Yamasaki, and Tami Miyashiro (1955) The polymorphism of cordierite and indialite. *Am. Jour. Sci.*, **253**, 185–208.

Muir, I. D. (1951) The clinopyroxenes of the Skaergaard intrusion, eastern Greenland. *Mineral. Mag.*, **29**, 690–714.

Nagy, Bartholomew, and George T. Faust (1956) Serpentines; natural mixtures of chrysolite and antigorite. *Am. Mineral*, **41**, 817–838.

Ostrovsky, I. A. (1946) *Acad. Sci. U.S.S.R., Belyankin Jubilee Vol.*, 505–512.

Parker, Ronald B. (1961) Rapid determination of the approximate composition of amphiboles and pyroxenes. *Am. Mineral*, **46**, 892–900.

Pecora, William T., and Joseph J. Fahey (1950) The lazulite-scorzalite isomorphous series. *Am. Mineral,* **35**, 1–18.

Poldervaart, Arie (1950) Correlation of physical properties and chemical composition in the plagioclase, olivine, and orthopyroxene series. *Am. Mineral,* **35**, 1067–1079.

———— and H. H. Hess (1951) Pyroxenes in the crystallization of basaltic magma. *Jour. Geol.,* **59**, 472–489.

Posnjak, E., and H. E. Merwin (1922) The system Fe_2O_3-SO_3-H_2O. *Am. Chem. Soc., Jour.,* **44**, 1970.

Rabbitt, John C. (1950) A new study of the anthophyllite series. *Am. Mineral,* **35**, 263–323.

Rapp, George, Jr. (1960) Name clinozoisite should be dropped (abs.). *Bull. Geol. Soc. Am.,* **71**, 2039.

Reed, Frank S., and John L. Mergner (1953) Preparation of rock thin sections. *Am. Mineral,* **38**, 1184–1203.

Rinne, Friedrich W. B., and Max Berek (1953) "Anleitung zu optischen Untersuchungen mit dem Polarisationmikroskop," 2d ed. E. Schweizerbart, Stuttgart.

Rogers, A. F. (1906) The determination of minerals in crushed fragments by means of the polarizing microscope. *Columbia School Mines, Quart.,* **27**, 340–359.

Rosenblum, S. (1956) Improved techniques for staining potash feldspars. *Am. Mineral,* **41**, 662–664.

Rosenfeld, John L. (1950) Determination of all principal indices of refraction on difficultly oriented minerals by direct measurement. *Am. Mineral,* **35**, 902–905.

Ross, Clarence S. (1950) The dark field stereoscopic microscope for mineralogic studies. *Am. Mineral,* **35**, 906–910.

———— (1962) Microlites in glassy volcanic rocks. *Am. Mineral,* **47**, 723–740.

———— and S. B. Hendricks (1945) Minerals of the montmorillonite group, their origin and relation to soils and clays. *U.S. Geol. Surv. Prof. Paper,* **205-B.**

———— and P. F. Kerr (1930) The kaolin minerals. *U.S. Geol. Surv. Prof. Paper,* **165-E.**

Sabine, P. A. (1950) The optical properties and composition of the acmitic pyroxenes. *Mineral. Mag.,* **29**, 113–125.

Sahama, Th. G. (1953) Mineralogy of the humite group. *Ann. Acad. Sci. Fenn. Ser. A III, Geol.-Geogr.,* **31.**

———— (1953) Mineralogy and petrology of a lava flow from Mt. Nyiragongo, Belgian Congo. *Ann. Acad. Sci. Fenn. Ser. A III, Geol.-Geogr.,* **35.**

————and Kai Hytönen (1958) Calcium-bearing magnesium-iron olivines. *Am. Mineral,* **43**, 862–871.

Schaller, W. T., and E. P. Henderson (1932) Mineralogy of drill cores from the potash field of New Mexico and Texas. *U.S. Geol. Surv. Bull.*, **833.**

Schmidt, E. (1919) Die Winkel der krystallographischen Axen der Plagioklase. *Chem. Erde*, **1,** 351–406.

Schreyer, W., and H. S. Yoder, Jr. (1960) Hydrous Mg-cordierite. *Ann. Rpt. Direct. Geophys. Lab.*, **1959–1960,** 91–94.

Schroeder van der Kolk, J. C. (1900) "Tabellen zur mikroskopischen Bestimmung der Mineralien nach ihren Brechungsindex." Wiesbaden.

Schuster, Max (1880) Ueber die optische Orientierung der Plagioklase. *Tscherm. Min. Petr. Mitt.*, **3,** 117–284.

Seki, Yotaro, and Masao Yamasaki (1957) Aluminian ferroanthophyllite from the Kitakami Mountainland, Northeastern Japan. *Am. Mineral*, **42,** 506–520.

Shannon, Earl V. (1924) The mineralogy and petrology of intrusive Triassic diabase at Goose Creek, Loudoun County, Virginia. *Proc. U.S. Natl. Mus.*, **66** (2), 1–86.

Shaw, Dennis M. (1960) The geochemistry of scapolite. Part I. Previous work and general mineralogy. *Jour. Petrology*, **1,** 218–260.

Skinner, Brian J. (1956) Physical properties and end-members of the garnet group. *Am. Mineral*, **41,** 428–436.

Slawson, Chester B. (1934) An objective with a variable diaphragm. *Am. Mineral*, **19,** 24–28.

———— and A. B. Peck (1936) The determination of the refractive indices of minerals by the immersion method. *Am. Mineral*, **21,** 523–528.

Smith, Harold T. V. (1937) Simplified graphic method of determining approximate axial angle from refractive indices of biaxial minerals. *Am. Mineral*, **22,** 675–681.

Smith, J. R. (1958) The optical properties of heated plagioclases. *Am. Mineral*, **43,** 1179–1194.

Smith, J. V. (1958) Effect of temperature, structural state and composition of the albite, pericline and acline-A twins of plagioclase feldspars. *Am. Mineral*, **43,** 546–551.

Sun, Ming-Shan (1957) The nature of iddingsite in some basaltic rocks of New Mexico. *Am. Mineral*, **42,** 525–533.

Sundius, N. (1931) On the triclinic manganiferous pyroxenes. *Am. Mineral*, **16,** 411–429, 488–518.

Swart, B. (1950) Morphological aspects of the Bokkeveld Series at Wuppertal, Cape Province. *Ann. Univ. Stellenbosch*, **26,** Ser. A (3–11), 413–480.

Thiele, Erich (1940) Die Beziehung der chemischen Zusammensetzung zu den physikalischen-optischen Eigenschaften in einigen Mineralien des Kontakts. *Chem. Erde*, **13,** 64–91.

Tilley, C. E. (1951) The zoned contact-skarns of the Broadford area, Skye: A study of boron-fluorine metasomatism in dolomites. *Mineral. Mag.,* **29,** 621–666.

Tobi, A. C. (1956) A chart for measurement of optic axial angles. *Am. Mineral,* **41,** 516–519.

Toler, L. G., and John Hower (1959) Determination of mixed layering in glauconites by index of refraction. *Am. Mineral,* **44,** 1314–1318.

Tomita, T. (1934) Variations in optical properties, according to chemical composition, in the pyroxenes. *Jour. Shanghai Sci. Inst. Sec. 2, Geol.,* **1,** 41–58.

Topkaya, Mehmed (1950) Recherches sur les silicates authigènes dans les roches sédimentaires. *Thes. Univ. Lausanne, Fac. Sci.*

Tröger, W. E. (1952) "Tabellen zur optischen Bestimmung der gesteinsbildenden Minerale." E. Schweizerbart'sche Verlags., Stuttgart.

Tunell, George (1953) Two definitions of positive and negative extinction angles in the plagioclase feldspars: one leading to consistency and clarity, the other to inconsistency and confusion. *Am. Mineral,* **38,** 404–411.

Tuttle, O. F. (1952a) Optical studies on alkali feldspars. *Am. Jour. Sci., Bowen Vol.,* 553–567.

———— (1952b) Origin of the contrasting mineralogy of extrusive and plutonic salic rocks. *Jour. Geol.,* **60,** 107–124.

Venkatesh, V. (1954) Twinning in cordierite. *Am. Mineral,* **39,** 636–646.

Wahlstrom, Ernest E. (1960) "Optical Crystallography," 3d ed. John Wiley & Sons, Inc., New York.

Walker, F., and A. Poldervaart (1949) The Karoo dolerites of the Union of South Africa. *Bull. Geol. Soc. Am.,* **60,** 591–705.

Ward, G. W. (1931) A chemical and optical study of the black tourmalines. *Am. Mineral,* **16,** 145–190.

Wayland, Russell G. (1942) Composition, specific gravity and refractive indices of rhodochrosite; rhodoschrosite from Butte, Montana. *Am. Mineral,* **27,** 614–628.

Weaver, C. E. (1956) The distribution and identification of mixed-layer clays in sedimentary rocks. *Am. Mineral,* **41,** 202–221.

Wilcox, Ray E. (1964) Immersion liquids of relatively strong dispersion in the low refractive index range (1.46–1.52). *Am. Mineral,* **49,** 683–688.

Wilshire, H. G. (1958) Alteration of olivine and orthopyroxene in basic lavas and shallow intrusions. *Am. Mineral,* **43,** 120–147.

Winchell, A. N. (1924) Studies in the amphibole group. *Am. Jour. Sci.,* **7,** 287–310.

———— (1937) "Elements of Optical Mineralogy." Part I, 5th ed. John Wiley & Sons, Inc., New York.

———— and Horace Winchell (1951) "Elements of Optical Mineralogy." Part II, 4th ed. John Wiley & Sons, Inc., New York.

Winchell, Horace (1946) A chart for measurement of interference figures. *Am. Mineral*, **31**, 43–50.

———— (1957) Large angular aperture and useful interference figures. *Am. Mineral*, **42**, 570–572.

Winchell, N. H., and A. N. Winchell (1909) "Elements of Optical Mineralogy." D. Van Nostrand Company, Inc., Princeton, N.J.

Wolfe, C. W., and Virginia Franklin (1949) Refractive indices of high index liquids by the prism method on the two-circle goniometer. *Am. Mineral*, **34**, 893–895.

Wright, Fred E. (1907) The measurement of the optic axial angle of the minerals in the thin section. *Am. Jour. Sci.*, **24**, 317–369.

———— (1911) "The Methods of Petrographic-microscopic Research." *Carnegie Inst. Publ. 158.*

———— (1951) Computation of the optic axial angle from the three principal refractive indices. *Am. Mineral*, **36**, 543–556.

Wülfing, E. A., and O. Mügge (1925) "Mikroskopische Physiographie der Mineralien und Gesteine," 5th ed. E. Schweizerbart'sche Verlags., Stuttgart.

INDEX

Mineral species described in detail are set in **boldface** type as are the pages covering the detailed description. Mineral species partly described are set in *italics*.